TROPICAL WETLAND M

Ashgate Studies in Environmental Policy and Practice

Series Editor: Adrian McDonald, University of Leeds, UK

Based on the Avebury Studies in Green Research series, this wide-ranging series still covers all aspects of research into environmental change and development. It will now focus primarily on environmental policy, management and implications (such as effects on agriculture, lifestyle, health etc), and includes both innovative theoretical research and international practical case studies.

Also in the series

At the Margins of Planning
Offshore Wind Farms in the United Kingdom
Stephen A. Jay
ISBN 978 0 7546 7196 1

Contentious Geographies
Environmental Knowledge, Meaning, Scale
Edited by Michael K. Goodman, Maxwell T. Boykoff and Kyle T. Evered
ISBN 978 0 7546 4971 7

Environment and Society
Sustainability, Policy and the Citizen
Stewart Barr
ISBN 978 0 7546 4343 2

Multi-Stakeholder Platforms for Integrated Water Management
Edited by Jeroen Warner
ISBN 978 0 7546 7065 0

Protected Areas and Regional Development in Europe
Towards a New Model for the 21st Century
Edited by Ingo Mose
ISBN 978 0 7546 4801 7

Energy and Culture
Perspectives on the Power to Work
Edited by Brendan Dooley
ISBN 978 0 7546 4514 6

Tropical Wetland Management
The South-American Pantanal
and the International Experience

ANTONIO AUGUSTO ROSSOTTO IORIS
University of Aberdeen, UK

Routledge
Taylor & Francis Group

LONDON AND NEW YORK

First published 2012 by Ashgate Publishing

2 Park Square, Milton Park, Abingdon, Oxon OX14 4RN
711 Third Avenue, New York, NY 10017, USA

Routledge is an imprint of the Taylor & Francis Group, an informa business

First issued in paperback 2016

British Library Cataloguing in Publication Data
Tropical Wetland Management: The South-American Pantanal and the International
    Experience. – (Ashgate Studies in Environmental Policy and Practice)
    1. Wetland management – Pantanal. 2. Nature conservation – Pantanal. 3. Ecosystem
    management – Pantanal. 4. Wetland management.
    I. Series II. Ioris, Antonio Augusto Rossotto.
    333.9'1816'098171–dc23

Library of Congress Cataloging-in-Publication Data
Ioris, Antonio Augusto Rossotto.
    Tropical Wetland Management: The South-American Pantanal and the International
    Experience / by Antonio Augusto Rossotto Ioris.
        p.   cm – (Ashgate Studies in Environmental Policy and Practice)
    Includes bibliographical references and index.
    1. Marsh ecology – Pantanal. 2. Marsh conservation – Pantanal. 3. Wetland
    conservation – Pantanal. 4. Environmental protection – Pantanal. 5. Environmental
    policy – Pantanal. 6. Wetland management – Tropics. I. Title.
    QH117.I67 2012
    577.68098–dc23   2012012008

ISBN 978-1-4094-1878-8 (hbk)
ISBN 978-1-138-25261-5 (pbk)

Contents

List of Figures

List of Tables

Notes on Contributors

Robert Bruce Boler is an Ecologist and Senior Project Manager at the South Florida Natural Resource Centre, Everglades National Park. He has worked on different facets of Everglades restoration for the past 15 years, including work with the US Environmental Protection Agency. He is currently the principal author of the Final Environmental Impact Statement and manager for the Tamiami Trail Modifications: Next Steps project.

Tom Ball holds a PhD in Ecology from the University of Cambridge and is an Academic Fellow in Sustainable Flood Management at the University of Dundee and associated with the UNESCO Centre for Water Law, Policy and Science Centre. Qualified as both an ecologist and a lawyer, his main interests are on ecological and hydrological science and societal response to environmental hazards, as well as projects on virtual water and GIS applications.

Mpaphi C Bonyongo is an ecologist by training. He is currently a senior research scholar at Okavango Research Institute, University of Botswana. He received his PhD from Bristol University in Britain. His current research interests are in herbivore ecology.

Synara A O Broch is an Environmental Supervisor Environment at the Institute of Environment of Mato Grosso do Sul (IMASUL) and an assistant lecturer at the Federal University of Mato Grosso do Sul (UFMS). She has a degree in Civil Engineering from the Vale do Rio dos Sinos University and a PhD in Sustainable Development from the University of Brasília (UnB).

Débora F Calheiros is a Senior Researcher at Embrapa Pantanal. Calheiros holds a MSc in Civil Engineering and a PhD in Sciences from the University of São Paulo. Recent work has been dedicated to study the hydro-ecological processes and ecotoxicology of Pantanal and Upper Paraguay River Basin through long-term studies of watershed management.

Joffre Castro received a MSc in Water Resources Engineering at University of Kansas and a PhD in Geochemistry from the University of South Carolina. He currently works as a senior scientist with the South Florida Natural Resources Centre. His main areas of interest include water-quality restoration projects and probabilistic risk assessments. He most recently collaborated with Florida International University on the ban of the insecticide endosulfan.

Eben Chonguiça is the Chief Executive officer of the Permanent Okavango River Basin Water Commission (OKACOM) with offices in Maun. He holds a PhD in Physical Geography from Uppsala University, Sweden.

Catia Nunes da Cunha is a professor at the Federal University of Mato Grosso (UFMT) and member of the Pantanal Ecological Research Group (NEPA) and of the National Institute of Science and Technology in Wetlands (INAU) in the same university. She obtained her doctoral degree at UFSCAR (Brazil) and had post-doctoral training at the Max Planck Institute for Limnology (Germany).

Eliana F G C Dores is an Assistant Professor of Chemistry at the Federal University of Mato Grosso (UFMT), Brazil. She obtained a PhD in Chemistry at the State University of São Paulo (Brazil) and is the coordinator of the MSc programme on Water Resources at UFMT. Her work is related to the environmental dynamics of pesticides and consequences to the water environment.

Vic C Engel received a M.S. in Systems Ecology from the University of Florida and a PhD in Earth and Environmental Science from Columbia University in New York. Since 2003 he has worked as an Ecohydrologist at the South Florida Natural Resources Centre, Everglades National Park. His research focuses on the interactions between plant communities and hydrologic cycles, mangrove forest carbon cycling, nutrient dynamics, and sea level rise.

Daniela M Figueiredo is a limnologist, director and technical expert of Aquanálise, a company based in Cuiabá (Brazil) that deals with water and wastewater analyses and consultancy. She received her MSc in Ecology and Biodiversity Conservation from the Federal University of Mato Grosso and DSc in Ecology and Natural Resources from the Federal University of São Carlos. She is also a teacher at the Mato Grosso Science and Technology Department and collaborator teacher at the Federal University of Mato Grosso in the Water Research Masters.

Sandra G Gabas is an Associate Professor at the Federal University of Mato Grosso do Sul. She is a geologist and has a PhD in Civil Engineering from the University of São Paulo. Her current research focuses on groundwater contamination.

Pierre Girard is an Associate Professor at the Department of Botany and Ecology of the Bioscience Institute of the Federal University of Mato Grosso, Brazil. He is also a researcher at the Pantanal Research Centre where he leads action-research involving networks of local actors on environmental issues and climate change. He has a PhD on isotope hydrology from Quebec University.

Fábio V Gonçalves has a degree in Civil Engineering from the University for Development of State and the Region of the Pantanal (UNIDERP), MSc in Environmental Technologies from the Federal University of Mato Grosso do Sul

(UFMS) and PhD in Civil Engineering at Instituto Superior Técnico de Lisboa (Technical University of Lisbon – TU/Lisbon). He is an Assistant Professor in the Department of Hydraulic and Transports of the Exact Science and Technologic Centre of the Federal University of Mato Grosso do Sul (CCET/UFMS). Main experience is in water resources and renewable energy production in water distribution system.

Carlos N Ide is an Associate Professor in the Department of Hydraulic and Transports of the Exact Science and Technologic Centre of the Federal University of Mato Grosso do Sul (CCET/UFMS). He has a degree in Civil Engineering from the State University of Mato Grosso (UEMT) and PhD in Water Resources and Environmental Sanitation from the Federal University of Rio Grande do Sul (UFRGS). Has experience in water resources and sanitary engineering.

Antonio A R Ioris is a lecturer in the School of Geosciences at the University of Aberdeen and a research fellow at the Aberdeen Centre for Environmental Sustainability. Recent academic publications include work on institutional reforms, socio-spatial conflicts and environmental justice in Latin America and Europe. He is the coordinator of the Pantanal Research Network established in 2008 between South American, North American and European scientists.

Wolfgang J Junk is an academic with a long experience in the Pantanal, in the Amazon and other wetland systems. He is a senior researcher and team leader at the Max Planck Institute for Limnology. Since 2007 Junk is an emeritus professor in Germany, while also working in Brazil at the State University of Amazonas (UEA), at the National Amazon Research Institute (INPA) and at the National Institute of Science and Technology in Wetlands (INAU).

Donald L Kgathi is an Associate Professor of research at Okavango Research Institute, University of Botswana. He holds a PhD in Development Studies from the University of East Anglia. His current research interests are in economics of rural livelihoods and farming systems, adaptation to shocks and climate change, socio-economic impacts of biofuel production and use, and economic valuation of natural resources.

C Naidu Kurugundla is a Principal Botanist in the Department of Water Affairs, Botswana responsible for the protection of wetlands from the alien invasive weeds and pollution. Kurugundla holds PhD in Botany and Pollution from the University of Madras. Before taking up the present assignment, he served as an Associate Professor in Botany at the University of Madras, India.

Giancarlo Lastoria is an Associate Professor at the Federal University of Mato Grosso do Sul (UFMS). He is a geologist with a PhD in geosciences and environmental geology from the State University of São Paulo (UNESP).

Reinaldo Lourival works on biodiversity conservation and has a PhD from the University of Queensland. As the Conservancy's Latin American Region, Reinaldo Lourival is a Science Advisor who develops and implements conservation planning, monitoring, and research activities across the region.

Lapologang Magole is a research scholar at the Okavango Research Institute, University of Botswana. She holds a PhD in Development Studies from the University of East Anglia. Her research work and interest are in the areas of natural resources governance (land and water) and rural development planning.

Joseph E Mbaiwa is an Associate Professor (tourism studies) at Okavango Research Institute, University of Botswana. His research interests are on tourism development, rural livelihoods and conservation. He holds a PhD in tourism sciences from Texas A&M University, USA.

Carol Mitchell received a MA and PhD from the Department of Biology at Princeton University, specializing in Ecology, Evolution and Behaviour. She currently works for the US National Park Service, as Deputy Director for Science at the South Florida Natural Resources Centre, Everglades National Park, managing science and natural resources programmes that support the restoration and protection of the Everglades habitats.

Natalie Mladenov is a Research Scientist at the University of Colorado and has studied aquatic processes in the Okavango Delta since 2000. She holds a PhD in Civil and Environmental Engineering from the University of Colorado at Boulder, USA. Her expertise is in water resources and aquatic biogeochemistry.

Gagoitseope Mmopelwa is a senior research scholar at Okavango Research Institute, University of Botswana. He holds a PhD in Agricultural Economics from the University of Pretoria. His areas of specialization are economic valuation of natural resources and conservation, livelihood systems and adaptation to climate change.

Nkobi M Moleele is leader of the BIOKAVANGO, a biodiversity conservation programme for the Okavango Delta funded by the GEF, UNDP and the Government of Botswana, and implemented by the Okavango Research Institute of the University of Botswana. He holds a PhD in Physical Geography (Ecology) from Stockholm University, Sweden.

Sibangani Mosojane is the Biodiversity Coordinator at the BIOKAVANGO project where he is mainstreaming biodiversity management objectives into the key landscape production sectors of the Okavango Delta. He was previously the District Wildlife Coordinator in the Department of Wildlife and National Parks. Mosojane holds a MSc in Conservation Ecology and Planning from the University of Pretoria.

Guilherme M Mourão is a researcher at the Pantanal Research Centre of Embrapa, in Corumbá, and holds a PhD in Ecology from the Instituto Nacional de Pesquisas da Amazônia.

Michael Murray-Hudson is an ecologist and holds the position of Research Scholar at Okavango Research Institute, University of Botswana. He received a PhD from the University of Florida. His research interest is in vegetation ecology particularly in wetlands.

Márcia D de Oliveira is a researcher at the Pantanal Research Centre of Embrapa, in Corumbá, and holds a PhD in Ecology from the Federal University of Minas Gerais. Her mains research interests are in wetland aquatic ecology and the biogeochemical characteristics of rivers and floodplain in the Pantanal wetland, transport of nutrients and suspended solids.

Carlos R Padovani is a researcher at the Pantanal Research Centre of Embrapa, in Corumbá, and has a PhD from the University of São Paulo.

Adriano R Paz is an Assistant Professor at the Department of Civil and Environmental Engineering of the Federal University of Paraíba (Brazil). He is a civil engineer by training and received his MSc and DSc degrees in water resources and environmental sanitation from the Institute of Hydraulic Researches of the Federal University of Rio Grande do Sul.

Leonard Pearlstine is a Landscape Ecologist at South Florida Natural Resources Centre, Everglades National Park, and has 25 years of experience on multi-disciplinary and applied research in natural resource management. He has degrees in Economic Zoology (BSc, Clemson University), Environmental Planning (MSc University of South Carolina) and Geomatics (PhD, University of Florida). He leads the Everglades National Park ecological modelling team.

Hugh P Possingham completed his DPhil at Oxford University and is a professor at the University of Queensland. He was elected to The Australian Academy of Science in 2005 and now sits on their council. Professor Possingham is currently an ARC Federation fellow and Director of a Commonwealth Environment Research Facility. Possingham has a variety of public roles including past Chair of the federal government Biological Diversity Advisory Committee.

Lars Ramberg is the former Director of the Okavango Research Institute (formerly the Harry Oppenheimer Okavango Research Centre), University of Botswana and the Institute's founder. He holds a PhD in Limnology from Uppsala University, Sweden. His research focuses mainly on the interplay between terrestrial and aquatic ecosystems, functional relations between land use and water use and their effects on hydrology, and chain effects on ecology and human responses.

Maria L Ribeiro is an Associate Professor at the Federal University of Mato Grosso do Sul. She has a degree in Chemical Engineering from the Federal University of São Carlos (UFSCar), MSc in Physical Chemistry from the University of São Paulo (USP) and PhD in Water Resources and Environmental Sanitation from the Federal University of Rio Grande do Sul (UFRGS). She has experience in sanitary engineering, water chemistry and wastewater treatment.

Susan Ringrose is the current Director of the Okavango Research Institute and a Professor of the University of Botswana. She holds a PhD from the University of London, England in geomorphology and remote sensing. Outside of these topics, Professor Ringrose also has expertise in transboundary water issues, land use change, and paleo-environmental research.

Thomas Safford is an assistant professor in the Department of Sociology at the University of New Hampshire. His research focuses on analyzing inter-organizational relationships and understanding the roles different public and private sector organizations play in environmental management. Safford holds a PhD in Development Sociology from Cornell University.

Dilip Shinde is an Ecologist with the South Florida Natural Resources Centre, Everglades National Park. His work focuses primarily on the Comprehensive Everglades Restoration Plan. He received a PhD in Soil and Water Science at University of Florida, Gainesville.

Jhonatan B Silva is Laboratory Technician at the Federal University of Mato Grosso do Sul (UFMS). He has a degree in Environmental Engineering, MSc in Environmental Technologies and is currently a PhD student at the Federal University of Mato Grosso do Sul (UFMS).

Christopher F Souza is an Assistant Professor at the Technology Centre of the Federal University of Alagoas (Brazil). He got his BEng in Civil Engineering and obtained MSc and DSc degrees in Water Resources and Environmental Sanitation from the Institute of Hydraulic Researches of the Federal University of Rio Grande do Sul, with work on urban drainage and environmental flows.

Jorge L Steffen is an Associate Professor at the Federal University of Mato Grosso do Sul. He has a degree in Civil Engineering from the Federal University of Mato Grosso (UFMT), a MSc in Water Resources and sanitation from the Federal University of Rio Grande do Sul (UFRGS) and a PhD in Hydraulics and Drainage from the University of São Paulo (USP). He works in the area of water modelling, transport phenomena, hydraulics and hydrology.

Humberto C do Val is an agronomist graduated from the State University of São Paulo (UNESP). He is currently a MSc student at the Federal University of Mato Grosso do Sul (UFMS).

Luiz A A do Val is an Assistant Professor at the Department of Hydraulic and Transports of the Federal University of Mato Grosso do Sul (UFMS). He is a degree in Civil Engineering from the Federal University of Uberlandia and master in Transportation Engineering from the Military Institute of Engineering (IME). He is currently a PhD student at the Federal University of Mato Grosso do Sul (UFMS).

Cornelis Vanderpost is an Associate Professor at Okavango Research Institute, University of Botswana. He is a geographer with a PhD from the University of Utrecht, The Netherlands. He has research interests in the social, ecological and geographic dimensions of nature conservation and the role of conservation in rural development.

Andrew Vinten is a principal scientist in the Catchment Management Group at the James Hutton Institute (formerly Macaulay Institute). He has over 25 years experience in research on impacts of land use on water quality. His current work aims to provide research that support policy on the management of water resources and water quality, focused by an understanding of stakeholder needs and economic cost.

Matt Watts is a winner of the Eureka Prize and his research interests are in systematic conservation planning, decision support systems, geographical information systems, systems analysis and design, and computer programming.

Piotr Wolski is an Associate Professor at the Okavango Research Institute, University of Botswana. He has been studying hydrology of the Okavango Delta since 2000. Piotr holds a PhD in Earth Sciences from Free University of Amsterdam.

Preface

The main objective of this edited book is to provide a critical update of recent scientific development and politico-institutional experiences related to the conservation of the Pantanal and compare it with other wetland areas of international importance. The different chapters are written by experts in bio-physical and socio-ecological sciences working in South American, European, African and North American wetlands. Most of the authors are active participants in the Pantanal International Network, an interdisciplinary initiative that started in 2008 between Scottish and Brazilian academics (initially supported by The Leverhulme Trust) and now includes academics from many other countries and numerous research organisations. The Pantanal International Network has facilitated the communication and the establishment of several joint research initiatives with very successful results.

One of the main paradoxes of science and policy-making today is the fact that, although the intricacy and the magnitude of environmental impacts are increasingly recognised by government and society, the reactions to those problems have been typically fragmented and inadequate, particularly in the case of tropical wetlands. Cases of environmental degradation and social conflicts continue to defy most contemporary responses, which are still largely based on techno-bureaucratic approaches and market-driven solutions. Therefore, it is important to meticulously enquire into the causes of environmental degradation, the asymmetric distribution of opportunities and the unfair sharing of negative impacts. Interdisciplinary work, like the one that permeates this book, should emphasise the historical and geographical currency of environmental problems, especially the double exploitation of nature and society. Special attention must also be dedicated to the limits of mainstream environmental management and to the politicised nature of technical assessments and policy implementation.

Considering the socio-natural complexity of tropical wetlands, this publication aims to fill an obvious gap in the international literature. Despite the fact that the Pantanal and comparable wetland systems have attracted significant international and academic interest, there are very limited publications that comprehensively offer a cross-disciplinary discussion of the multiple aspects of wetland management. The challenge that the authors of this book faced was to interconnect the many ecological, socio-economics, institutional and political issues that influence the conservation and sustainable use of wetlands. This was not a trivial goal, but the final results clearly demonstrate the complexity and relevance of wetlands, not only for the local ecosystems and social groups, but for the entire global society.

Antonio Augusto Rossotto Ioris
University of Aberdeen, UK

Chapter 1

Introduction and Overview of the Book

Andrew Vinten

The Pantanal is a 140,000 km^2 complex of savannah wetlands dependent on a natural flood pulse, draining the highland savannah forest plateau at the heart of South America. The biodiversity, the low intensity cattle ranching and abundant fisheries are threatened by damming for hydro-electricity and water supply, urban growth, and agricultural intensification in the high plateau area surrounding the wetlands.

I first visited the Pantanal in 2006, after an International Water Association conference on Diffuse Pollution in Belo Horizonte, Brazil. There I was presenting Scottish plans for research to support implementation of the EU Water Framework Directive, with respect to improving water quality in Scottish rivers, lakes, coastal and ground waters. Travelling by boat in the dry season along the sluggish channels of the inappropriately name Rio Claro, in the Poconé area, with my water quality sunglasses on, I could see a system enriched by nutrients leading to intensive algal growth. I could see the multitude of caiman alligators and oxygen starved fish congregating in ever declining pools along the Transpantaneira highway. Although these processes were clearly part of a natural cycle of flood and drought, I started thinking about the impact of the city of Cuiabá, which discharges around 70% of its sewage into the Cuiabá River, and of the pressure on the ecology of this system generated by this large input of organic pollution.

On return to the UK I contacted Antonio Ioris, a scientist with wide experience of both the governance of the Pantanal waters as a Water Resources Manager for the Brazilian Ministry of the Environment and of the Water Framework Directive through his employment with the Scottish Environment Protection Agency. He was at that time a university lecturer and senior scholar with the Aberdeen Centre for Environmental Sustainability (a partnership between Aberdeen University and the James Hutton Institute), and together we conceived a plan to develop a network of Brazilian, Scottish and other scientists to build understanding, promote research and inform governance of the Cuiabá river basin. This is the largest of the 6 main tributaries of the Paraguay River which feed the Pantanal basin. This plan was put to the Leverhulme Trust who kindly funded a network project for three years. During this time a series of three workshops has been held (two in Cuiabá, one in Scotland) which have sought to build understanding of the interaction between the human actors and the functioning of this huge ecosystem. After the first of these workshops I again visited the Pantanal, this time along the Cuiabá River and in the Lake Chaccorroré area. With my hydro-electric sunglasses on, I noted the concerns of the local fishermen and others that the natural flood cycle, on which the lake and associated wetland ecology depended, were severely compromised by the Manso dam. This is built on the Cuiabazinho River upstream of Cuiabá to provide water supply, hydro power and flood control for the city and elsewhere in Brazil. I also spoke to river communities concerned about the impact of the urban inputs and recreational fishing on their livelihoods. At the second workshop at the Macaulay Institute in Scotland (no sunglasses necessary!) the network team conceived a group of research projects to take forward characterisation and inform management of the Cuiabá River. This led to one of the team, Peter Zeilhofer, a researcher at the Federal University of Mato Grosso, gaining Brazilian funding for a project (PRONEX) to "quantify pollution by nutrients and pesticides in the Planalto-Depression system of the Pantanal, and identify impacts on diverse sections of society".

On my third visit to the Pantanal, at the height of the flood season in March 2011, I visited the vast plateau which drains into the Pantanal. I got some idea of the scale of agricultural production which has developed in the last three decades in this area and the consequences for soil erosion, sediment and pesticide transport into the Pantanal. Accompanied by Carlos Padovani, of Embrapa Corumbá, I had the privilege to undertake a 5 day descent of 600 km of the Cuiabá River by open boat in order to sample the water quality and collect data on the pollution impacts and the self-cleaning capacity of the river and associated wetlands. With my soil erosion sunglasses on, I noted that the turbidity of the river remained virtually unchanged while the river remained within its channel, downstream of Cuiabá, but that the water hyacinth and other marginal vegetation within the wetland, once the river was not longer canalised, were able to filter out the colloidal material in the water over the course of about a 100 km section of the river between Rio Mutum and Porto Cercado. This transect also showed (changing sunglasses) that while effects of the organic pollution inputs from the city of Cuiabá were clearly evident downstream in the form of a sag in dissolved oxygen in the water, these were small compared with the impact of organic matter brought in from the natural process of decomposition of waterlogged vegetation in flood areas. This temporary natural process, known locally as *dequada* or *decoada*, led to a steady decline of the oxygen content of the river Cuiaba from 8 mg/L to near zero at the confluence with the Paraguay River. I noted the pressure on the National Park by recreational fishing companies travelling 150 km upstream of Corumbá into the Cuiaba River in order to provide clients (many with sunglasses) with aerated fishing grounds.

These experiences have shown me that in order to understand the processes occurring, to grasp their significance, and to analyse the way people impact and are affected by these processes, one not only needs several pairs of sunglasses, but also the inputs from many pairs of eyes with many different perspectives. Only then can the tasks of characterisation, process understanding, management, valuation and ultimately, achieving sustainable governance of this captivating wetland, and its surrounding catchment area, be addressed. This book brings together the expertise not of those with a cursory, superficial, monochromatic view of the Pantanal (or of those with a few pairs of newly acquired sunglasses), but of those who have spent a lifetime living in, researching, characterising, campaigning and promoting sustainable management for the area, with a passion and enthusiasm, a determination and a vision, that are both contagious and humbling.

In addition, perspectives from experienced catchment managers from similar ecosystems elsewhere in the world (the Okavango, the Everglades, and in Europe) are brought to the table to enable a sharpened perspective on what may be helpful for the management of the Pantanal wetland. The Brazilian and State government agencies have a progressive approach to water policy, and a clear readiness to compare experiences with water managers and researchers elsewhere in the world. It is to be hoped that this multi-disciplinary compilation of perspectives and experience from the Pantanal and elsewhere, will prove a significant resource to facilitate the translation of knowledge into effective governance.

In Chapter 2, "The *Pantaneiros*, Perceptions and Conflicts about the Environment in the Pantanal", Pierre Girard explores the diverse perceptions of the Pantanal among the various stakeholders (river dwellers, land owners, intensive ranchers, cowboys, tourists, NGOs, local and national governments), and shows how historical developments have led to conflicting situations between them. A central dichotomy in his argument is the difference between conservation as an activity and as a way of life, and he argues that some of the key groups in the Pantanal have been excluded from the debate about its future, for example in the establishment of the national park. He argues for the involvement of academics to increase the awareness of traditional groups, such as fisherman and peons (farm workers), and to enable their voice to be heard.

In Chapter 3, "Hydro-ecological Processes and Anthropogenic Impacts on the Pantanal Wetland" Débora Calheiros, Márcia Divina de Oliveira and Carlos Padovani explore the anthropogenic impacts on the ecological functioning of the Pantanal. The pressures on the ecology considered include deforestation, hydropower dams, intensive agriculture, industrial and urban development, wildfires, mining, recreational and commercial fishing, and climate change. These pressures need to be considered against a backdrop of the natural ecological attributes, one of which is the predominance of heterotrophy in the river system, especially in the flood season. This leads to special effects, such as the *decoada* process, when O_2 is severely depleted and CO_2 enhanced in the water column. It also leads to a food web which is dominated by detritivory in the fish.

Understanding the impact of the anthropogenic pressures requires a quantitative appreciation of the hydrological elements – precipitation, evapotranspiration, soil water, groundwater and surface water, and human management – and their interaction. In Chapter 4 "Availability, Uses and Management of Water in the Brazilian Pantanal", Daniela Maimoni de Figueiredo, Eliana Freire Gaspar de Carvalho Dores, Adriano Rolim Paz and Christopher Freire Souza describe theses key hydrological features of the Pantanal, and also the dynamics of the annual flooding cycle and its impact on water quality through the *decoada* process.

The soil, land and geological properties of the region and the impacts of their management on water quality (nutrients, sediment, pesticides) are further elaborated in Chapter 5 "Soil and Water Conservation in the Upper Paraguay River Basin: Examples from Mato Grosso do Sul, Brazil" by Carlos Ide and colleagues. The problems of inappropriate agricultural use in the highlands of the region are graphically illustrated with respect to the Taquari catchment which suffers from the effects of very large sediment loads as well as dangerous concentrations and loads of pesticides. The sediment has led to major alternations in the morphology of the river, including the formation of a permanently flooded area where formerly productive farmland occurred.

In Chapter 6 "Systematic Zoning Applied to Biosphere Reserves: Protecting the Pantanal Wetland Heritage" by Reinaldo Lourival and colleagues, the need for systematic planning of the Pantanal Biosphere Reserve (PBR) to integrate qualitative and quantitative tradeoffs between objectives within and between zones while providing quantitative assessment of target achievement (i.e. biodiversity, socio-cultural and economic sustainability) is discussed. Principles of comprehensiveness, adequacy, representation and efficiency of care are integrated into the Biosphere Reserve model. The authors show how spatial variability of ecological, social and economic processes can be used within a spatially explicit tool for public engagement into the decision process, whereby objectives and goals are laid out and compromises are negotiated explicitly. Their analysis, using the Marxan with Zones spatial planning software, shows that the current design of the PBR not only under-represents the biodiversity objectives but also ignores the other two essential components of a BR, the protection of local cultures and the economic sustainability of the region. There is potential, using such systematic planning software, for socio-cultural objectives to be explicitly optimized and represented across all zones in the same way biodiversity and economic objectives can.

In Chapter 7, "Organizational Complexity and Stakeholder Engagement in the Management of the Pantanal Wetland", Thomas Safford notes that organizational networks play an important role in shaping the relationships between public sector actors and stakeholder groups engaged in these collaborative management activities. The nature of these interactions is often a key factor determining whether these planning efforts achieve their objectives. Analysing interviews with >200 organizational actors involved in environmental planning in the Pantanal region between 1998-2002 Safford found that tactics used by business and production groups as well as among international environmental NGOs were based on a highly rationalized belief that political lobbying is the most effective mechanism for ensuring these actors' interests are addressed within natural resource management programmes. He argues that managers might be better served by acknowledging the political nature of management efforts and illustrating to politically engaged actors that contributing to multi-party planning does not necessarily undermine their interests. Findings from this study point to strained inter-organizational relationships as the social forces that are inhibiting the formation of shared beliefs and understandings.

In Chapter 8 "Reassessing Development: Pantanal's History, Dilemmas and Prospects" by Antonio Ioris, the author re-assesses the past and present social and ecological impacts being brought to bear on the Pantanal and concludes that both currently and in the past, it is the same overall model of development that has been responsible for the double exploitation of nature and society alike. He contends that it is unhelpful to maintain a contrast between the traditional, low impact agriculture and the destructive elements of modern forms of production. He traces the continuum of the history of the Pantanal from the earliest intervention by European settlers in the region, the development of ranching and gold extraction, through the Paraguay war (1864-1870) which was largely associated with the territorial control of the Pantanal region, through to more modern development programmes formulated by the Brazilian State for the Pantanal since the 1970s. There has been a transition from one elitist economic system, dominated by large Pantanal farmers, to another elitist system, now controlled by industrial interests and plantation farmers. For the majority of the population in the cities and in the countryside the recent modernization process has offered scarce improvements in living standards and in terms of political opportunities. Newly formed decision-making forums have been dominated by the same political groups that always traditionally controlled economic and social opportunities. He illustrates this graphically in the context of the Taquari River where the disruption caused by deforestation and soil erosion in the upstream of the river basin ended up impacting the only area of small farms in the Pantanal floodplain. He also notes that long-term ranching has had direct impacts on fauna and flora (due to changes in land use, the spread of diseases and competition for forage between cattle and undomesticated herbivores, and has also facilitated the activities of external fishermen, hunters and poachers. It means that the floodplain was largely preserved after nearly three centuries only because of the low profitability of conventional cattle production.

In Chapter 9 "Management and Sustainable Development in the Okavango Delta" by Cornelis Vanderpost and colleagues relate the impacts of recent pressures (mainly tourism, irrigation and power generation) on one of the world's most pristine tropical wetlands. Tourism contributes positively through development of national policies, and community based tourism, but can cause problems such as the growth in number of elephants leading to conflict with farmers. Invasive plants and climate change impacts are major concerns. The Okavango Delta Management Plan formulates the limits for sustainable development in the Botswana portion of the basin. It is important to recognize that many wetland communities depend on agricultural production (crops and livestock) and fishing, hunting and gathering. Conservation efforts in the basin should not only concentrate on making these production systems sustainable, but also promote ways to involve residents in community-based and commercial eco- tourism activities.

In Chapter 10 "Wetlands and the Water Environment Europe in the First Decade of the Water Framework Directive: Are Expectations Being Matched by Delivery?" Tom Ball discusses the experience of regulating and restoring wetlands in Europe. This is highly relevant, as Brazil has modelled much of its water policy approach on the EU Water Framework Directive. This policy framework is one of the first to bring the ecosystem approach explicitly into environmental policy. It is based upon identifying the characteristics of good ecological status of water, characterising the status of regulated water bodies, then devising a programme of measures to raise water bodies to good status based on a six year river basin planning cycle. However, there is limited evidence that European member states are developing formal worked strategies for wetlands as part of Directive implementation. This is partly due to the difficulties of defining good ecological status for wetlands in physico-chemical, ecological and morphological terms, and how wetland status relates to linked water body status. It is also due to the problems of identifying restoration measures and mechanisms to finance these measures. Agri-environment schemes are the only land management schemes that currently further this purpose. A recent (2010) major thrust in Scottish policy, coming on the back of the 2009 river basin plans, is the funding of restoration projects to address the impacts of the past activities identified above, principally channel straightening, loss of bed and bank vegetation, the increase in invasive species, and the removal of barriers to fish migration. A key challenge is in delivering on objectives that require proper valuation of ecosystem services, an approach that is still in its infancy.

In Chapter 11, "Managing the Everglades and Wetlands of North America", Victor C. Engel and colleagues focus on the fundamental characteristics of the Everglades ecosystem and how they have changed as a result of human impacts, such as drainage, national park establishment, road construction and flood control, and nutrient enrichment, over the last 120 years. The US institutional framework which governs wetland management and restoration is then described, along with results of research to guide restoration activities in the Everglades. The approach to restoration includes benefit-cost analysis and incremental cost analysis to examine the costs of a project in comparison to the potential ecological benefits. Some examples of metrics which may be used to assess benefits include the area of increased spawning habitat for anadromous fish, number of stream kilometres restored to provide fish habitat, increases in number of breeding birds, and increases in the population of target species. Restoration projects considered include removing barriers to flow, and restoring hydroperiods. These are incorporated within a Comprehensive Everglades Restoration Plan, whose primary ecological benefit is establishment of the natural sheetflow of water through the system. This has a key role in shaping the ridge-slough structure of the landscape. Dynamic hydrological, water quality and vegetational models are helping to inform this plan.

The challenges of adaptive water management in the Pantanal at individual, stakeholder group, local, regional and national government levels, across diverse perspectives, are immense. By bringing together the multi-disciplinary expertise and experience of a wide range of academics and practitioners, Antonio Ioris, the editor of this important and timely publication, has provided a valuable and accessible reference. This will surely help to inform development of appropriate policies for this seemingly endless (yet finite), pure (yet polluted), peaceful (yet contested) and extraordinarily diverse ecosystem.

Chapter 2

The *Pantaneiros*, Perceptions and Conflicts about the Environment in the Pantanal

Pierre Girard

Introduction

A few decades ago the Pantanal was largely ignored by the world press and media. The focus on the Amazon region and on its controversial development was eclipsing this region where nothing of interest attracted the world's eyes and was even overlooked by the Brazilians, Bolivians and Paraguayans. Since then, the Pantanal was designated a Brazilian National Heritage Site in 1988 by the Federal Constitution and then a World Heritage Site in 2000 by UNESCO. It is only recently that the international community has turned the Pantanal into an object of global significance based on the region's contribution to global biodiversity and climate stability. From this point of view, the Pantanal became a few years ago a 'global public good' underscoring the idea that the whole of mankind has a direct stake in the region, a right to see it preserved and a duty to make sure this happens (Carter et al. 2004, Charnoz 2010). The proponents of this preservation discourse are mainly international, national (Brazil, Bolivia, Paraguay) and local NGOs, but also international agencies such as UNDP, OAS, and part of the scientific community from the countries sharing the floodplain and from abroad.

Traditional actors in the Pantanal were mainly ranchers, their employees, federal and state governments and governmental agencies and Indian Nations (Earthwatch 2004, Junk and Nunes da Cunha 2005, Silva and Girard 2004). These were primarily concerned with local development and landownership questions. In the seventies and eighties, the Brazilian government prompted numerous development plans such as PRODEPAN, POLOCENTRO, POLOALCOOL, etc, drawing newcomers to establish themselves in the areas surrounding the Pantanal (Silva and Girard 2004). They implemented mechanized agriculture and cattle ranching that rapidly started to compete with the floodplain cattle production. These newcomers did not see themselves as actors on the Pantanal scene. They were largely preoccupied with their own survival and success in turning the Brazilian Cerrado (savannah) into a storehouse for their country and the rest of world. However, as the concern about Pantanal grew, most studies showed that threats to its ecological integrity were not originating within the floodplain but in the surroundings, turning the newcomers into important actors for the Pantanal (Silva and Girard 2004) even though they still do not see themselves as such.

These recent historical developments set the stage for several conflicting situations. First, as the cattle production in the surrounding areas is more competitive than in the floodplain, it threatens the Pantanal ranchers (some of the key stakeholders). But these face competition not only from the plateau beef producers: many sell their land to foreigners, mainly from South and South East Brazil, that challenge, within the Pantanal, the traditional ranchers with new production methods, including machinery, new pastures and engineering works to change water flow locally (Wantzen et al. 2008).

Secondly, the legitimacy of traditional ranchers is contended by the conservationism of some international and national NGOs, some of which have bought land that was previously used for ranching and turned it into strict conservation preserves. NGOs have also adopted strategies for conservation based on the traditional ranching methods that are evidently less aggressive to the environment than those commonly adopted by the foreigners buying land in the Pantanal. The development of a Pantanal brand which would certify that Pantanal beef from traditional farms is more environmentally friendly is at the core of these strategies. However, efforts have been going on for several years and still no brands have emerged. One failure in attempting to create such a brand – the *vitelo pantaneiro* ('Pantanal calf') – within the scope of the Pantanal Natural Regional Park project, is well documented (Charnoz 2010). Even though this approach by NGOs is clearly not confrontational, its lack of success sheds light on the uneasy relationship between environmental NGOs and traditional ranchers.

Third, there is a conflict between international NGOs, environmental scientists, international agencies and the Brazilian Federal and State governments. The Pantanal as a "global public good" (Charnoz 2010) discourse does not appeal to the Brazilian authorities. In the recent past, for Brazilian governments, economic development, not environmental conservation, has been a priority. The position of the Federal and State governments is that NGOs are welcome as long as the rules about how to use natural resources emanate from the democratically elected authorities. In recent years however, international NGOs and agencies as well as many environmental scientists have shifted their discourse saying that conservation and development goals need to converge.

The idea that maintaining traditional ranching might be the solution to promote environmental conservation in the Pantanal is a relatively new construction. This new concept keeps the traditional ranchers, especially large landowners, in a power position. In the case of the Pantanal Natural Regional Park episode, this did not warrant effective control over the events and success for this particular group and even less for the Pantanal society as a whole (Charnoz 2010, Vargas 2006).

The new context of globalization puts the cattle ranching of the floodplain in an ever more competitive scenario. Defining cattle ranchers, especially large ones, as the protectors of the Pantanal by sticking to tradition is unlikely to be enough to warrant economic success and sustainability in the Pantanal. New models for cattle ranching and beef or meat production are necessary, as are the emergence of endogenous activities based on knowledge and perception of the local environment and know-how of local populations, traditional or not.

Tourism has been growing in the Pantanal in the last quarter of a century. Unlike new modes of beef production, which are unlikely to upset the structure of Pantanal society profoundly, it is quite possible that tourism will. Growing tourism implies that the influx of visitors will increase, that new infrastructures – not oriented by the cattle producers – will be built and moreover that a larger segment of Pantanal society will come into frequent contact with non-*pantaneiros* and their world view. Tourism, as far as sustainability of the Pantanal is concerned, might well be salvation or perdition.

A better understanding of who the *pantaneiros* are, how the public and the academics perceive them and the Pantanal and how they interact with the world surrounding the Pantanal is required to better understand the mounting conservation challenges. The following pages aim at providing definitions of who the *pantaneiros* are, showing the differences between landowners, peons (i.e. farm workers) and *ribeirinhos* (i.e. those who live in the river banks). This chapter also intends to describe various perceptions of the Pantanal and *pantaneiros*: in the electronic media, in Brazilian literature and in the formal academy. It also strives to understand how the *pantaneiros* relate to other actors – *non-pantaneiro* – involved in the Pantanal and what fractures in *pantaneiro* society conflicts with *non-pantaneiros* might expose.

Pantaneiros

The simplest way to define *pantaneiro* is the person who lives in the Pantanal. For the casual observer it is clear that there is a peculiar way of living and perceiving the world in the Pantanal. Work processes, means of subsistence, life style are adapted to the local environmental conditions in such a way that it distinguishes them from any other in the surrounding areas (Corrêa Filho 1955). However, the Pantanal is vast and there are in fact many landscapes and peoples within it. The *pantaneiros* from Poconé in the northern part of the Pantanal share similarities with those of Nhecolândia farther South, but nonetheless are from different ethnic origins and history. In the following description an overview is given. For simplicity's sake, we use the term outsiders for all those who are not *pantaneiros*.

The question of identity for the *pantaneiros* is recent. In the past, due to the geographic isolation of the Pantanal, they were not readily exposed to foreign elements and there was no need to define them. In the sixties and seventies, there was an immigration wave. Migrants came and established themselves mainly on the surrounding plateaus, but also in the Pantanal, attracted by fiscal incentives from the government. In Poconé, for example, the encounter with other cultures and ways of life prompted many persons identified with the tradition to re-think it and to reconstruct it since, from this moment, it could readily be asked: what does it mean to be a *pantaneiro*? (cf. Campos Filho 1998).

Usually the *pantaneiros* are associated with cattle ranching. Livestock has been the main economic activity in the Brazilian Pantanal since 1737 (Borges 1991). However, Pantanal society is also composed of native peoples – mainly Guatós and Bororos (Campos Filho 1998, Silva and Silva 1995) – and of communities of fishermen living mainly along the Paraguay and Cuiabá riverbanks. Nevertheless, cattle ranching is still the most important activity in the Pantanal from economic and land use perspectives. Historically, the region was divided into large farms that organized the social relations. Most people lived – and still live – in rural areas. However, some small cities like Poconé, Barão de Melgaço and Corumbá are centuries' old urban nuclei. With the exception of the large land owners and cattle-ranchers, who today spend a significant amount of their time in the cities, it is mostly on the cattle ranching farms that men and women live as janitors, peons, cooks, maids, etc.

Farmers and Peons

In these farms the *pantaneiros* are mainly *caboclos*, i.e. mixes of natives of the Pareci, Guató, Chiquito and other Bolivians tribes with Afro-descendents and Spanish or Portuguese migrants and their descendents (Corrêa Filho 1946). The culture that they have established was linked to a subsistence economy based on minimal solutions barely sufficient to keep individuals alive (Candido 1964). This extreme simplicity in the way of living is still a characteristic of the farm dwellers in the Pantanal, independent of their social position, from farmers, small ranchers, labourers to foremen (Rossetto 2009).

Banducci Junior (1995) states that until the first half of the 20th century the farmers (the landowners) and their employees (today the so called peons, the cowboys or farm workers in general) shared a common lifestyle and common world vision. Until that time the main difference was land ownership. The peons did not own land in the Pantanal. However, it was fairly common that they raised small herds on the farmers' land (Campos Filho 1998).

The descendants of these farmers and their families used to live on the same farms as their fathers grew up on to avoid expenditure on judicial separation of the land. Their employees lived, for many generations, nearby or even in their houses, all gathered under the law of the owner (Corrêa Filho 1946). For a long time, the custom of protecting the integrity of heritage guided the transmission of family property, an element that facilitated the reproduction of cultural norms and values (Rossetto 2009).

In the Poconé area, until the sixties, peons were called *camaradas* (comrades). Campos Filho (1998) affirms that oral sources stated that life was extremely hard. Hard times create a strong team spirit and close cooperation (Vayda and McCay 1975). In the vastness of the Pantanal, without any schedules to keep to, at times sleeping in the wild, without food, facing danger from animals and the elements, there was a need for good companions, willing and ready to go ahead with the work to be done.

In the past, the different social classes coexisted with more harmony and thought that their social system was contradiction free. The *camaradas* obeyed the landowner, as he (normally landowners are male) was the patriarchal figure of authority. However the landowner, his sons and the peons did the work as equals, a habit that was maintained until the migration of the landowners to the city. With time, the way of life of landowners and peons differentiated. Many landowners started to show signs of their wealth and also new ideas introduced by their education. The peons lost the trust that their bosses once had in them and this was substituted by regulation of work and a lowering of wages (Campos Filho 1998).

In practice the families of farmers cannot maintain the traditional patterns. Changes due to globalization and the consequent introduction of modern techniques in cattle ranching, as well as inter-generational conflicts result in the reconfiguration of their way of life. As time passes their identity is changing and is becoming similar to cattle ranchers throughout Brazil. Less and less farming in the Pantanal can be considered as an example of sustainable use of the natural environment even though, in the representations of the media and of the Brazilian people in general, the Pantanal farmers are associated with a rustic and 'all natural' form of cattle-ranching (Rossetto 2009).

In spite of their knowledge of the unique Pantanal environment, the introduction of new productive techniques, such as the planting of fodder and changes in livestock management, challenges the very existence of the peons. New skills are needed, for example to drive tractors, to repair machinery such as crawler tractors, seeders and mechanical trimmers. Peons are told to qualify for carrying out the new functions or leave and migrate to urban areas (Rossetto 2009).

Not only is their existence threatened by modernization and globalization, but one of their most entrenched values can also impair them on the work market. According to Campos Filho (1998), freedom of action is valued by the *pantaneiro*. In the past, peons acted without necessarily communicating this to the landowners (who often implicitly agreed). They could set fires, kill an ox or hunt if they deemed it necessary. Today, when peons are contracted by outsiders for other services, like in a conservation action or by a hotel or other jobs linked to tourism, their relationship with their employers might be difficult because they do not readily accept invasion of their freedom of action by practices that differ from their own.

Today most farmers live in the cities around the floodplain. On many large farms peons still live in the Pantanal caring for the cattle. In most images of the Pantanal it is them who can be seen riding horses and driving cattle. According to Charnoz (2010), they are the ones maintaining the traditions, the 'soul' of the *pantaneiro*.

As such, the peons constitute great public relations for the farmers when they need attention from authorities or even on the international scene. The peons carry the idyllic image of life in the open and a symbiotic connection with nature that are alive in the media and in the minds of the Brazilian public and many others and that are so useful to attract the attention of politicians and donors on the national and international scene (Charnoz, 2010).

Fishermen

There is a regional characterization that distinguishes *pantaneiros* from *ribeirinhos*. These are the people living by the riverside, with greater identification to the water than to the land. Fishing, rather than ranching, is their main activity even though they practice some agriculture, sometimes on permanently dry land and sometimes on land subject to inundation (Silva and Silva 1995).

The fishermen are people of mixed blood descended from Indians such as Guatós, Borroros, Paiaguas, etc, and Afro-descendents (Silva and Silva 1995, Charnoz 2010). Many move from one place to another depending on the flood stages (the point at which the surface of a river, creek, or other body of water has risen to a sufficient level to cause damage or affects use of man-made structures) and intensities (i.e. probabilistic description of the frequency of high, modest and low flood). They do not fish on an industrial scale thereby having limited impact on the environment. These people are largely ignored by local public policies (Charnoz 2010).

In the Northern Pantanal, on the margins of the Cuiabá River, the river dwellers share their time between farming and fishery. In general, they plant beans, rice, maize, cassava and sugar cane to produce brown sugar. The predominance of one or other activity mainly depends on the availability – or not – of land, its size, its carrying capacity (i.e. the maximum, equilibrium number of organisms of a particular species that can be supported indefinitely in a given environment) and also the available workforce. The exclusivity of fishing activity is directly related to lack of access to farmland.

Fruit production and handicrafts also contribute to their income. At the beginning of the rainy season from October, with the fruiting of the cashew, women are engaged in the manufacture of sweets, giving them an extra source of income. In November and December, when mango – very abundant in the region – fruits, many families on the Cuiabá River sell the fruit to raise extra income. In some other localities older people weave nets, produce cassava flour, crochet, pestles, bowls and clay pots (Silva and Silva 1995).

The Perception of the Pantanal and the *Pantaneiros*

The media portray a vision of the Pantanal that is spread with the globalization of economy and culture and is based on images of sanctuary and paradise. This vision is disseminated by the governments, by international NGOs and by the tourist industry.

It is also interesting to take a look at how the *pantaneiros* are pictured in Brazilian literature. This reveals how most Brazilian and many foreigners conceive and, even more, what their expectations of the *pantaneiros* and their home landscapes are.

The Pantanal on the Internet

To form an idea of how the electronic media present the Pantanal, Girard and Vargas (2008) used Google and entered 'Pantanal' as a search expression. It was hypothesized that a user not knowing the Pantanal and wanting to obtain information about it would do that spontaneously. This was performed on 24 August 2007.

The search returned sites from tourist firms, government and international NGOs. The image and discourse within these sites clearly express the paradise concept that is disseminated by these institutions. The dominant themes were extracted from the images and text observed at each site. Three themes prevail: landscapes, indigenous fauna and culture. On the sites where fauna was the principal theme, birds (toucans, jabiru storks, macaws, and herons), jaguars, alligators (jacarés), and capybaras predominate. Nonetheless, anyone who visits the Pantanal a few times would know that the animals he is most likely to see are jacarés and beef cows – animals not frequently shown in the images on the internet. Also, from the Internet one would think that the Pantanal is always wet, when in fact the high waters last for only three to five months. The 'virtual' ignores the drought: it is either non-existent or does not have any appeal.

The words most commonly used to describe the Pantanal were: 'immense', 'zoo', 'exuberant', 'diverse', 'ecological heritage', 'experience', 'the world's largest flood plain', 'adventure', 'strong emotions', 'eco-adventure', 'world heritage', 'fishing', 'ecotourism', 'paradise' and 'sanctuary'. Almost no mention is made of the Pantanal being the homeland of the *pantaneiros* and that, even though it can be all that these descriptors convey, it is also a dwelling place where life happens day-to-day. But the sites barely show people at all. The *pantaneiro* is ignored, and the sites found on Google suggest a virgin land, suitable for adventure, for the unexpected, to dazzle. A place almost untouched, to be preserved and maintained, a true paradise. The Pantanal viewed on Google consist of "uniform landscapes, aesthetically harmonized, with fields, forests, trees, flowers, lakes, rivers, perfectly ordered [...] resembling the Garden of Eden" (Vargas and Heeman 2003).

The Pantanal and Pantaneiros *in the Literature*

The most renowned writer from the Pantanal is poet Manoel Wenceslau Leite de Barros. He was born in 1916 in Cuiabá and has spent a large part of his life close to the Pantanal, mainly in its Southern part. Another well-known account of the Pantanal is by João Guimarães Rosa, the famous author of *Grande Sertões Veredas* one of the most acclaimed Brazilian novels both in Brazil and internationally.

Manoel de Barros

Manuel de Barros has won many awards for his work, including the Jabuti Prize, the most important literary award in Brazil. He is considered by many as the greatest living poet from Brazil. The poet Carlos Drummond de Andrade recognized Manoel de Barros as the principal poet of Brazil. In 1998 the poet was rewarded with the National prize of Literature of the Ministry of Culture from Brazil, for his work. He has lived in Corumbá, a small city in the heart of Southern Pantanal and he is now in Campo Grande, in the upper area surrounding the Pantanal (the Planalto). A large portion of his work is related to the Pantanal landscape, the way time passes in the floodplain and to the *pantaneiros*. Some excerpts of his work are presented to illustrate his perception of the land and its people (note that the translations are only indicative and were made by the author).

This first poem briefly describes the *pantaneiro*:

> "He has the force of the spring the *Pantaneiro*
> He lives in the state of a tree
> And he has to be a continuation of the waters"

> "Tem força de minadouro o *Pantaneiro*
> Vive em estado de árvore
> E há de ser uma continuação das águas" (in Campos 2007)

The main impression is that the *pantaneiro* is part of nature, he is not really dissociated from the landscape – he is its continuation. As well, the *pantaneiro* is strong – he has force – but he is not a brute, but is rather, slow, patient but efficient: water can dig holes in the rock as tree roots can break it. The poet creates an image of the *pantaneiro* in complete harmony with nature, even compelling him: "...he has to be the continuation of the waters".

In this second excerpt – the first paragraph of the poem *Mundo Renovado* (Renewed World) from his book "*Livro de Pré-Coisas*" (Book of Pre-Things) (Barros 2007: 29), the poet gives an account of the Pantanal landscape.

> "In the Pantanal no one can pass a ruler. Especially when it rains. The ruler is the existence of limit. And the Pantanal has no limits."

> "No Pantanal ninguém pode passar régua. Sobremuito quando chove. A régua é existidura de limite. E o Pantanal não tem limites."

These short sentences explain that for whoever lives in the Pantanal, the landscape is limitless, especially when inundated. And man – the *pantaneiro* – has to abide by this: "no one can pass a ruler." The poet presents the Pantanal as infinite, no beginning, no end, somewhat like a definition of the divine, reinforcing the image of a land outside the current of history, some place that has no rule, especially no rule from men.

In *Lides de Campear* (Works of Cowboying) (Barros 2007: 33–34), the author states that he will search in the dictionary for the meaning of *pantaneiro*. The definition he finds:

> "It is said of, or that whoever works little, spending his time chatting."

> "Diz-se de, ou aquele que trabalha pouco, passando o tempo a conversar."

But then, he ponders the nature of this work:

> "Nature determines a lot. For our work that is of riding horse is always an uncapped mouth. Always a challenge. An inherent stubbornness. Like the nighthawk does.
>
> In the cattle drive, which is a monotonous task for many hours, sometimes for whole days – it is in the use of chants and retellings that the *pantaneiro* finds his being. In the exchange of prose or riding, he dreams over the fences. It's really a work in the wide, where the *pantaneiro* can invent, transcend, and go out of orbit by the imagination.
>
> But at the time of the rounding up stallions to geld them, the *pantaneiro* pulls his weight evenly. In sunlight or in the cold zero."

> "Natureza determina muito. Pois sendo a lida nossa de a cavalo, é sempre um destampo de boca. Sempre um desafiar. Um porfiar inerente. Como faz o bacurau.
>
> No conduzir de um gado, que é tarefa monótona, de horas inteiras, as vezes de dias inteiros – é no uso de cantos e recontos que o *pantaneiro* encontra seu ser. Na troca de prosa ou de montada, ele sonha por cima das cercas. É mesmo um trabalho na larga, onde o *pantaneiro* pode inventar, transcender, desorbitar pela imaginação.
>
> Mas na hora do pega-pra-capar, *pantaneiro* puxa na força por igual. No lampino do sol ou no zero do frio."

The poet at first seems to make little of the *pantaneiro*: he appears not to be doing much: he only rides horse all day long guiding the cattle. But, as de Barros says, it is not as easy as it seems: "Always a challenge". He also hints at the character of the *pantaneiro*: "...inherent stubbornness". But the main point is that the *pantaneiro* use these long expanses of time to chat and by doing so finds himself and defines his own world vision. As such, he is like the ultimate poet leaving the earth by imagination. This is again an idyllic image of the *pantaneiro*.

João Guimarães Rosa

João Guimarães Rosa (1908–1967) was a Brazilian novelist, considered by many to be one of the greatest, born in the 20th century. His best-known work is the novel *Grande Sertão: Veredas* (The Devil to Pay in the Backlands).

In one of his stories under the title *Entremeio com Vaqueiro Mariano* (Inset with Cowboy Mariano), which was written after he visited Manoel de Barros in the Pantanal, Rosa captures the extreme images of the imbrications man-cattle-landscape. In this passage the cowboys, leading cattle through the Pantanal during the drought, are trapped by fire and Mariano described how they finally escaped. Note that *vaqueiro* is an equivalent of peon.

> "—... It was another violent race, the road was a narrow alley, fire here, fire there. A fire jaguar, tall and bearded, one could even see the healthy grass bent the body to escape from it ... I smelled burning flesh. My face could not stand that heat, which aggravated. Smoke coming in, us crying. I no longer had spit to swallow, my mouth turned inside out, sick. Piece of fire was flying, falling on the cattle, and making them scream their worst, suffering. Ash flew up lifted by the foot of the cattle, really. The thunder of the herd was beating in my ear: I'm dead, I'm dead ... and the rough sound that fire boiled was what was the most evil, to figure what it was, the rush...
>
> — ...Well, then, God descended from heaven and sat me on the saddle: the fire had been sleeping behind because of a blessed ground of swamps, and we went in another peaceful wide, the fire labouring away. The fresh air rained. I took off my hat. It was a relief. Not that we, all the cattle even, we had been born..."
> (Guimarães Rosa 2001: 124 -125).

> "—...Foi outra corrida *friçosa*, o caminho era um beco apertado, fogo de cá, fogo de lá. Um fogo onça, alto e barbado, que até se via o capim são dobrar o corpo p'ra fugir dele... Senti o cheiro de carne queimada. Minha cara não agüentava mais aquele calor, que agravava. Fumaça entrando, a gente chorando. Não tinha mais cuspe no engolir, minha boca ascava virada do avesso. Voava pedaço de fogo, caindo em boi, e fazendo eles berrarem pior, sofrente. Voava cinza até levantada pelo pé do boi, mesmo. O trupo da boiada batia no meu ouvido: *tou morto, tou morto*... E o barulho bronco que o fogo fervia é que era o mais maligno, p'ra dar idéia dele, das pressas...
>
> — ...Enfim, aí, Deus desceu do Céu e me sentou na sela: o fogo tinha dormido p'ra trás, por causa de um bento chão de brejos, e entramos em outro largo sossegado, a queimada lavorando por longe. O ar choveu fresco. Tirei meu chapéu. Foi um descanso. Não que nós, os bois todos até, a gente tinha nascido..."

In this excerpt, fire challenges the life of the cowboys. Guimarães Rosa uses his talent to give us the tale of how they escape, using Mariano's voice. Mariano uses details and images to convey the danger and final relief he felt. This relates to the last excerpt by Manoel de Barros, about the *pantaneiros* telling themselves stories through the long hours driving the animals.

This piece also illustrates the relationship between men and cattle. Even though the cowboys risked their lives, they stood alongside the animals. This seems so natural that Mariano does not even mention anything about leaving the herd on its own. Men are teaming up with the animals, and even more, they give them their feelings: "...we, all the cattle even, we had been born..."

The account by Guimarães Rosa is as idyllic about the *pantaneiro* as was de Barros' one. Mariano is a great storyteller, just like Guimarães Rosa himself, he is, like the others peons, a natural man, utterly courageous and intimately linked to nature.

The images of *pantaneiros* depicted in the literature are quite similar to the images of the Pantanal in the electronic media. There, the Pantanal is an idyllic paradise, free of humans. Here the Pantanal is almost divine, and the *pantaneiro* is a man who is almost an extension of nature. Ironically, the Pantanal and the *pantaneiros* depicted both by the media and by literature – fixed in a somewhat mythical time – do not seem fit for the challenges of the twenty-first century.

The Pantanal and *Pantaneiros* for the Academy

Scholars have attempted to capture the descriptions and representations that *pantaneiros* produce of the Pantanal. Almost 20 years ago, sociolinguist Albana Nogueira indicated that the wetland ecological system cannot be restricted only to the study of birdlife and flora, "much more important is the man who lives in it." She also emphasized that the daily contact with the environment provided *pantaneiros* with the ability to read from nature's most subtle changes, a fundamental knowledge for human survival in the natural region (Castelnou et al. 2003).

The flood pulse occurring in the Pantanal is also responsible for the forms of human life that have, throughout its history, adapted to the coming and going of flood and drought. This is the main element shaping the unique 'Pantanal culture'. The peculiar life-cycles of the fauna and flora, as well as the vastness of the landscape, are other elements of this culture which have produced its technologies, particularly regarding cattle breeding. Finally, relative isolation from large urban centres has helped differentiate the Pantanal culture (Vargas 2006).

Depending on the place they live and their main activity – fishing or cattle ranching – *pantaneiros* evolve different descriptions of their landscape. What strikes the observer is the quantity of words *pantaneiros* have to talk about their environment. These words are not only descriptive but many times refers to the potential uses for a specific environment. Studies about the perception of landscape are incipient and have not yet evolved a single classification scheme based on environmental perception of the *pantaneiros*. Depending on the studies, terms and classification schemes vary a great deal.

For example, in a small community of fishermen (around 260 people), Galdino (2006) performed a study on the landscape as perceived by the inhabitants of Cuiabá Mirim, located on the banks of the Cuiabá River in the Northern portion of the Pantanal. The people of Cuiabá Mirim spontaneously named 17 landscape types. In a reunion within the community they re-arranged their list, excluding some redundancy into 14 descriptors of the landscape that they organized into three categories. Two of them were named by themselves: '*baixada*' and '*alto*', while they left the third unnamed. This type was named by Galdino (2006) as '*aquático*'. '*Alto*' or highland, which refer to areas that are always dry or that flood only during very high and infrequent floods. '*Baixadas*', or lowlands refers to landscapes that usually flood and '*aquático*' or aquatic landscapes are those where water is always present such as rivers, canals, lakes, etc.

Some names were used more frequently than others: '*rio*' (river), '*brejo*' (swamp), '*baía*' (a part river that opens like a lake or a lake linked to a river), '*firme*' (land that is almost never inundated), and '*lagoa*' (lake – usually not linked to the river network during the drought). In this case study, the predominance of water-related landscapes in the frequently used names, reflects the interdependency of these fishermen with their environment (Galdino 2006).

The multiplicity of uses, accessibility and subsistence relationships, are the mechanisms constructing the traditional knowledge of the landscape types. Subsistence is a key relationship between the *pantaneiro* and the environment. The *pantaneiros* of Cuiabá-Mirim develop various subsistence activities such as agriculture, mining, gathering, fishing, hunting, farming and small-scale manufacturing of houses, canoes and fishing tackle. The reduction of subsistence activities, caused by commercial fishing and tourism, the loss of access to some areas of landscape, and the reduction of various forms of land use due to the decrease in traditional practices such as hunting, manufacture and fishing are the factors that most threaten the knowledge of landscape types (Galdino 2006).

The various forms of use of the multiple types of landscape associated with traditional knowledge about seasonality, is a form of adaptive management. The adaptive management recognizes that environmental conditions will always change, requiring societies to adjust and or develop strategies to cope with these changes (Berkes et al. 2000). This form of management makes the people who practice it more ecological and socially resilient (Berkes and Folkes 1998).

For the communities linked to cattle ranching, for example San Pedro de Joselândia, the representation of space differs, resulting in a particular geography. They acknowledge '*o rio*' (the river) as a place to fish for their own consumption or to sell the fish, but also see it as a transport route. The lowlands close to the river is an extension named '*o Pantanal*', where the cattle can graze during the drought. On somewhat higher grounds, monospecific stands of *Vochysia divergens* can grow. This tree is commonly call '*cambará*' and the landscape is named '*o Cambarazal*' in which the livestock is established at the beginning of the rainy season when the water starts to rise. '*O Cambarazal*' is also where wood and timber can be found. Sometimes expanses of this landscape can be cleared for cultivation that can serve the farms and livestock at the beginning of the rains and this is also a source of timber. Further away from the river, '*o firme*' (the firm – with reference to the ground that is always firm as opposed to the soft ground once inundated), is found. This land, which is almost never inundated, is divided by fences enclosing pasture for livestock during the floods. It can also be used for agriculture (Castro et al. 2006).

Species diversity is part of the pace of life of marshland communities. Galdino (2006), in her work on the house and landscape in the Pantanal, illustrates how the shrub and tree species are used in construction by *pantaneiros*. Even today, in the community of Cuiabá Mirim, most of the timber upper structure of the houses comes from the landscape itself.

Finally, for the *pantaneiro*, the Pantanal is a social fabric. This tissue is composed of persons residing or working in the Pantanal, sharing the habits and values of the local culture. They share in a common history, are subjected to the rules of the social interactions inherent to this culture. This social fabric is composed of social networks and kinship, associated to knowledge networks and knowledge disseminated and perpetuated in the community.

Interactions between *Pantaneiros* and Outsiders

Even though the term *pantaneiros* is used to designate someone who was born in the Pantanal or lives according to the local lifestyle, this does mean that the *pantaneiros* are a uniform group always sharing common views and interests. Conflicts in relation to economic issues, land and use of resources can expose times when *pantaneiro* act as a homogeneous group and when they do not. Two case studies illustrate this quite well. The first is about the use of the resources in an area in the northern portion of the Pantanal and the second is about the experience of the Pantanal Regional Park set in the southern part of the Pantanal.

Conflicts in the Chacororé and Siá Mariana Area

Fagundes Silveira (2001) studied a conflict that developed about the use of the resources and land around two lakes – Chacororé and Siá Mariana – located close to the city of Barão de Melgaço in Mato Grosso state. Until about 1980 land use in this area was marked by the consolidation of urban nuclei and consolidation of traditional communities living in rural areas. Starting from 1980, the Chacororé-Siá Mariana system started to be used intensively for commercial fishing. The strengthening of economic interests intensified the circulation of boats in the area and caused sediment movements and perturbations in the local ecosystem. This brought up conflicts about the uses of these lakes.

By the end of the 1990s, mansions, summer houses and hotels for tourists had been constructed in the Siá Mariana floodplain. Also, the intense navigation on the canals linking the two lakes to the Cuiabá River increased the flow from

Chacororé Lake to the Cuiabá River resulting in a diminution of the Chacororé area during the drought. The stakeholders involved, local traditional communities, the State and Federal environmental agencies, NGOs and media, reacted differently to this transformation of the landscape. The situation prompted an intervention by the Mato Grosso State Environmental Agency which allowed the construction of dykes to contain the Chacororé waters. However the Federal Environmental Agency, as well as civil society were opposed to the dykes and the case went to court for a solution.

Fagundes Silveira (2001) used Stakeholders Analysis (Grimble et al. 1995) to identify the stakeholders in the Chacororé-Siá Mariana system and assess how they perceive, compete over and share this system. He identified 22 stakeholders that were grouped in four levels along a continuum of institutional levels regarding the use of the lakes' system resources ranging from local to global (Table 2.1).

Table 2.1   Stakeholders of the Chacororé-Siá-Mariana Lakes System

Continuum level	Stakeholders	Environmental interest
Global/International	Multilateral agencies; future generations	Conservation; sustainable development; development of tourism
National	Ministry of Environment; Federal environmental agency; Public prosecutor; Federal police	Conservation; resources protection; tourism development; supervision
Regional	State environmental agency; State ministries; Forest police; Media	Conservation; resources protection; tourism development; supervision
Local (surroundings)	Tourist guides; researchers; local NGOs; dwellers association; municipalities	Conservation; access to resources; tourism development; resources protection; economic gain
Local (lakes area)	Professional, amateur and subsistence fishermen; hotels and houses owners; boat drivers; national and international tourists; housewives; traditional communities; farmers; bait collectors.	Conservation; economic gain; food; interaction with natures and wildlife; access to resources; tourism development; land for farming; attractive sites

Source: Modified from Fagundes Silveira (2001)

Note that with respect to this classification, *pantaneiros* are encountered in the two local classes.

Interests with respect to the use of the Chacoroé-Siá Mariana lake system vary along the continuum of levels. At the local levels, the stakeholders have a more practical and immediate interest in the resource motivated by subsistence and economic gain. Stakeholders from regional to global levels show greater care for sustainability and ecosystem maintenance. At intermediate levels, resources protection and supervision, as well as development of tourism are common goals. At the national and international levels long term objectives such as sustainable development and conservation are central (Fagundes Silveira 2001).

Conservation is one interest that was expressed by the stakeholders at all levels. However the meaning of conservation varies along the continuum. At the global end, conservation is an activity while at the local end it is a way of life. These two modes for conservation are different and result in different definitions of problems and solutions. Usually, conservation as a way of life is an attitude of the local stakeholders – *pantaneiros* – such as the fishermen, traditional communities, inhabitants, traditional farmers, local commerce, etc. These actors use the resources in the context of their own existence and of long-term maintenance of the various ecosystems components. This attitude prompts a moderate use of the system in such a way that long term negative impacts have to be taken into account when considering short term political or economic gains (Showers 2000).

Conservation as an activity implies a series of rules to be enforced such as limiting access to the river, floodplain or headwaters, environmental cleansing, monitoring of water uses and mitigation of the consequences of these uses; removal of dykes or other flood control structures; reforestation of riparian forest and re-introduction of aquatic species. Conservation as an activity is usually driven from the outside of the system being considered by governments and/ or other conservationists (international/national NGOs, international agencies, tourist industry) and, depending on how it is proposed and performed, there can be opposition or cooperation by the local stakeholders (Boon 1992, Showers 2000).

In the Chacororé-Siá Mariana lakes system Fagundes Silveira (2001) demonstrates that for the *pantaneiros* traditional communities, the local dwellers including fishermen and small farmers, conservation is an attitude. At the international, National and State levels conservation is an activity. As such it can bring together or polarize local communities which can reject conservation actions or objectives if they are imposed instead of supporting local initiatives.

The various interests manifested by the stakeholders of the Chacororé-Siá Mariana system generated conflicts and partnerships between local and higher levels. At stake are the use of fish and the landscape of the lake system. For example, the system dwellers – *pantaneiros* – say that they are much more supervised than the tourists and amateur fishermen. However, these declare themselves more heavily supervised than the local fishermen, while both consider themselves as the guardians of the ecosystems, accusing each other of predatory fishing. It is worth mentioning that predatory fishing can be one of the causes of declining fish population, especially the most prized species, locally called *pacu* (*Piaractus mesopotamicus*) and *pintado* (*Pseudoplatystoma ssp*), regrouping several catfish species (Mateus and Penha 2007, Peixer and Petrere Jr 2007). The use of the levees and their riparian forests evidence other conflicting attitudes. The inhabitants infer misuse of the areas by tourists and amateur fishermen who camp there. They, in turn accuse the inhabitants of using them for housing and small scale agriculture.

At another level the attitude of NGOs, striving to increase the awareness of other stakeholders about the maintenance of quantity/quality of water in the lakes and the conservation of fish stocks and wildlife, conflicts with the influx of tourists and amateur fishermen who call for the opening of *jacaré* hunting as well as for an increase in the number of boats equipped with high powered motors that they use for fishing. NGOs also ask whether predatory fishing can be attributed as much to professional (*pantaneiros*) as to amateur fishermen.

Finally, conflicts between the local and higher levels resulted when the State and Federal environmental agencies try to enforce the fishing regulation, which in the words of local stakeholders should be: "more efficient and done by more responsible agents". In spite of these diverging positions, the stakeholders in the lake system agree regarding the necessity to limit the speed and engine size of boats, reforest the riparian areas of the Cuiabá River, control silting, produce better information about the lakes system, have better environmental laws, adequately manage garbage and have an environmental education programme (Fagundes Silveira 2001).

In these conflicts about the use of the Chacororé and Siá Mariana lakes system, the *pantaneiros* – here mainly fishermen – acted, within the scope of the described conflicts, as a homogenous group, that is no conflicts were detected among the *pantaneiros*. This is not always the case as will be seen next.

The Pantanal Natural Regional Park

In 2002 the Pantanal Natural Regional Park (PNRP) was created. This park had no equivalents in the Brazilian system of parks and other nature conservation designations under the National System of Conservation Units (SNUC in Brazil). This new conservation regulation was created in the style of the French Regional Natural Parks. In France these parks differ from the National Parks, which were created to protect the natural system. The Regional Parks were created in 1967 to promote the preservation of nature, the development of traditional activities – particularly agriculture – and establish a tourist infrastructure (Vargas 2006, Billaud 1984). The PNRP was thus established under the assumption that traditional farming in the Pantanal had maintained the landscape and that insuring the permanence of traditional farmers and their activities would contribute to the conservation of the biome.

The PNRP was initiated by a group of large farmers from Southern Pantanal in the Mato Grosso do Sul State, a federal senator and the local state of Mato Grosso do Sul, supported by the French Government. Each actor had its own motive for joining this initiative. The coalition of landowners saw it as a way to counter the influence of conservationist NGOs and as a response to the growth of competitors on the beef market. The local and national politicians in part responded to the anxiety of the large landowners, but also saw the PNRP as an instrument to contain the influence of the conservationist forces emphasizing the global public good aspect of the Pantanal far more than the regional development aspect. Finally the French Government had the will to export its expertise and exert influence in the Pantanal, a natural area of global significance (Charnoz 2010, Vargas 2006).

To implement the PNRP, its initiators created the Park Pantanal Institute (*Instituto Parque Pantanal*), a civil association that was recognized by the Brazilian Government for its public relevance. Under the statutes of IPP the PNRP was constituted by free initiative of the rural landowners and encompassed only private land, not necessarily contiguous (Vargas 2006). At the time of its foundation, the total area of the PNRP was about 20 per cent of the Southern Pantanal.

The creation of the PNRP was controversial. NGOs and other environmental organizations were not receptive: they attacked the PNRP, stating it was not legal because it did not fit any of the designations of the national system. According to Charnoz (2010) the cause of the NGOs reluctance regarding the PNRP was their lack of trust in the local farmers, who were hardly perceived to be good maintainers of the environment. As well, opposition was noted within the Mato Grosso do Sul State government, among researchers active in the Pantanal and producers not included in the PNRP (Vargas 2006).

The establishment of the Natural Park also showed some fractures in the social fabric of the *pantaneiros*. The PNRP project defined 'traditional community' in the Pantanal mainly as large landowners and cattle ranchers, leaving aside other important groups such as the fishermen and the peons. According to Charnoz (2010), 'a specific "system of signification", centered on landowners, was indeed used throughout the project to identify and locate the "pantaneira community"'. Arguably, an exclusionary exercise in structural power took place through a biased delineation of: "who can speak in the name of the community"; "which actors are the most 'traditional' ones in the region"; and "who embodies the soul of the Pantanal".

It is interesting to note that "traditional farming", the selling argument of the PNRP, is, in the mind of the public, Brazilian or international, associated mainly with the traditional activities developed by the peons. Farmers mostly live in the cities while their farms are operated by their employees. Tradition as a means to conserve the Pantanal ecosystems is maintained, and can arguably be transmitted, by the local traditional communities, the fishermen and the peons' families. However these groups do not have the social capital to be heard in a project such as the PNRP, or as a matter of fact in any other significant forums dealing about the Pantanal governance. Only the large landowners and cattle ranchers have a sufficient level of organization and social capital to achieve this. This is a significant drawback for participative Pantanal governance. The very groups that in the 'public eye' represent the Pantanal, its soul, values, and tradition do not have an active political voice and are easily overcome by more powerful groups active in the Pantanal.

Conclusion

The Pantanal and *pantaneiros* lived fairly isolated from the rest of the world until the beginning of the seventies. During this period *pantaneiros* evolved a culture of their own basically linked to cattle ranching, even though communities of fishermen also gathered mainly on the margins of the Paraguay and Cuiabá rivers. With the arrival of immigrants from southern Brazil, new environmental and economic pressures started to develop and are still increasing (Silva and Girard 2004, Wantzen et al. 2008). Mainly because of its preserved biodiversity, the Pantanal gained importance on the national and international scenes. In the meantime, cattle ranching establishing itself on the surrounding plateaus caused the economic decline of cattle-ranchers in the Pantanal.

This new scenario raised several questions about the identity of the *pantaneiros* and the governance of the Pantanal. At the same time both Pantanal and the *pantaneiros* started to emerge in the media, in the minds of Brazilians and on the international scene. On the one hand the Pantanal is depicted as a natural paradise, almost free of the presence of man, an image driven by the tourist industry and also by conservationist NGOs, both interested in maintaining the Pantanal as close to this ideal as possible for their own purposes. On the other hand the *pantaneiro* is perceived by society as a primordial man, almost an extension of nature. This image has been used by the local large landowners to promote the idea that it is traditional cattle ranching that has preserved the Pantanal's biodiversity in the last two centuries.

The use of these idyllic images is prompting a false debate wherein most of the Pantanal population is excluded. Whether or not traditional populations preserved the natural status of their homeland has turned into a sterile academic discussion unable to lead to any real decision-making (Brown 2003, Fraser et al. 2006, Gerhart 2010). What seems crucial at this moment is to increase the participation of *pantaneiros*, especially those who historically do not have voice in public debates about the governance of their own land, that is, the peons and the fishing communities. This seems essential, principally in their own interest. These groups, with little or no formal education and wealth, do not readily voice their concerns, even though the new economic and environmental situation is affecting them: peons may lose their jobs as they do not readily have new skills that are more and more necessary to work on farms and fishing pressure is growing in the Pantanal while there is indication that fish populations are declining, a potentially disastrous situation for the fishing communities. The Academy can find a role here: using its skills to increase awareness of the fishermen and peons, using its prestige to make their voice heard.

Acronyms

AFD - Agence Française de Développement (French Development Agency)
IIED - International Institute for Environment and Development

IPP - Instituto Parque Pantanal (Pantanal Park Institute)

NERA - Núcleo de Estudos, Pesquisas e Projetos de Reforma Agrária (Nucleus of study, research and projects for the agrarian reform – this entity is linked to the University of the State of São Paulo - UNESP)

NGO - Non Governmental Organization

NUPAUB - Núcleo de Apoio à Pesquisa sobre Populações Humanas e Áreas Úmidas Brasileiras (Nucleus to support research on Brazilian human populations and wetlands - linked to USP)

OAS - Organization of American States

PNRP - Pantanal Natural Regional Park

POLOCENTRO - Programa de Desenvolvimento dos Cerrados (Cerrado Development Programme)

PROÁLCOOL - Programa Brasileiro de Álcool (Brazilian Alcohol Programme)

PRODEPAN - Programa de Desenvolvimento do Pantanal (Pantanal Development Programme)

SNUC - Sistema Nacional de Unidades de Conservação (National Conservation Units System)

UNDP - United Nations Development Programme

UNESCO - United Nations Educational, Scientific and Cultural Organization

USP - University of São Paulo

References

Campos, M.C.A. 2007. Manoel de Barros: O Demiurgo das Terras Encharcadas – Educação pela Vivência do Chão. PhD Thesis. São Paulo: University of São Paulo.

Banducci Junior, A. 1995. Sociedade e Natureza no Pensamento *Pantaneiro*: Representação de Mundo e o Sobrenatural entre os Peões das Fazendas de Gado na "Nhecolândia" (Corumbá/MS). MSc Dissertation. São Paulo: University of São Paulo.

Barros, M. 2007. *Livro de Pré-Coisas: Roteiro para uma Excursão Poética no Pantanal*. 5th edition. Rio de Janeiro: Record.

Berkes, F. and Folke, C. 1998. *Linking Social and Ecological Systems: Management Practices and Social Mechanisms for Building Resilience*. Cambridge: Cambridge University Press.

Berkes, F., Colding, J., and Folke, C. 2000. Rediscovery of traditional ecological knowledge as Adaptive Management. *Ecological Applications*, 10(5), 1251-1262.

Billaud, J-P. 1984. *Marais Poitevin: Rencontres de la Terre et de l'Eau*. Paris: L'Harmattan.

Boon, P.J. 1992. Essential elements in the case for river conservation, in *River Conservation and Management*, edited by P.J. Boon, P. Calow and G. Petts. Chichester: John Wiley, 11-33.

Borges, F.T.M. 1991. Do Extrativismo à Pecuária: Algumas Observações sobre a História Econômica de Mato Grosso (1870 a 1930). MSc Dissertation. São Paulo: University of São Paulo.

Brown, K. 2003. Three challenges for a real people-centred conservation. *Global Ecology and Biogeography*, 12(2), 89–92.

Campos Filho, L.V.S. 1998. *Tradição e Ruptura: Subsídios ao Planejamento Conservacionista, Direcionado à Pecuária e ao Turismo, no Pantanal de Poconé-MT*. Cuiabá: Biosciences Institute, Federal University of Mato Grosso.

Candido, A. 1964. *Os Parceiros do Rio Bonito: Estudo sobre o Caipira Paulista e as Transformações dos seus Meios de Vida*. Rio de Janeiro: Livraria José Olympio Editora.

Carter R., Kim, J.A., Chambers, W.B., Teixeira, P. and Girard, P. 2004. *Interlinkages Approach for Wetland Management: The Case of The Pantanal Wetland*. Yokohama: United Nations University Institute of Advanced Studies.

Castelnou, A., Floriani, D., Vargas, I.A. and Dias, J.B. 2003. Sustentabilidade socioambiental no Pantanal Mato-grossense e seu espaço vernáculo como referência. *Desenvolvimento e Meio Ambiente*, 7, 43-70.

Castro, S.P., Barrozo, J.C., Castro, C.A., and Almeida, R.A. 2006. *Identificação das Unidades Produtivas Familiares em seus Espaços de Ocupação no Município de Barão de Melgaço. O caso de Joselândia*. Cuiabá: Federal University of Mato Grosso.

Charnoz, O. 2010. *Community participation in the Pantanal, Brazil*. Working Paper No. 93. Paris: AFD.

Corrêa Filho, V. 1946. *Pantanais Matogrossenses (Devassamento e Ocupação)*. Rio de Janeiro: Brazilian Institute of Geography and Statistics, National Council of Geography.

Corrêa Filho, V. 1955. *Fazendas de Gado no Pantanal Mato-grossense*. Rio de Janeiro: Ministry of Agriculture.

Earthwatch. 2004. *Pantanal Conservation Research Initiative: Annual Report 2003*. Maynard: Earthwatch.

Fagundes Silveira, J.M. 2001. Aplicação do Método Stakeholder Analysis no Sistema de Baías Chacororé-Siá Mariana, Pantanal de Mato Grosso. MSc Dissertation. Cuiabá: Federal University of Mato Grosso.

Fraser, D.J., Coon, T., Prince, M.R., Dion, R. and Bernatchez, L. 2006. Integrating traditional and evolutionary knowledge in biodiversity conservation: a population level case study. *Ecology and Society*, 11(2), 4, Available at: www.ecologyandsociety.org/vol11/iss2/art4.

Galdino, Y.S.N. 2006. A Casa e a Paisagem Pantaneira Percebida pela Comunidade Tradicional Cuiabá Mirim, Pantanal de Mato Grosso. MSc Dissertation. Cuiabá: Federal University of Mato Grosso.

Gerhart, C. 2010. Researchers and their discursive tactics in the debate on traditional populations and protecting biodiversity. *Desenvolvimento e Meio Ambiente*, 21, 43-67.

Girard, P. and Vargas, I.A. 2008. Tourism, development and knowledge in Pantanal: possible dialogues and partnerships. *Desenvolvimento e Meio Ambiente*, 18, 61-76.

Grimble, R. J., Chan, M-K., Aglionoby, J. and Quan, J. 1995. *Trees and Trade-offs: A Stakeholder Approach to Natural Resource Management*. London: IIED Gatekeeper Series 52. International Institute for Environment and Development.

Guimarães Rosa, J. 2001. *Estas Estórias*. 5th edition. Rio de Janeiro: Nova Fronteira.

Junk, W.J. and Nunes da Cunha, C. 2005. Pantanal: a large South American wetland at a crossroads. *Ecological. Engineering*, 24, 391-401.

Mateus, L.A.F. and Penha, J.M.F. 2007. Avaliação dos estoques pesqueiros de quatro espécies de grandes bagres (Siluriformes, Pimelodidae) na bacia do rio Cuiabá, Pantanal norte, Brasil, utilizando alguns Pontos de Referência Biológicos. *Revista Brasileira de Zoologia*, 24(1), 144-150.

Peixer, J. and Petrere Jr., M. 2007. Hook selectivity of the pacu *Piaractus mesopotamicus* (Holmberg, 1887) in the Pantanal, the state of Mato Grosso do Sul, Brazil. *Brazilian Journal of Biology*, 67(2), 339-345.

Rossetto, O.C. 2009. Sustentabilidade ambiental do Pantanal Mato-Grossense: interfaces entre cultura, economia e globalização. *Revista NERA*, 15, 88-105.

Showers K.B. 2000. Popular participation in river conservation, in Global Perspectives, in *River Conservation: Science, Policy and Practice*, edited by P.J. Boon, B.R. Davies and G.E. Petts. Chischester: John Wiley, 459-474.

Silva, C.J. and Girard, P. 2004. New challenges in the management of the Brazilian Pantanal and catchment area. *Wetlands Ecology and Management*, 12, 553–561.

Silva C.J. and Silva F.A. 1995. *No Rítmo das Águas do Pantanal*. São Paulo: NUBAUP.

Vargas, I. A. 2006. Território, Identidade, Paisagem e Governança no Pantanal Mato-grossense: Um Caleidoscópio da Sustentabilidade Complexa. PhD Thesis. Curitiba: Federal University of Paraná.

Vargas, I.A. and Heemann, A. 2003. Sentir o paraíso no Pantanal: reflexões sobre percepção e valoração ambientais. *Desenvolvimento e Meio Ambiente*, 7, 135-148.

Vayda, A.P. and McCay, B.F. 1975. New directions in ecology and ecological anthropology. *Annual Review of Anthropology*, 4, 293-306.

Wantzen, K.M, Cunha, C.N., Junk, W.J., Girard, P., Rossetto, O.C., Penha, J.M, Couto, E.G., Becker, M., Priante, G., Tomas, W.M., Santos, S.A., Marta, J., Domingos, I., Sonoda, F., Curvo, M. and Callil, C. 2008. Towards a sustainable management concept for ecosystem services of the Pantanal wetland. *Ecohydroloy and Hydrobiology*, 8(2-4), 115-138.

Chapter 3

Hydro-ecological Processes and Anthropogenic Impacts on the Ecosystem Services of the Pantanal Wetland

Débora F Calheiros, Márcia D de Oliveira, Carlos R Padovani

Introduction

The human future depends on environmental conservation to ensure the maintenance of the dynamic character of ecosystem functions or services, particularly those related to the supply of water. All life, terrestrial and aquatic, ranging from microbes to vertebrates, including humans, depends on and is shaped by water and watershed processes. Nonetheless, the degradation and loss of wetlands, in particular, is happening faster than that of other ecosystems and, as a consequence, the status of freshwater wetland species is also deteriorating faster, as pointed out by the Millennium Ecosystem Assessment (MEA 2005). The disruption of the structure and functioning of an ecosystem, such as water flow, matter and energy transfer, production and decomposition processes (nutrient cycling) can have dire consequences for nature and society (Junk et al. 1989, Neiff 1997, Falkenmark et al. 2007). In many ecosystems considerable effort goes into ensuring crop and energy production, but often at the expense of other important services, such as fisheries, freshwater supply and flood mitigation (Welcomme 1995, Postel and Richter 2003, Palmer 2010). At the same time ecosystem rehabilitation is normally costly, if at all feasible, given that many changes can be practically irreversible (Buijse et al. 2005, Poff et al. 1997, 2010). As a result the ecological role of riverine systems should be emphasised and clearly demonstrated to the general public and government leaders (Thorp et al. 2010). The Water Decade declared by the United Nations for the period 2005–2015, emphases this main concern about the conservation of wetlands (UN 2010).

A short definition of Ecosystem Services is 'the benefits people obtain from ecosystems' or, in a more detailed definition, 'the quantifiable or qualitative benefits of ecosystem functioning to the overall environment, including the products, services, and other benefits that humans receive from natural, regulated, or otherwise perturbed ecosystems'. Benefits are frequently separated into 'supporting services' (e.g. soil formation, biogeochemical cycling, production, habitat or refugia, and biodiversity), 'regulating services' (e.g. regulation of water quality, climate, floods, drought, land degradation and erosion, as well as biological processes such as pollination, pests, and diseases), 'provisioning services' (direct or indirect food for humans, fresh water, wood and fibre, and fuel) and 'cultural services' (e.g. aesthetic, spiritual, educational, recreational, and other nonmaterial benefits) (cf. MEA 2005, Thorp et al. 2010, Palmer and Richardson 2009). There is also the concept of the 'wise use of wetlands', as defined by the Ramsar Convention – the Convention on Wetlands of International Importance – which entails "the maintenance of their ecological character, achieved through the implementation of ecosystem approaches, within the context of sustainable development" (Ramsar 2005).

Agricultural systems, in particular, depend on ecological processes and services provided by ecosystems that are crucial for supporting and enhancing human well-being. Despite this, intensive agricultural management has caused profound changes in land cover, watercourses, and aquifers, contributing to ecosystem degradation and undermining the provision of a wide range of ecosystem function. As a rule, agriculture farms, or agroecosystems, have been managed as disconnected from the wider landscape, with scant regard for maintaining the ecological components and processes. Irrigation, drainage, extensive clearing of vegetation and the use of agrochemicals (fertilizers and pesticides) have altered the quantity and quality of water. The resultant modifications of water flows and water quality have had major ecological, economic, and social consequences, including effects on human health and loss of services provided by ecosystems, such as fisheries (Welcomme 1995, MEA 2005, Dudgeon et al. 2006, Palmer and Richardson 2009). Similarly, river regulation (i.e. the introduction of reservoires, water abstraction, irrigation, waterways, etc.) can cause the modification of the flow regime beyond the ecological thresholds, leading to changes in the structure and function of ecosystems and the loss of ecosystem services due to the alteration of the dynamic equilibrium between physical and biological features. Those changes can occur suddenly, although normally they represent the cumulative outcome of a gradual decline in biodiversity and reduced ecological resilience (Petts 1990, Welcomme 1995, Sparks 1995, Bunn and Arthington 2002, Hamilton 2002a,b, Postel and Richter 2003, Dudgeon et al. 2006). Besides, the invasion of exotic species has the potential to deeply alter the linkages between species through the food chain and modify access to food and habitat (Dudgeon et al. 2006, Oliveira et al. 2010a,b, 2011). The combination of threats is harmful for the healthy function and structure of wetlands, because it affects hydrodynamics and harms complex biota interrelations.

This chapter will deal with environmental pressures in the context of the South American Pantanal, where the repeated failure to tackle the loss and degradation of ecosystems has seriously undermined the progress toward achieving the Millennium Development Goals of reducing poverty, combating hunger, and increasing environmental sustainability (MEA 2005, Falkenmark et al. 2007). The main aim is to briefly describe the hydro-ecological functioning of the Pantanal wetland and discuss the detrimental influences of the sort of use of natural resources and the associated anthropogenic impacts on the ecosystems services (water, soil, nutrient cycling and biodiversity). An improved understanding of river–floodplain interactions is certainly necessary for the conservation of the Pantanal and particularly for the management of the Paraguay River, its main watercourse. That is even more urgent due to the number of infrastructure projects being implemented or under consideration, besides the increasing in the abnormal sedimentation processes capable to put in risk the maintenance of the hydro-ecological dynamics (Galdino et al. 2005, Calheiros et al. 2009). The current discussion has serious implications for policy-making in relation to environmental and socio-economic measures, especially the need to secure the welfare and livelihood of the communities that live in the Pantanal, who have the right, as stated in the Brazilian Constitution, to a healthy and ecologically balanced environment (Brazil 1988). The regional experience has wider significance for the management and conservation of tropical wetlands in other parts of the world, and particularly in the Paraguay-Parana River System as a whole (La Plata Basin), downriver from the Upper Paraguay River Basin.

The Pantanal: An Overview

The Pantanal Wetland is an extensive floodplain area in the centre of South America that belongs to the Parana-Paraguay River System (Figure 3.1). It is the largest continental freshwater wetland in the world, recognized as a National Heritage site by the Brazilian Federal Constitution (Brazil 1988) and Biosphere Reserve by the United Nations, and also as a World Natural Heritage site (the Pantanal Matogrossense National Park) (UNESCO 2000a,b). It is also considered an important biome under the aforementioned Ramsar Convention, which promotes the wise and sustainable use of wetlands. It should be noted that in the Pantanal alone there are three Ramsar Sites specified as waterfowl habitat: the Pantanal Matogrossense National Park, SESC Pantanal Private Reserve and the Negro River Private Reserve (Ramsar 2010). The environmental features of the Pantanal contain a rich biodiversity, but also populations of endangered species of large mammals and birds, which have almost disappeared in other parts of Brazil. For instance, the large blue macaw (*Anodorhynchus hyacinthinus*), the giant otter (*Pteronura brasiliensis*), the jaguar (*Panthera onca*) and the marsh deer (*Blastocerus dichotomus*) are some endangered species that can still be seen easily in the Pantanal (Campos and Magnusson 1995, Harris et al. 2005, Britski et al. 2007, Leuchtenberger and Mourão 2008, Mourão et al. 2000, 2010).

On the one hand, the livelihood of the '*pantaneiros*' and indigenous people in the floodplain closely depend on the environmental health of the wetland (see Chapter 2 in this book). Ecosystem services sustain the traditional economic activities of the region as extensive cattle ranching, subsistence, commercial and recreational fisheries, as well as cultural and ecological tourism (Brazil 1997, 2006, Calheiros 2007, Calheiros et al. 2000, 2009). On the other hand, the use of natural resources, mainly in the surrounding plateaus or highlands, has taking place in

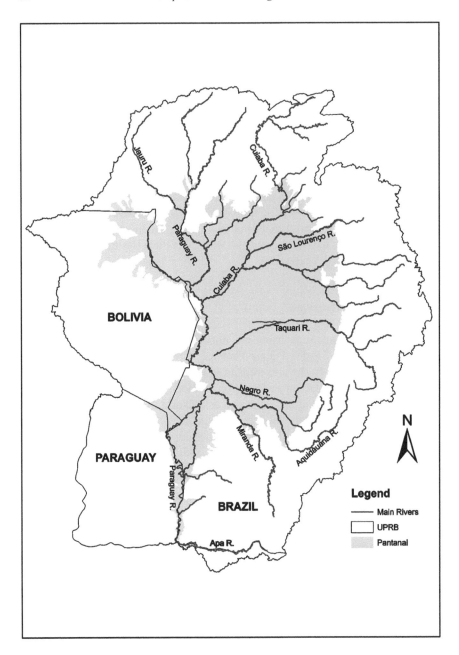

Figure 3.1   The Pantanal Wetland in the Upper Paraguay River Basin (modified from Padovani 2010)

unsustainable ways since the 1970s. In more recent years the intensification of farming activities, mainly cattle farming, has been growing on floodplain areas. Mounting agricultural pressures on the Pantanal change the ecological equilibrium, which is aggravated by the expansion of agro-industrial plants and hydropower schemes as well as by the increasing concentration of urban populations in the surrounding highlands, as in the main metropolitan area of the region, Cuiabá City, with around 780,000 inhabitants (Brazil 2010). An assessment of the structure of poverty in Brazil (Barros et al. 2000) emphasises that, at the dawn of the 21st century, the country is not a poor country, but extremely unequal. Brazilian society has abundant availability of resources and scarcity is not an excuse for the current levels of poverty. However, economic growth has not generated satisfactory results, but perpetuated inequality. This socio-economic situation is not different in the Pantanal region, but the relative conservation of its natural resources still offers benefits to the poor riverine and traditional peoples since it guarantees the access to food and water with high quality.

At the same time, the threats to the conservation of the Pantanal system, mainly the unwise uses of land and aquatic resources, are reducing its resilience and biocapacity, resulting in serious current and latent environmental and social problems. Hence, it is essential to focus on promoting environmental conservation before valuable ecosystem services are irreversibly lost. It is necessary to resort to economically and environmentally friendly development alternatives (Alho et al. 1988, Lourival et al. 1999, Da Silva 2000, Hamilton 2002b, UN-IAS 2004, Harris et al. 2005, Junk and Nunes da Cunha 2005, Wantzen et al. 2008, Calheiros et al. 2009). The regeneration of ecosystem services in ecologically degraded areas costs more than preventing their loss in the first place and promoting poverty alleviation (ESPA-AA 2008). In addition, according to Poff et al. (2010) and Palmer (2010), the challenge facing river scientists is to define ecosystem needs clearly enough to guide policy formulation and management actions that strive to balance competing demands and visions as conservation versus the abusive use of natural resources.

The Hydro-ecology of the Pantanal Wetland

The Upper Paraguay River Basin (UPRB) covers about 360,000 km^2 and comprises two main areas, namely, the surrounding plateaus and the Pantanal floodplain. The Pantanal spreads for about 140,000 km^2 and is shared by Brazil (80%), Bolivia (19%) and Paraguay (1%); see Figure 3.1. The Paraguay River is its main drainage channel and has large tributaries, such as the Jauru, Cuiabá, Taquari, and Miranda Rivers. The highest flow of the Paraguay River (2,950 m^3.s^{-1}) occurs between June and August and the lowest (1,900 m^3.s^{-1}) between December and January (Brazil 2006). There are four distinct hydrological phases: rising, flood, falling and dry period. The annual and multi-annual cycles of flooding (Figure 3.2) are the most important hydrological phenomena in the UPRB and define the ecological functioning of the Pantanal (Hamilton 2002a,b, Mourão et al. 2002a, b, Junk and Nunes da Cunha 2005). Practically all aquatic and terrestrial life within the Pantanal, including that of human groups, depends upon the timing and magnitude of floods (Hamilton 2002b, Calheiros et al. 2000, Mourão et al. 2010, Oliveira and Calheiros 2000, Oliveira et al. 2010a, Andrade 2011).

In hydro-ecological terms, the Pantanal is a tropical wetland subject to a predictable monomodal flood pulse, which controls the structure and function of the floodplain ecosystems. The annual and multi-annual variability affects the biological and ecological processes with different intensities, coverage and time scales (Hamilton et al. 1997, Calheiros and Hamilton 1998, Junk 2000, Oliveira and Calheiros 2000, Hamilton 2002a; Calheiros 2003, Junk and Nunes da Cunha 2005, Oliveira et al. 2010 a,b, 2011, Padovani 2010). The Paraguay River drains the Pantanal wetland, a vast complex of internal deltas structured as a mosaic of coalesced alluvial fans built by each in-flowing river (Hamilton 2002a, Assine and Soares 2004, Galdino et al. 2005) which delays and potentially modifies runoff from the drainage basin. The floodplain is characterised by a very low gradient (Brazil 1997) from east to west (30-50 cm.km^{-1}) and an even lower gradient from north to south (3-15 cm.km^{-1}). During the flow passage through the Pantanal floodplain, about 90 per cent of the water returns to the atmosphere, contributing considerably to the regional water and heat balance (Brazil 1997, Hamilton 2002 a,b, Junk and Nunes da Cunha 2005). Waters throughout most of the floodplain tend to flow in a characteristic direction, albeit slowly (often 2–10 cm.s^{-1}, cf. Hamilton et al. 1995) because of the low gradients and the resistance offered by the dense vegetation. The Paraguay River shows stronger floodplain effects than most other rivers, because of the high degree of river–floodplain contact area, which can reach up to 20 km wide (Hamilton et al. 1997, Hamilton 2002a, Brazil 2006).

The floodplain, therefore, functions as a large 'reservoir' that stores water from the plateau during the rainy season and delivers it slowly to the lower sections of the Paraguay River, buffering the magnitude of the floods. The average and maximum inundated areas are 34,880 km^2 and 130,920 km^2, respectively (Hamilton et al. 2002). The rainy season begins between September and December, increases between January and March, promoting the flood wave movement from the north through the south Pantanal, inundating as much as 70 per cent of the floodplain until July, taking about six months to pass through the Brazilian territory. Significant portions of the Pantanal floodplain are submerged from four to eight months each year by water depths from a few centimetres to more than two meters (Brazil 1997). The low water season is typically from September to December, whereas the highest water level occurs around late March in North Pantanal, and late July in South Pantanal, where the Paraguay River water levels have been recorded for more than a century at the Ladário Station (Figure 3.2).

Because of the low slope gradients of the Pantanal floodplain, backwater effects control river water levels and thus the floodplain inundation (Brazil 1997, Hamilton 1999, 2002a). It influences the water levels not only in the river channels but also in adjacent floodplains and tributary courses. Water levels of the tributaries are controlled by the Paraguay River near their confluences. The peak

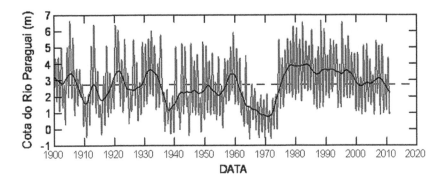

Figure 3.2   Daily Levels of Paraguay River Measured at Ladário Station since 01 January 1900 (in grey); the black line shows the inter-annual variations; the horizontal intermittent line represents the historical average = 2.72 meters. Modified from Mourão et al. 2002b)

discharge of these tributaries tends to occur earlier than that of the Paraguay River, given that the rise of the Paraguay impounds the lower courses of the tributaries, decreasing flow velocity and sometimes even reversing their flow temporarily. As a consequence, the highest water levels in the lower courses of them tributaries do not necessarily correspond with their peak discharges (Hamilton 2002a). This backwater effect often extends for considerable distances upriver, increasing the inundated area, the level and the duration of the floods.

Such geomorphologic and hydrologic characteristics confer to the Pantanal a large diversity of habitats that enhance species diversity (Mourão et al. 2000, Da Silva 2000, Junk and Nunes da Cunha 2005). Although species diversity is not particularly high and endemism is practically absent, the region is notable for its extraordinary concentration and abundance of wildlife, being considered the richest single wetland site for birds in the world (Harris et al. 2005). It provides a habitat to an estimated number of 658 species of birds, as well as over 190 species of mammals, 50 reptiles, 1,132 species of butterfly and about 270 fish species. It also provides a wintering ground for a large number of migratory birds (Campos and Magnusson 1995, Tomas et al. 2001, UNU-IAS 2004, Mamede and Alho 2006, Mourão et al. 2010, Britski et al. 2007).

The Pantanal wetland provides numerous environmental goods and services to the inhabitants of the region. Consistent with many authors (Hamilton et al. 1997, Hamilton 1999, 2002a,b, Junk and Nunes da Cunha 2005, UNU-IAS 2004, Calheiros 2003, Calheiros et al. 2009, Oliveira et al. 2010a,b, 2011), the scientific comprehension of rivers–floodplain interactions is still very limited, especially in relation to the role of 'floodplain effects' (i.e. the hydro-ecological processes resulting of the interaction between aquatic and terrestrial ecosystems). A very particular example of the influence of its hydro-ecological processes is a natural phenomenon, locally called '*dequada*', or more grammatically correct '*decoada*': during the beginning of the rising hydrological phase, the water–land interactions promote a series of transformations in the limnological characteristics of the watercourses owing to the decomposition of the newly submerged organic material (mostly terrestrial vegetation comprised of grasses and dead or senescent macrophytes). The '*decoada*' is characterised by colour changes in the water due to dissolved organic carbon, decline of the dissolved oxygen concentration and pH values, and increase of the electrical conductivity and concentrations of carbon dioxide, methane and nutrients such as carbon, nitrogen and phosphorus (Hamilton et al. 1995, 1997, Calheiros and Hamilton 1998, Calheiros et al. 2000, Oliveira and Calheiros 2000, Calheiros 2003, Oliveira et al. 2010a,b, 2011, Bastviken et al. 2010). Depending on the magnitude of the changes in water quality during the '*decoada*', massive fish kills of the order of tens of thousands of tons can occur as a result of anoxia and high levels of carbon dioxide. Significant fish kills have been observed in the floodplain portion of the large rivers of the UPRB such as the Cuiabá, Taquari, Negro, Miranda and Paraguay Rivers (Hamilton et al. 1997, Calheiros and Hamilton 1998, Calheiros et al. 2000).

The overall flood pulse maintains the floodplain in a permanent condition of primary succession, with typical high rates of production and decomposition. The dry season is characterised as a period of stress for aquatic species but at same time a source of food to piscivore animals. In turn, the '*decoada*' phenomenon is also a form of stress to the aquatic environment and a source of nutrients and energy, both to aquatic and terrestrial environments and species, especially to birds and other piscivore animals, such as giant otter and caiman. Thus this phenomenon acts like a second regulatory factor on the structure and dynamics of aquatic communities including fish, phytoplankton, zooplankton, bacterioplankton and benthic animals (Calheiros and Hamilton 1998, Calheiros et al. 2000, Oliveira and Calheiros 2000, Calheiros 2003, Oliveira et al. 2010a,b 2011, Andrade 2011). '*Decoada*' can be considered as a a complex process that synthesises the hydro-ecological functioning of the Pantanal extensive floodplain (Calheiros and Hamilton 1998).

Data are scarce about total and net phytoplankton or macrophyte photosynthetic rates in the rivers, marginal lakes and floodplains of the Pantanal system. However, is possible to infer that the P:R relationship, the ratio between primary production (P) and heterotrophic metabolism (R= respiration), has a tendency towards values lower than the unity during most of the seasonal cycle, since the respiratory rates would supplant by far the photosynthetic ones (Hamilton et al. 1995, Calheiros 2003, Bastviken et al. 2010). In general, aquatic environments are heterotrophic systems because of the low P/R (< 1) values (Battin et al. 2008, Finlay 2001).

Calheiros (2003) studied the stable isotopic flow of carbon and nitrogen in the Pantanal, and concluded that it is difficult to determine which primary energy source (algae or vascular plants) is more important to support its aquatic food chain. The high degree of river–floodplain interaction in the Pantanal contributes principally with a type of allochthonous biomass that is easily decomposed, such as grasses, the majority being C_3 grasses, derived from the previous dry season.

As a result of the decomposition of such amounts of allochthonous submerged organic matter during the '*decoada*' phase, the free CO_2 and CH_4 produced have an important role as carbon sources for autochthonous (photo and chemoautotrophic) primary production. The relative abundance of detritivorous fish, essential for productivity of fish species in the upper trophic levels of large tropical rivers (Lowe-McConnell 1987, Catella and Petrere 1996), suggest an expressive role of the decomposition route by producing methane for the methanotrophic bacteria as well carbon dioxide for algal sources (Calheiros 2003). Catella and Petrere (1996) also highlighted that the detritus, as an item of fish diet, is a strategy to shorten food chains, increasing the efficiency of energy assimilation and resulting in high detritivorous fish biomass. Besides algae, the detritus is a rich source of energy due to the presence of bacteria, protozoans and organic compounds (Wotton 1990).

The aforementioned phenomenon of '*decoada*' is indicative of the promptness and magnitude of the decomposition rate: the wave of *decoada's* water coming from the northern Pantanal can move approximately 300 km through the southern part along the Paraguay River in a matter of weeks, in years of high floods, and can continue for up to three months. So both trophic systems, autotrophic and heterotrophic, are related and complementary along all the hydrologic cycle in the Pantanal wetland, and directly associated to the flood pulse/biogeochemistry dynamics (Calheiros 2003). Thus, it is possible to estimate that any modification in the hydrodynamics of the system has elevated potential to alter the flux of matter and energy in the aquatic environments of Pantanal, which means: potential to alter the magnitude and duration of the 'decoada' phenomenon. Similarly we can estimate that changes in the hydrodynamics will also have an effect on renovation and primary production of the pasture grasses in the extensive floodplain able to undergo inundation and, consequently, on the offer of energy to the herbivores, including cattle. In the medium term, it is also possible to expect negative effects in the safeguarding of the carnivorous fauna, depending on the degree of alteration of the system hydrodynamic by the anthropogenic impacts, affecting the biodiversity which means, in an ultimate analysis, would affecting the ecological functioning of the Pantanal system as a whole.

Main Anthropogenic Impacts

The previous pages showed the complexity and vulnerability of the Pantanal system, which is highly dependent on seasonal and interannual variability. The consequence is that any long-term alteration of the flood pulse is likely to result in fundamental ecological changes and also influence the living conditions of the local human population (Calheiros et al. 2000, 2009, Hamilton 2002a,b, Junk and Nunes da Cunha 2005). Even so, in the last four decades, the Upper Paraguay River Basin has suffered from significant anthropogenic impacts related mostly to cattle ranching and agricultural land uses in the surrounded highlands, which have caused: (a) uncontrolled deforestation to produce agricultural commodities, including cattle meat, soybean, cotton, corn, and sugar cane, as well as charcoal; (b) increasing erosion and land degradation, causing sedimentation and settlement of water bodies; (c) increasing contamination by heavy metals, pesticides, fertilizers, domestic and agro-industrial effluents in surface and groundwater sources; (d) mining (especially of diamonds and gold) particularly in the spring and headwater areas with such negative consequences as erosion, sedimentation and mercury contamination; (e) abusive use of burning of the native vegetation cover as a way to deforest or for pasture management; (f) introduction of exotic grass, molluscs and fish species; (g) large development projects: the progress of waterway implementation by means of different small procedures (dredging, ports, etc.) not as a single project, and the implantation of the whole forecasted number of hydroelectric power dams in the rivers that feed into the Pantanal wetland, and finally (h) the increasing damage to the environmental services. We now discuss the socio-ecological outcomes of the main anthropogenic impacts.

Agricultural land uses

It has been estimated that around 85 per cent of the Pantanal floodplain (Figure 3.3) still retains its original vegetation cover (Monitoring 2010) The Pantanal includes the open water areas (lakes) and permanently or frequently flooded areas that are not a target for deforestation. If these areas were excluded from the Pantanal's total area, the percentage of area converted to agricultural land uses within the Pantanal will increase considerably. Another aspect is that the areas of higher elevation in the Pantanal that include most of woody vegetation (forests and savannahs with tall trees called "cerradão") are the main areas under high pressure to be converted into pastures of exotic African grassland species. Examining the map of converted areas within the Pantanal, most are near the northern, eastern and southern borders, where the flood regimes are less intense and the access by roads is easier (Padovani et al. 2008). Frequently the woody vegetation extracted from the Pantanal has been exported as charcoal to support the metals industry of other states such as Minas Gerais.

For the last 40 years the original savannah vegetation in the plateaus has been converted to pasture and crops in around 50–80 per cent of its area (Monitoring 2010). Consequently, the majority of the rivers that bring water to the Pantanal floodplain have now suffered substantial impacts due to the deforestation and the consequent increase of erosion and sedimentation (Godoy et al. 2002, 2005, Galdino et al. 2005), mainly in the headwater areas, as discussed further in later chapters of this book. However, with the prospect of reaching 80% of deforestation in each property, which is authorised under the law (excluding the Permanent Protection Areas of riverine forest and headwaters), allows an easy replacement of native vegetation cover with exotic pasture and other uses. In order to stop that, it is certainly required the approval of more stringent conservation legislation, what is currently being described as the 'Pantanal law'. An additional impact caused by agriculture is the chronic contamination by pesticides and fertilizers. The pesticides in use include DDT, a substance that is banned in Brazil, λ-cyhalothrin and atrazine (Galdino et al. 2005, Miranda et al. 2008, Dores and Calheiros 2008).

In the savannah areas fire events are common and strongly influence the vegetation structure and, consequently, the animals, with both positive and negative effects, depending on its occurrence on different intensities and/or different time scales. Fire is considered a major natural determinant of savannah vegetation and it is also used as a management tool during the dry season. However, wildfires and human-induced fires applied to pasture management, when employed in an indiscriminate way, represent an important additional stress. The extent of fire effects on animal communities generally depends on the extent of change in habitat structure and species composition caused by fire. The long-term impact of fire on the distribution and abundance of the various organisms is not yet fully understood (Silveira et al. 1999, Brown and Smith 2000, Prada and Marinho-Filho 2004). In turn, the response of individual plants to burning involve morphological and physiological changes, while at the community level, there are changes in the dynamics of the association between species. Fire has complex effects on vegetation structure, since burning affects tolerant and sensitive species that have different sites of preference on the environment (Brown and Smith 2000). An uncontrolled fire in areas with pasture biomass can spread to the savannah, causing large and uncontrolled fires and, consequently, negative effects (Rodrigues et al. 2002, Crispim and Branco 2002). The open structure, dense herbaceous stratum, high primary productivity and strong phenology seasonality all promote seasonal fires. The distribution of fire density in the Pantanal is heterogeneous and is associated with vegetation classes. Besides the human influence, flood regime and climatic factors has been pointed out as strong factors contributing to the occurrence of fires in the Pantanal (Padovani et al. 2008).

Exotic Species Invasion

Invasive exotic species have been considered an important cause of biodiversity loss (Clavero and García-Berthou 2005), mainly by the modification of environments of native species (Simberloff 2010). In the Pantanal, species such as cattle (*Bos taurus*) and pig (*Sus scrofa*) have been present since early European settlement and have had unknown effects (Mourão et al. 2002a). To upport cattle, native pasture grasses have been replaced with the African *Brachiaria* genus, although on small scales (Crispim and Branco 2002), but with high tendency to increase. Aquatic exotic *Brachiaria subquadripara* and *Panicum repens* were also introduced in the Pantanal (Pott and Pott 2004). These species decreased macrophyte diversity in small habitats in the Upper Paraná River floodplain (Thomaz et al. 2009). At the same time, native species can also become invasive (Simberloff 2010), depending on the water regime, changing from grassland to woodland landscapes, as observed in the Pantanal wetland (Santos et al. 2006).

In terms of the impacts of other species, in the Upper Paraguay River Basin the legislation only allows aquaculture of native species. Even so, *tucunaré*

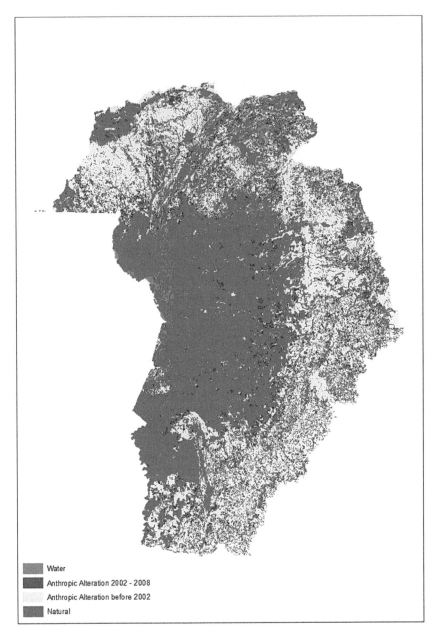

**Figure 3.3   Anthropogenic Alteration of Natural Vegetation Cover and
            Land Use of Upper Paraguay River Basin in the Brazilian
            Territory, between 2002 and 2008.**

Source: Monitoring (2010)

(*Cichla* cf. *monoculus*) and the *tambaqui* (*Colossoma macropomum*) (Marques and Resende 2005, Rezende E.K. pers. commun.), from the Amazon basin, were introduced in the Pantanal and can potentially alter the structure of fish communities. Four species of invasive freshwater molluscs were also recorded in the UPRB (*Limnoperna fortunei*, *Corbicula fluminea*, *Corbicula largillierti* and *Melanoides tuberculatus*) that might negatively affect native species by competing for food and space as well as alter the characteristics of water, and change the structure of native biota. Some of these species are widely distributed in the basins of the Cuiabá and Miranda rivers (Callil and Mansur 2002, Oliveira and Mansur, unpublished), but their impacts have not been measured yet.

A special threat is related to the presence of the golden mussel, *Limnoperna fortunei*, which has already caused major ecological impacts in Brazil. *L. fortunei* is native to China and Southeast Asia and was introduced into the Parana River system after 1991. The upstream introduction in the UPRB was facilitated by regular barge traffic along the Paraguay-Parana waterway, which may carry the organisms in the ships' hulls. The dispersion between the Paraguay River and tributaries has been slower due to the lower traffic of boats (Oliveira et al. 2010b). The current occurrence area of this mussel is restricted to the Paraguay River main channel, connected lakes, and the lower portions of tributaries such as Cuiabá, Miranda and Apa Rivers. In the Paraguay River the species populations have become established in waters of lower pH and lower calcium concentrations compared to the native environments (Oliveira et al. 2011). With this high tolerance *L. fortunei* is able to establish in most of the rivers of UPRB. Natural characteristics of the Paraguay River floodplain such as the '*decoda*' phenomenon, with depletion events of dissolved oxygen during rising water, approaching 0.0 mg.L^{-1}, pH 5.0, and free CO_2 above 100 mg.L^{-1}, as well as water temperatures around 30.0°C have limited the establishment *of L. fortunei*. Besides, the natural water level fluctuation of the Paraguay River, that is about 5 m between low and high water, also causes mortality of mussels that are exposed to the air during the low water phase (Oliveira et al. 2010a).

Regarding such controlling factors related to the natural hydro-ecological functioning of the wetland, it is possible to predict that future anthropogenic alterations of the hydro-ecological river functioning will also incraase the suitability for invasion by *L. fortunei* and other exotic species. It should be noted that impoundments (Johnson et al. 2008), and other planned hydrodynamic alteration projects increase the chances of *L. fortunei* establishment and multiplication.

The Paraguay-Parana Waterway

The Paraguay-Parana waterway has long been used by indigenous people and colonisers and continues to be used today by local inhabitants. The natural seasonal variability of water levels in the Paraguay River impedes navigation by boats and barge trains with deep draft during only around three to four months, during the low water phase. Nonetheless, navigation by larger craft is generally possible during around eight months of the year, making the river one of the best natural navigable routes. In 1996, with the economic diversification and intensive production in the region, the agro-businesses and the mining sectors advocated engineering works to improve the navigability of the Paraguay River and facilitate year-round barge transport of commodities in order to reduce the transportation costs of soybeans and minerals shipped to the Atlantic Ocean via Asunción and Buenos Aires. Such a major navigation project has been under discussion for several decades, and this would entail structural alteration of the Paraguay River channel to make it deeper at low water (Bucher et al. 1993, Hamilton 1999, 2002a,b, Gottgens et al. 1991) which would lead to large-scale, irreversible wetland degradation (Lourival et al. 1999, Junk and Nunes da Cunha 2005, Harris et al. 2005). The potential impacts of navigation projects within the Pantanal may be grouped as follows: (a) degradation of the river environment; (b) degradation of riparian areas; and (c) alteration of the flood pulse and therefore of the river–floodplain interaction and exchanges of water, materials, and aquatic animals (Hamilton 2002a). This ambitious intergovernmental project involves the five countries of La Plata Basin, Brazil, Bolivia, Paraguay, Uruguay and Argentina, which have been officially considering the range of possible engineering interventions that could severely alter the natural hydrological dynamics of the Pantanal region with potential environmental impact on adjacent flood plains (Bucher et al. 1993, Hamilton 1999, 2002a,b, Gottgens et al. 2001). Such modifications aim to facilitate year-round and day-round navigation by the largest and heaviest barge trains through the Pantanal (WWF 2001, UNU-IAS 2004). At the present time, according to Araújo Jr. (2009) and IIRSA (2010), the project still persists, although based on a new approach with fewer engineering interventions, but with investments of US$ 422 million. For the year 2011 the IIRSA is planning to elaborate a further Action Strategic Plan on navigation for the period 2012–2022.

The Taquari River Sedimentation

The increased sedimentation conditions in the majority of the rivers that form the Pantanal wetland, although in different magnitudes, is related to the natural erosion and sedimentation processes characteristic of a floodplain as well as the growing land use activities. An example of alteration of natural flow regimes in the UPRB is the relative accelerated modification in the inundation pattern of the Taquari River, one of the main tributaries of the Paraguay River. In its sub-basin, since 1970, erosion and silting up has been increasing (Godoy et al. 2002, 2005) and has turned the river system into a more unstable braiding system, although mobility and instability is a natural characteristic of rivers with alluvial fans (Assine and Soares 2004). Such conditions are considered a major environmental and social problem in the region due to the abnormal sedimentation that has caused a permanent inundation area (around 5,000 to 8,000 km^2), instead of a natural periodic inundation, and also altered the river's main flow from the south (Nhecolândia subregion) to the north (Paiaguás subregion). The consequences are the impacts on biodiversity, decline of the fish populations and also the decline of the area for cattle breeding (Godoy et al. 2002, 2005, Galdino et al. 2005, Jongman 2005).

The increasing land use (cattle ranching and crop plantations) in the highlands surrounding the Pantanal, has been pointed out as the main or the only cause of river bed sedimentation in the Pantanal. But it is important to highlight that the soils and topography prone to erosion in the highlands and a flat plain with very shallow slopes in the Pantanal, extracted from the plateau and naturally deposited in the floodplain millions of tons of sediments in the last thousands of years, forming one of the major alluvial fans in the world, the Taquari mega fan, covering around 50,000 km² (Galdino et al. 2005). The Taquari highland basin shows strong erosion processes of different types from laminar erosion to huge gully erosions that form canyons. The major shape and erosion processes of this landscape are natural processes and the land use has actually promoted the increase in their magnitude. However, how many of these processes are natural and how much is caused by land use is still unknown. The predominance of erosive sandy soils and high slope conditions promote land slices erosion. The laminar erosion is the most evident erosion process in pasture areas suffering overgrazing and excessive cattle trampling. In a long term approach, gully erosion seems to be much more important, but in recent decades the laminar erosion increased a lot due the cattle management influence. Crop plantations are less important to the erosion processes because the soils are less erosive and the topography is gentle. The soil granulometry of the croplands presents more silt and clay content than sandy soils characteristic of pasture land use. Observing the suspended sediment concentrations, solid discharge and the sediment at the river bed, fine sand from pastures soils predominates as much as 80 per cent of the sediment load at the border between the highlands and the Pantanal (Padovani et al. 2005). Another point is that much of the sediment eroded in the highlands is trapped in the highlands river beds and, according to Souza (1998), will take at least 100 years to reach the Pantanal.

The increase in rainfall in recent decades (Soriano and Galdino, 2005), another important factor causing the increase in the rain erosivity (i.e. the power of rain to break and transport the soil as sediments), is contemporary to the erosion processes caused by land use. It means that is very difficult to separate the contribution of each source of erosion: if due to the mega alluvial fan natural geomorphological processes, to the rising of pluvial precipitation and to the influence of land use for pastures or crops. However, urgent remediation and recuperation measures should be done, as well as new and carefully measures of land use practices should be implemented.

Hydroelectric Schemes

Considerable alterations of the hydrodynamics of the main tributaries and consequently the flood pulse modification of the whole Pantanal system are expected after the establishment of hydroelectric power plants (Girard 2002,

Calheiros et al. 2009). Considering that the flood pulse is the main ecological driving force (Junk et al., 1989) for a floodplain wetland, according to Poff et al. (1997) there are five critical components of the flow regime that regulate ecological processes in river ecosystems: the magnitude, frequency, duration, timing, and rate of change of hydrologic conditions. However, river regulation through dams and dikes can significantly alter these processes. Consistent with MEA (2005), the degree of fragmentation of rivers can be used as a good indicator to infer the likely condition of wetlands. Dams play a major role in fragmenting and modifying aquatic habitats, transforming lotic (running water) ecosystems into lentic (standing water) and semi-lentic ecosystems, altering the flow of matter and energy, and establishing barriers to migratory species movement. According to the World Commission on Dams (WCD 2000), impoundments fundamentally alter rivers and the use of natural resources, frequently entailing a reallocation of benefits from local riparian users to new groups of beneficiaries at a regional or national level. WCD (2000) also concludes that whilst dams have made significant contributions to economic development, in too many cases an unacceptable and often unnecessary price has been paid to secure these benefits. It traced many of the challenges to deficient decision-making processes that were not transparent and participatory, particularly with regards to negatively affected people.

While some dams in the United States (268 out of 80,000) are being decommissioned, the demand and untapped potential for these structures is still high in the developing world, particularly in Asia and South America; in the La Plata Basin, which is one of the most threatened basins, currently 27 large dams (> 60 meters) are planned or under construction. In the UPRB/Pantanal region, the first large dam was installed in the Manso River, a tributary of the Cuiabá River in 1999. The Manso Dam has a reservoir with a large surface area (387 km^2), generation capacity of 220 MW (Hamilton 2002a,b, Da Silva 2000, Junk 2000) and also regulates the seasonal flooding that affects the city of Cuiabá. Junk and Nunes da Cunha (2005) mention changes in hydrology caused by the large Manso reservoir, which began to affect flora, fauna and also fishermen and cattle ranchers along the Cuiabá River inside the Pantanal. UNU-IAS (2004) also forecast that lower and shorter flood peak in the Cuiabá River only related to the Manso Dam could have profound ecological impacts in the northern Pantanal. Junk and Nunes da Cunha (2005), alluding to Girard (2002), emphasise the possibility of the number of reservoirs in the catchment area to rise in the future from nine to 31, three of them being of large size on the rivers Correntes (176 MW), Itiquira (156 MW) and Jauru (110 MW).

In the Brazilian section of the Upper Paraguay River Basin, there are 38 reservoirs already in operation (7 large hydroelectric plants, 23 small hydroelectric plants and 8 hydroelectricity generating centrals), 3 small schemes under construction, 63 small schemes with licence application and 31 under consideration. That totals 135 projects (Calheiros et al. 2009, Brazil 2012) with 73 per cent being small schemes (≤ 30 MW), and the majority (75%) planned to be constructed in the northern part of the basin responsible for around 70 per cent of the water of the whole system (Brazil 1997, 2006) (Figure 3.4). The majority of the small projects are localized and/or planned for the same river, resulting in significant synergic impacts, compared in magnitude to those associated with huge dams. The physical barrier of such small dams, even operating at "run-of-river" mode, and with a reservoir area that should be less than 3 km², also may change nutrients and suspension matter fluxes and loading and therefore the cycling of nutrients downstream, affecting the floodplain area as well (Kingsford 2000). By examining multiple low-head dams in the Fox River there is clear evidence that these small structures may adversely affect many biotic and abiotic components of rivers and streams on local and landscape scales (Santucci et al. 2005). On the other hand, the presence of a physical barrier also prevents the movement of migratory fish species during the spawning season, affecting fish production in the medium and long term (Fernandes et al. 2009, Suzuki et al. 2009).

This pursuit of electricity production is an initiative by the Brazilian Federal and State governments to increase energy production for the whole country through investments by the private sector. The small schemes especially have attracted high business interest in recent years, because of their easy access to financial credits and simplified planning permission. However, the cumulative effects on the hydro-ecological functioning because of changing water-flow dynamics are predictable but completely unknown, except for the Manso Dam (Zeilhofer and Moura 2009). At any rate, it is possible to anticipate a disturbing scenario related to the high potential of all these development projects cumulatively to change the regime of seasonal and interannual flooding throughout the Pantanal floodplain (Girard 2002, Calheiros et al. 2009) and particularly threaten the protection of the main conservation areas and Ramsar Sites of the biome.

Based on these arguments and according to the Ramsar Convention Guidelines for Contracting Parties, relating to the maintenance of natural water regimes to maintain wetlands, and therefore to maintain traditional riverine communities (Ramsar 2007), it would be possible to include the two northern Pantanal Ramsar sites (The Pantanal Mato-Grossense National Park and the Private Reserve SESC Pantanal) in the Montreaux Record, a list of endangered wetlands. The Montreux Record, established by Ramsar Resolution 4.8 in 1990, is the "record of Ramsar Sites where changes in ecological character have occurred, are occurring or are likely to occur as a result of technological developments, pollution or other human interference". Resolution 5.4 (Ramsar 1993) specified that the purpose of the Montreux Record is to identify priority sites for positive national and international conservation attention, including applications of the Ramsar Advisory Mission and allocation of resources available under financial mechanisms.

Calheiros et al. (2009), based on the guidelines formulated during the workshop 'Influences of hydroelectric power dams in the hydro-ecological functioning of the Pantanal, Brazil', conducted during the VIII INTECOL (International Wetlands Conference, which took place in Cuiabá in 20–25 July 2008 (Cuiabá 2008), discuss the negative consequences of hydropower dams and recommend a series of actions to avoid and minimize the impacts. As stated in the guidelines,

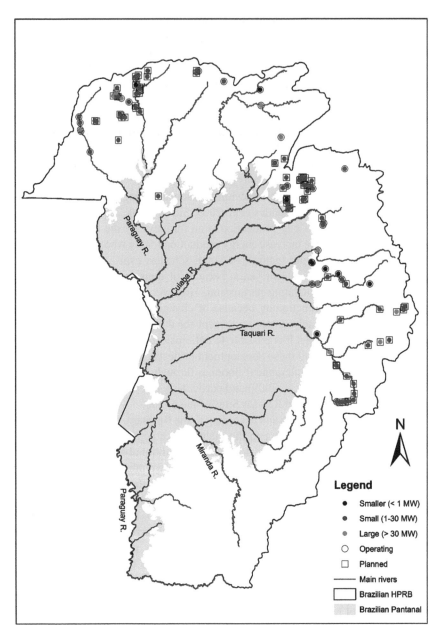

**Figure 3.4   Present and Planned Hydroelectric Power Dams to the Upper
               Paraguay River Basin in the Brazilian Territory, totaling 135 plants**

Source: ANEEL (Brazil 2012)

the Brazilian decision-makers of the electric power and of the environmental and water resources sectors of the State (Mato Grosso and Mato Grosso do Sul) and Federal governments must secure the protection of the fragile ecosystem of the Pantanal regarding the hydrological changes and systematically adopt the methodology of Environmental Strategic Assessment to help identify the impacts of hydroelectric generation in the UPRB and the sub-basins that must be kept free of dams, to ensure fishery, as well as recommend the necessity of following the ecological flow as a tool of regulating the operation of the reservoirs already built.

Conclusions

Taking into account its hydro-ecological characteristics and various forms of impact, we can conclude that the long-term conservation of the Pantanal is seriously endangered. The synergic impacts of the anthropogenic pressures on the ecosystem health and services in the plateau and floodplain areas that form UPRB have pushed the whole system into a critical level of conservation. The current pattern of development in the region is based on short-term, profit-oriented agriculture and on extensive cattle ranching, on large-scale road construction and flood control, and stimulation of industrial production by the construction of large hydroelectric power plants, canalisation of the Paraguay River for large ship traffic, and other measures to improve infrastructure. This model of development is likely to continue to affect one of the largest and best-preserved wetland systems in the world, and the economic return is questionable (Junk and Nunes da Cunha 2005). Despite the multiple pressures on and manifold threats to the stability and sustainable management of the Pantanal, the scientific knowledge, although still modest, has been contributing to the formulation of public policies and the enforcement of environmental regulation. However, such knowledge is totally disregarded by the decision makers in the time of settling on about the kind of development should be followed. On the other hand, detailed knowledge of the ecosystems and about the major ecological, economic and social factors involved is still necessary to better understand the dimension of influences of those anthropogenic impacts on the traditional economic activities (cattle ranching, fishery and tourism) and on the natural hydro-ecological functioning of the Pantanal floodplain.

An integrated approach to land, water, and ecosystems at the river basin or catchment level is urgently needed to increase multiple benefits and to mitigate detrimental impacts among ecosystem services. This approach , based also on the Precautionary Principle, would require a better understanding of the functioning of the ecosystems and how they generate multiple ecosystem services and of the value of maintaining biodiversity, habitat heterogeneity, and landscape connectivity in agricultural landscapes. Actions based on science and traditional knowledge can also play a relevant role in the conservation and sustainable management of the complex Pantanal ecosystem (UNU-IAS 2004). Therefore, it is ultimately necessary to rethink the development of the Pantanal region, based on integrated and participative planning and management, taking into account the whole catchment area. Changes in crops and cattle production practices also need to be considered in the UPRB, following the environmental legislation related to protecting riparian forests and water systems, putting together a strong program of recovering degraded areas, as well as shifting the agribusiness view to agro-ecological production systems in the surrounding plateaus and floodplain areas. If the wise use of natural resources is not put into practice, the hydro-ecological processes of the Pantanal wetland will be seriously damaged, losing ecosystem services in quantity and quality. All these impacts together are interfering negatively in its ecological processes, resulting in current and potential environmental (habitat loss, decreasing of species richness and abundance, species extinction) and social problems.

Acknowledgements

We would like to deeply thank all Embrapa Pantanal staff for their technical support, in particular the librarians Viviane Solano and Marilisi Cunha, as well as the financial agencies LTER/CNPq (Site 2, #520056/98-1), FINEP/CT-HIDRO (001/2004) and Embrapa. Also Stephen Hamilton, from Michigan State University, in the USA, and Andrew Vinten, from the James Hutton Institute in Scotland, for their important suggestions and commentaries.

References

Alho, C.J.R., Lacher JR., T.E. and Gonçalves, H.C. 1988. Environmental degradation in the Pantanal ecosystem. *BioScience,* 38, 164-171.

Andrade, M.H.S. 2011. O Fenômeno da "decoada" no Pantanal do Rio Paraguai, Corumbá/MS: Alterações dos Parâmetros Limnológicos e Efeitos sobre os Macroinvertebrados Bentônicos. PhD Thesis. Instituto de Biociências, Universidade de São Paulo: São Paulo.

Araújo Jr, J.T. 2009. *Infraestrutura e Integração Regional: O Papel do IIRSA.* Rio de Janeiro: Centro de Estudos de Integração e Desenvolvimento – CINDES. Available at: http://www.iirsa.org/BancoMedios/Documentos%20PDF/oe_rio09_cindes_tavares.pdf [acessed: 25 February 2011].

Assine, M.L. and Soares, P.C. 2004. Quaternary of the Pantanal, west-central Brazil. *Quaternary International*, 114, 23–34.

Barros, R.P., Henriques, R. and Mendonça, R. 2000. Desigualdade e pobreza no Brasil: retrato de uma estabilidade inaceitável. *Revista Brasileira de Ciências Sociais*, 15(42), 123-142.

Bastviken, D., Santoro, A.L., Marotta, H., Pinho, L.Q.; Calheiros, D.F., Crill, P. and Enrich-Prast, A. 2010. Methane emissions from Pantanal, South

America, during the low water season: toward More comprehensive sampling. *Environmental Science Technology*, 44(14), 5450-5455.

Battin, T.J., Kaplan, L. A. and Findlay, S. 2008. Biophysical controls on organic carbon fluxes in fluvial networks. *Nature Geoscience*, 1, 95-100.

Brazil. 1988. *Constituição da República Federativa do Brasil*, 1988. Available at: http://www.senado.gov.br/sf/legislacao/const [accessed: 15 September 2009].

Brazil. 1997. *Plano de conservação da Bacia do Alto Paraguai (Pantanal): PCBAP. Análise integrada e prognóstico da bacia do Alto Paraguai*. Diagnóstico dos Meios Físicos e Bióticos. v. 2, t. 3. Brasília, DF: PNMA/MMA.

Brazil. 2006. *Caderno da Região Hidrográfica do Paraguai – Plano Nacional de Recursos Hídricos*. Ministério do Meio Ambiente, Secretaria de Recursos Hídricos. Brasília: MMA. Available at: http://www.mma.gov.br/sitio/index. php?ido=publicacao.publicacoesPorSecretaria&idEstrutura=161 [accessed: 02 September 2009].

Brazil. 2012. *Acompanhamento de Registro dos Aproveitamentos Hidrelétricos*. SIGEL – Sistema Georreferenciado do Setor Elétrico Brasília: ANEEL. Available at: http://sigel.aneel.gov.br/ [accessed:2 February 2012].

Brazil. 2010. *XII Censo Demográfico*. Brasília: Instituto Brasileiro de Geografia e Estatística. Available at: http://www.censo2010.ibge.gov.br/primeiros_dados_ divulgados/index.php?uf=51 [accessed: 12 January 2011].

Britski, H.A., Silimon, K.Z.S. and Lopes, B.S. 2007. *Peixes do Pantanal - Manual de Identificação*. 2ª. ed. Brasília: Embrapa Informação Tecnológica.

Brown, J.K. and Smith, J.K. (eds.). 2000. Wildland fire in ecosystems: effects of fire on flora. Gen. Tech. Rep. RMRS-GTR-42-vol. 2. Ogden, UT: US Department of Agriculture, Forest Service, Rocky Mountain Research Station. Available at: http://www.fs.fed.us/rm/pubs/rmrs_gtr042_2.pdf [accessed: 24 February 2011].

Bucher, E.H., Bonetto, A., Boyle, T., Canevari, P., Castro, G., Huszar, P. and Stone, T. 1993. *Hidrovia: An Initial Environmental Examination of the Paraguay–Paraná Waterway*. Wetlands for the Americas Publ. no. 10, Manomet, Massachusetts.

Buijse, A.D., Klijn, F., Leuven, R.S.E.W. et al. 2005. Rehabilitation of large rivers: references, achievements and integration into river management. *Arch. Hydrobiol. Suppl.*, 155 (Large Rivers 15), (1-4), 715-738.

Bunn, S.E. and Arthington, A.H. 2002. Basic principles and ecological consequences of altered flow regimes for aquatic biodiversity. *Environmental Management*, 30(4), 492–507.

Calheiros, D.F. 2003. Influência do Pulso de Inundação na Composição Isotópica (δ^{13}C e δ^{15}N) das Fontes Primárias de Energia na Planície de Inundação do Rio Paraguai (Pantanal-MS). PhD Thesis. CENA/Universidade de São Paulo: Piracicaba-SP.

Calheiros, D.F. 2007. Baía do Castelo: local community participation in the Upper Paraguay River Basin, Pantanal Wetland, Mato Grosso do Sul State, in *Ramsar*

Handbooks for the Wise Use of Wetlands, 3rd Edition. Gland, Switzerland: Ramsar Convention Secretariat.

Calheiros, D.F., Arndt, E., Rodriguez, E.O. and Silva, M.C.A. 2009. *Influências de usinas hidrelétricas no funcionamento hidro-ecológico do Pantanal Mato-Grossense: Recomendações.* Corumbá: Embrapa Pantanal. Documentos, 102. Available at: http://www.cpap.embrapa.br/publicacoes/online/DOC102.pdf [accessed: 21 March 2010].

Calheiros, D.F. and Hamilton, S.K. 1998. Limnological conditions associated with natural fish kills in the Pantanal wetland of Brazil. *Verhandlungen - Internationale Vereinigung fur Theoretische und Angewandte Limnologie*, 26, 2189-2193.

Calheiros, D.F., Seidl, A.F. and Ferreira, C.J.A. 2000. Participatory research methods in environmental science: local and scientific knowledge of a limnological phenomenon in the Pantanal wetland of Brazil. *Journal of Applied Ecology*, 37, 684-696.

Callil, C.T. and Mansur, M.C.D. 2002. Corbiculidae in the Pantanal: history of invasion in southeast and central South America and biometrical data. *Amazon*, 17(1/2), 153-167.

Campos, Z. and Magnusson, W.E. 1995. Relationship between rainfall, nesting habitat and fecundity of *Caiman crocodilus yacare* in the Pantanal, Brazil. *Journal of Tropical Ecology*, 11, 351-358.

Catella, A.C. and Petrere, M. 1996. Feeding patterns in a fish community of baia da Onça, a floodplain lake of the Aquidauana River, Pantanal, Brazil. *Fisheries Management and Ecology*, 3, 229-237.

Clavero, M. and García-Berthou, E. 2005. Invasive species are a leading cause of animal extinctions. *Trends in Ecology and Evolution*, 20(3), 110-110.

Crispim, S.M.A. and Branco, O.D. 2002. *Aspectos gerais das Braquiárias e suas características na sub-região da Nhecolândia, Pantanal, MS.* Boletim de Pesquisa e Desenvolvimento, 33. Corumbá, MS: Embrapa Pantanal. Available at: http://www.cpap.embrapa.br/publicacoes/online/BP33.pdf [accessed: 25 January 2011].

Cuiabá. 2008. *Cuiabá Declaration on Wetlands: The State of Wetlands and their Role in a World of Global Climate Change.* International Wetlands Conference - INTECOL, 8. Cuiabá, Brazil, 20-25 July 2008, Available at: http://www.cppantanal.org.br/intecol/eng/sections.php?id_section=21 [accessed: 19 May 2009].

Da Silva, C.J. 2000. Ecological basis for the management of the Pantanal - Upper Paraguay River Basin, in *New approaches to river management*, edited by A.J.M. Smits, et al. Leiden, Netherlands: Backhuys Publishers, 97-117.

Dores, E.F.G.C. and Calheiros, D.F. 2008. Contaminação por agrotóxicos na bacia do rio Miranda, Pantanal (MS). *Revista Brasileira de Agroecologia*, 3(2 suppl.), 202-205.

ESPA-AA. 2008. Challenges to Managing Ecosystems Sustainably for Poverty Alleviation: Securing Well-Being in the Andes/Amazon. Situation Analysis prepared for the ESPA Program. Belém, Brazil: Amazon Initiative

Consortium. Available at: http://www.ecosystemsandpoverty.org/wp-content/uploads/2008/05/espa-aa-final-report-_small-version_.pdf [accessed: 6 April 2010].

Dudgeon, D., Arthington A.H., Gessner, M.O. et al. 2006. Freshwater biodiversity: importance, threats, status and conservation challenges. *Biological Reviews*, 81, 163–182.

Falkenmark, M., Finlayson, C. M., Gordon, L.J. et al. 2007. Agriculture, water, and ecosystems: avoiding the costs of going too far, in *Water for Food, Water for Life: A Comprehensive Assessment of Water Management in Agriculture*, edited by D. Molden. London: Earthscan, and Colombo: International Water Management Institute, 233-277.

Fernandes, R., Agostinho, A.A., Ferreira, E.A.; Pavanelli, C.S., Suzuki, H.I., Lima, D.P. and Gomes, L.C. 2009. Effects of the hydrological regime on the ichthyofauna of riverine environments of the Upper Paraná River Floodplain. *Brazilian Journal of Biology*, 69(2 suppl.), 669-680.

Finlay, J.C. 2001. Stable-carbon-isotope ratios of river biota: implications for energy flow in lotic food webs. *Ecology*, 82(4), 1052-1064.

Galdino, S., Vieira, L.M. and Pellegrin, L.A. 2005. *Impactos ambientais e socioeconômicos na Bacia do Rio Taquari – Pantanal.* Corumbá: Embrapa Pantanal. Available at: http://www.cpap.embrapa.br/publicacoes/online/Livro025.pdf [accessed: 21 June 2011].

Girard, P. 2002. *Efeito Cumulativo das Barragens no Pantanal: Mobilização para Conservação das Áreas Úmidas do Pantanal e Bacia do Araguaia.* Campo Grande, MS: Instituto Centro Vida. Available at: http://www.riosvivos.org.br/arquivos/site_noticias_758445261.pdf [accessed: 5 June 2011].

Godoy, J.M., Padovani, C.R., Guimarães J.R.D. et al. 2002. Evaluation of the siltation of River Taquari, Pantanal, Brazil, through ^{210}Pb geochronology of floodplain lake sediments. *J. Braz. Chem. Soc.,* 13(1), 71-77.

Godoy, J.M., Padovani, C.R., Vieira, L.M., Galdino, S. 2005. Geocronologia do assoreamento e níveis de mercúrio em lagos marginais do Rio Taquari no Pantanal, in *Impactos Ambientais e Socioeconômicos na Bacia do Rio Taquari – Pantanal,* edited by S. Galdino. L.M. Vieira. and L.A. Pellegrin. Corumbá: Embrapa Pantanal, 163-173.

Gottgens, J.F., Perry, J.E., Fortney, R.H., Meyer, J.E., Benedict, M. and Rood, B.E. 2001. The Paraguay- Paraná Hidrovía: Protecting the Pantanal with lessons from the past. *BioScience*, 51, 4, 301–308.

Hamilton, S.K. 1999. Potential effects of a major navigation project (Paraguay-Parana Hidrovia) on inundation in the Pantanal floodplains. *Regulated Rivers-Research & Management*, 15, 298-299.

Hamilton, S.K. 2002a. Hydrological controls of ecological structure and function in the Pantanal wetland (Brazil), in *The Ecohydrology of Southamerican Rivers and Wetlands*, edited by M. McClain, International Association of Hydrological Sciences. Special Publications 6, 133-158.

Hamilton, S.K. 2002b. Human impacts on hydrology in the Pantanal wetland of South America. *Water Science Technology*, 45, 35–44.

Hamilton, S.K., Sippel, S.J., Calheiros, D.F. and Melack, J.M. 1997. An anoxic event and other biogeochemical effects of the Pantanal wetland on the Paraguay River. *Limnology and Oceanography*, 42, 257-272.

Hamilton, S.K., Sippel, S.J. and Melack, J.M. 1995. Oxygen depletion and carbon dioxide and methane production in waters of the Pantanal wetland of Brazil. *Biogeochemistry*, 30, 115-141.

Hamilton, S.K., Sippel, S.J and Melack, J.M. 2002. Comparison of inundation patterns among major South American floodplains. *J. Geophys. Res.* [Atmos.] 107 (D20), LBA 5-1 to 5-14.

Harris, M.B., Tomas, W., Mourão, G., Da Silva, C.J., Guimarães, E., Sonoda, F. and Fachim, E. 2005. Safeguarding the Pantanal Wetlands: Threats and conservation initiatives. *Conservation Biology*, 19, 714-720.

IIRSA. 2010. Iniciativa para la Integración de la Infraestructura Regional Suramericana – IIRSA. Planificación Territorial Indicativa: Cartera de Proyectos IIRSA 2010. IV.5. Eje de la Hidrovía Paraguay-Paraná. Available at: http://www.iirsa.org/BancoMedios/Documentos%20PDF/lb10_seccion_iv_eje_hidrovia_paraguay_parana.pdf [accessed: 25 February 2011].

Johnson, P.T.J., Olden, J.D. and Vander Zanden, M.J. 2008. Dam invaders: impoundments facilitate biological invasions into freshwaters. *Frontiers in Ecology and the Environment*, 6 (7), 357-363.

Jongman, R.H.G. (ed.) 2005. Pantanal-Taquari: Tools for decision making in Integrated Water Management. ALTERRA Special Publication 2005/02 - Water for Food and Ecosystems Partners for Water 02.045. Wageningen: Alterra. Available at: http://www.cpap.embrapa.br/taquari/Taquari_final.pdf [accessed: 19 January 2011].

Junk, W.J. 2000. The Amazon and the Pantanal: a critical Comparison and lessons for the future, in *The Pantanal: Understanding and Preserving the World's largest Wetland*, edited by F.A. Swarts. St. Paul, Minnesota: Paragon House, 211–224.

Junk, W.J., Bayley, P.B. and Sparks, R.E. 1989. The flood pulse concept in river-floodplain systems. Proceedings of the International Large River Symposium (LARS) edited by D. Dodge. *Canadian Special Publication of Fisheries and Aquatic Sciences*, 106, 110–127.

Junk, W.J.; Nunes da Cunha, C. 2005. Pantanal: a large South American wetland at a crossroads. *Ecological Engineering*, 24, 391–401.

Kingsford, R.T. 2000. Ecological impacts of dams, water diversions and river management on floodplain wetlands in Australia. *Austral Ecology*, 25(2), 109–127.

Leuchtenberger, C. and Mourão, G. 2008. Social organization and territoriality of giant otters (Carnivora: Mustelidae) in a seasonally flooded savanna in Brazil. *Sociobiology*, 52(2), 257-270.

Lourival, R.F.F., Da Silva, C.J., Calheiros, D.F., Bezerra, M.A., et al. 1999. *Impactos da Hidrovia Paraná-Paraguai na Biodiversidade Pantaneira*: Proceedings of the 2nd Simpósio sobre Recursos Naturais e Sócioeconômicos do Pantanal. (Corumbá: EMBRAPA-CPAP), 518-534.

Lowe-McConnell, R.H. 1987. *Ecological studies in tropical fish communities*. Cambridge: Cambridge University Press.

Mamede, S.B. and Alho, C.J.R. 2006. Response of wild mammals to seasonal shrinking-and-expansion of habitats due to flooding regime of the Pantanal, *Brazilian Journal of Biology*, 66(4), 991-998.

Marques, D.K.S. and Resende, E.K. 2005. Distribuição do Tucunaré *Cichla cf. monoculus* (Osteichthyes, Cichlidae) no Pantanal. *Boletim de Pesquisa e Desenvolvimento*, 60. Corumbá, MS: Embrapa Pantanal. Available at: http://www. cpap.embrapa.br/publicacoes/online/BP60.pdf [accessed: 21 February 2011].

MEA. 2005. *Millennium Ecosystem Assessment. Ecosystems and Human Well-Being: Wetlands and Water - Synthesis*. Available at: http://www.maweb.org/ documents/document.358.aspx.pdf [accessed: 18 September 2009].

Miranda, K., Cunha, M.L.F., Dores, E.F.G.C. and Calheiros, D.F. 2008. Pesticide residues in river sediments from the Pantanal Wetland, Brazil *Journal of Environmental Science and Health, Part B: Pesticides, Food Contaminants, and Agricultural Wastes*, 43(8), 717 – 722.

Monitoring. 2010. *Monitoring alterations in vegetation cover and land use in the Upper Paraguay River Basin Brazilian Portion – Period of Analysis: 2002 to 2008*. Brasília: CI – Conservation International, ECOA - Ecologia e Ação, Fundação AVINA, Instituto SOS Pantanal, WWF- Brasil. Technical Report. Available at: http://www.wwf.org.br/informacoes/bliblioteca/publications_in_ english/?25202/Vegetation-and-Land-Use-Map-in-the-Upper-Paraguay-River [accessed: 22 April 2010].

Mourão, G.M., Coutinho, M.E., Mauro, R., Campos, Z., Tomás, W.M. and Magnusson, W. 2000. Aerial surveys of caiman, marsh deer and pampas deer in the Pantanal wetland of Brazil. *Biological Conservation*, 92, 175-183.

Mourão, G.M., Coutinho, M.E., Mauro, R.A., Tomás, W.M. and Magnusson, W. 2002a. Levantamentos aéreos de espécies introduzidas no Pantanal: porcos ferais (porco monteiro), gado bovino e búfalos. *Boletim de Pesquisa e Desenvolvimento*, 28. Corumbá, MS: Embrapa Pantanal. Available at: http:// www.cpap.embrapa.br/publicacoes/online/BP28.pdf [accessed: 21 February 2011].

Mourão, G.M., Oliveira, M.D., Calheiros, D.F., Padovani, C.R., Marques, E.J. and Uetanabaro, M. 2002b. O Pantanal Mato-Grossense, in *Os Sites e o Programa Brasileiro de Pesquisas Ecológicas de Longa Duração*, edited by U. Seedliger et al., Belo Horizonte: CNPq, 29-47.

Mourão, G., Tomas, W. and Campos, Z. 2010. How much can the number of jabiru stork (Ciconiidae) nests vary due to change of flood extension in a large Neotropical floodplain? *Zoologia*, 27(5), 751-756.

Neiff, J.J. 1997. El régimen de pulsos en ríos y grandes humedales de Sudamérica. En: Tópicos sobre grandes humedales sudamericanos, in *ORCYT-MAB*, edited by A.I. Malvarez and P. Kandus. Montevideo, Uruguay: UNESCO, 1-49. In Spanish.

Oliveira, M.D. and Calheiros, D.F. 2000. Flood pulse influence on phytoplankton communities of the south Pantanal floodplain, Brazil. *Hydrobiology*, 427, 102–112.

Oliveira, M.D., Calheiros, D.F., Jacobi, C.M. and Hamilton, S.K. 2011. Abiotic factors controlling the establishment and abundance of the invasive golden mussel *Limnoperna fortunei*. *Biological Invasions*, 13:717–729

Oliveira, M.D., Hamilton, S.K., Calheiros, D.F. and Jacobi, C.M. 2010a. Oxygen depletion events control the invasive golden mussel (*Limnoperna fortunei*) in a tropical floodplain. *Wetlands*, 30 (4), 705-716.

Oliveira, M.D., Hamilton, S.K., Calheiros, D.F., Jacobi, C.M. and Latini, R.O. 2010b. Modeling the potential distribution of the invasive golden mussel *Limnoperna fortunei* in the Upper Paraguay River system using limnological variables. *Brazilian Journal of Biology*, 70(3), 831-840.

Padovani, C.R. 2010. Dinâmica espaço-temporal das inundações do Pantanal. PhD Thesis. Piracicaba-SP: ESALQ/Universidade de São Paulo.

Padovani, C.R., Galdino, S. and Vieira, L.M. 2005. Dinâmica hidrológica e de sedimentação do Rio Taquari no Pantanal, in *Impactos Ambientais e Socioeconômicos na Bacia do Rio Taquari – Pantanal,* edited by S. Galdino. L.M. Vieira. and L.A. Pellegrin. Corumbá: Embrapa Pantanal, 153-162.

Padovani, C.R., Dias, F.A., Souza, G.F. and Calheiros, D.F. 2008. A remote sensing survey of vegetation disturbances in the Brazilian Pantanal wetland for the year 2004, in *Proceedings of 8th International Wetlands Conference - INTECOL*, Cuiabá.

Padovani, C.R., Cruz, M.L.L., Crispim, S.M.A. and Santos, S.A. 2008. Fire monitoring and spatial analysis for Brazilian Pantanal, in *Proceedings of 8th International Wetlands Conference - INTECOL*, Cuiabá.

Palmer, M.A. 2010. Beyond infrastructure. *Nature*, 467(7315), 534-535.

Palmer, M.A. and Richardson, D.C. 2009. Provisioning Services: A focus on fresh water, in *The Princeton Guide to Ecology*, edited by. S.A. Levin. Princeton: Princeton University Press, 625-633

Petts, G.E. 1990. Regulation of large rivers: problems and possibilities for environmentally-sound river development in South America. *Interciencia*, 15, 6, 388-395.

Poff, N.L., Allan, J.D., Bain, M.B., Karr, J.R., Prestegaard, K.L., Richter, B.D., Sparks, R.E. and Stromberg, J.C. 1997. The natural flow regime: a paradigm for river conservation and restoration. *BioScience*, 47, 769–784.

Poff, N.L., Richter, B.D., Arthington A.H. et al. 2010. The ecological limits of hydrologic alteration (ELOHA): a new framework for developing regional environmental flow standards. *Freshwater Biology*, 55(1), 147-170. Special Issue: Environmental Flows: Science and Management

Postel, S. and Richter, B. 2003. *Rivers for life: Managing water for people and nature*. Washington: Island Press.

Pott, A. and Pott, V.J. 2004. Features and conservation of the Brazilian Pantanal Wetland. *Wetlands Ecology and Management*, 12, 547–552.

Prada M. and Marinho-Filho, J. 2004. Effects of Fire on the Abundance of Xenarthrans in Mato Grosso, Brazil. *Austral Ecology*, 29, 568–573.

Ramsar. 1993. Ramsar Convention Secretariat. Resolution 5.4: The Record of Ramsar sites where changes in ecological character have occurred, are occurring, or are likely to occur ("Montreux Record"). Available at: http://www.ramsar.org/cda/en/ramsar-documents-resol-resolution-5-4-the/main/ramsar/1-31-107%5E23569_4000_0__ [accessed: 26 January 2011].

Ramsar. 2005. Ramsar Convention Secretariat. "Wetlands and water: supporting life, sustaining livelihoods" - 9th Meeting of the Conference of the Parties to the Convention on Wetlands (Ramsar, Iran, 1971). Kampala, Uganda, 8-15 November 2005.

Ramsar. 2007. Ramsar Convention Secretariat. *Participatory skills: Establishing and strengthening local communities' and indigenous people's participation in the management of wetlands*. Ramsar Handbooks for the Wise Use of Wetlands, 3rd Edition, vol. 5. Gland, Switzerland: Ramsar Convention Secretariat.

Ramsar. 2010. The annotated Ramsar List of Wetlands of International Importance – Brazil. Ramsar Convention Secretariat. Available at: http://www.ramsar.org/cda/en/ramsar-pubs-annolist-annotated-ramsar-16692/main/ramsar/1-30-168%5E16692_4000_0__ [accessed: 22 February 2011].

Rodrigues, C.A.G., Crispim, S.M.A. and Comastri Filho, J.A. 2002. Queima controlada no Pantanal. *Documentos, 35*. Corumbá, MS: Embrapa Pantanal. Available in: http://www.cpap.embrapa.br/publicacoes/online/DOC35.pdf [accessed: 25 February 2011].

Santos, S.A., Cunha; C.N., Tomás, W.M., Abreu, U.G.P. and Arieira, J. 2006. Plantas Invasoras no Pantanal: Como entender o problema e soluções de manejo por meio de diagnóstico participativo. *Boletim de Pesquisa e Desenvolvimento*, 66. Corumbá, MS: Embrapa Pantanal. Available at: http://www.cpap.embrapa.br/publicacoes/online/BP66.pdf [accessed: 15 February 2011].

Santucci Jr., V.J., Gephard, S.R., Pescitelli, S.M. 2005. Effects of Multiple Low-Head Dams on Fish, Macroinvertebrates, Habitat, and Water Quality in the Fox River, Illinois. *North American Journal of Fisheries Management*, 25(3), 975-992.

Silveira, L., Rodrigues, F. H.G., Jácomo, A.T.A. and Diniz Filho, J.A.F. 1999. Impact of wildfires on the megafauna of Emas National Park, central Brazil. *Oryx*, 33(2), 108-114.

Simberloff, D. 2010. Invasive species, in *Conservation Biology for all*, edited by N.S. Sodhi and P.R. Ehrlich. Oxford: Oxford University Press. Available at: http://www.dbs.nus.edu.sg/staff/details/sodhi/aConservation_Biology_for_All.pdf [accessed: 23 February 2011].

Soriano, B.M.A.; Galdino, S. 2005. Evolução da erosividade das chuvas na Bacia do Alto Taquari, in *Impactos Ambientais e Socioeconômicos na Bacia do Rio Taquari – Pantanal,* edited by S. Galdino. L.M. Vieira. and L.A. Pellegrin. Corumbá: Embrapa Pantanal, 119-124.

Souza, O.C. 1998. *Modern Geomorphic Processes along the Taquari River in the Pantanal: A Model for Development of a Humid Tropical Alluvial Fan.* Santa Barbara: University of California.

Sparks, R.E. 1995. Need for ecosystem management of large rivers and their floodplains. *BioScience*, 45(3), 168-182.

Suzuki, H.I., Agostinho, A.A., Bailly, D., Gimenes, M.F., Júlio-Júnior, H.F.and Gomes, L.C. 2009. Inter-annual variations in the abundance of young-of-the year of migratory fishes in the Upper Paraná River floodplain: relations with hydrographic attributes. *Brazilian Journal of Biology*, 69, (2suppl.), 649-660.

Thomaz, S.M., Carvalho, P., Padial, A.A. and Kobayashi, J.T. 2009. Temporal and spatial patterns of aquatic macrophyte diversity in the Upper Paraná River floodplain. *Brazilian Journal of Biology*, 69(2), 617-625.

Thorp, J.H., Flotemersch, J.E., Delong, M.D., Casper, A.F., Thoms, M.C., Ballantyne, F., Williams, B.S., O'neill, B.J. and Haase, C.S. 2010. Linking ecosystem services, rehabilitation, and river hydrogeomorphology. *Bioscience,* 60(1), 67-74.

Tomas, W., Salis, S.M., Silva, M.P. and Mourão, G. 2001. Marsh deer (*Blastocerus dichotomus*) distribution as a function of floods in the Pantanal wetland, Brazil. *Studies of Neotropical Fauna and Environment*, 36, 9-13.

UN. 2010. *United Nations International Decade for Action 'Water for Life' 2005-2015.* Available at: http://www.un.org/waterforlifedecade/background.html [accessed: 24 July 2009].

UNESCO. 2000a. *Biosphere Reserve Information - The Pantanal Biosphere Reserve.* MAB – Man and Biosphere Programme. Biosphere Reserves Directory. Available at: http://www.unesco.org/mabdb/br/brdir/directory/biores.asp?code=BRA+03&mode=all [accessed: 22 February 2011].

UNESCO. 2000b. *World Heritage by United Nations - Pantanal Conservation Complex on the World Heritage List.* Available at: http://whc.unesco.org/en/decisions/2428 [accessed: 22 February 2011].

UNU-IAS. 2004. *Inter-linkages Approach for Wetland Management: The Case of the Pantanal Wetland*: Proceedings of the Workshop Pantanal Wetland: Inter-linkages Approach for Wetland Management - best practices, awareness raising and capacity building (Porto Cercado, Mato Grosso, Brazil, 2003. United Nations University - Institute of Advanced Studies (UNU-IAS) and Federal University of Mato Grosso). Japan: UN-IAS. Report. Available at: http://www.ias.unu.edu/binaries2/Pantanal_Wetland_Report2004.pdf [accessed: 04 May 2008].

Wantzen, K.M., Nunes da Cunha, C., Junk, W.J., et al. 2008. Towards a sustainable management concept for ecosystem services of the Pantanal wetland. *Ecohydrology & Hydrobiology*, 8(2-4), 115-138.

WCD. 2000. *World Comission on Dams. Dams and Development: A New Framework for Decision-making.* London: Earthscan.

Welcomme, R.L. 1995. Relationships between fisheries and the integrity of river systems. *Regulated Rivers: Research & Management*, 11, 121–136.

Wotton, R.S. 1990. *The Biology of Particles in Aquatic Systems.* Boston: CRC Press.

WWF. 2001. Retrato da Navegação no Alto Rio Paraguai: Relatório da Expedição Técnica Realizada entre os dias 03 e 14 de novembro de 1999, no rio Paraguai, entre Cáceres (MT) e Porto Mutinho (MS). Brasília: WWF-Brasil. Available at: http://d3nehc6yl9qzo4.cloudfront.net/downloads/retrato_nav_wwf_brasil.pdf [accessed: 08 July 2011].

Zeilhofer, P. and Moura, R.M. 2009. Hydrological changes in the northern Pantanal caused by the Manso dam: Impact analysis and suggestions for mitigation. *Ecological Engineering*, 35. 105–117.

Chapter 4

Availability, Uses and Management of Water in the Brazilian Pantanal

Daniela M Figueiredo, Eliana F G C Dores, Adriano R Paz,
Christopher F Souza

Physical and Geographical context

The United Nations classifies its country members in six categories based on the average long-term *per capita* river water discharge, where Brazil is considered a country rich in water. However, in this classification, regional characteristics and different watersheds are not considered, since water is abundant in this country but unevenly distributed regarding its population (Rebouças 2004). Brazil stands out due to the large freshwater discharge of its rivers, whose water production, including the international Amazon Basin, represents 53 per cent of South America's and 12 per cent of total world surface water (Rebouças 1999).

In order to sustainably manage the water resources, Brazil adopted the watershed as the management unity. Twelve hydrographical regions (which comprise several watersheds or river basins) have been delimited in this country and are defined in the National Water Resources Plan (described in Figure 4.1; see more below). The Paraguay Hydrographic Region contains the Brazilian section of the Paraguay River Basin, which covers 1,095,000 km² of the South American territory with 34 per cent in Brazil and the remaining in Argentina, Bolivia and Paraguay (Figure 4.2).

The Upper Paraguay River Basin (UPRB) represents 4.28 per cent of the Brazilian territory and occupies part of Mato Grosso (48.2%) and Mato Grosso do Sul (51.8%) states. The UPRB is strategically important in the context of water resources management in Brazil, particularly for encompassing the Pantanal, one of the greatest wetland areas in the world, declared National Patrimony by the Brazilian Constitution of 1988, designated a Ramsar Site of International Importance by the Ramsar Convention on Wetlands in 1993 and a Biosphere Reserve by UNESCO in 2000 (Figueiredo and Salomão 2009). The UPRB in Brazil is highly heterogeneous from a geomorphological point of view, and is divided into several regions (Brasil 1997, Salomão et al. 2009), summarized in the three large features described below:

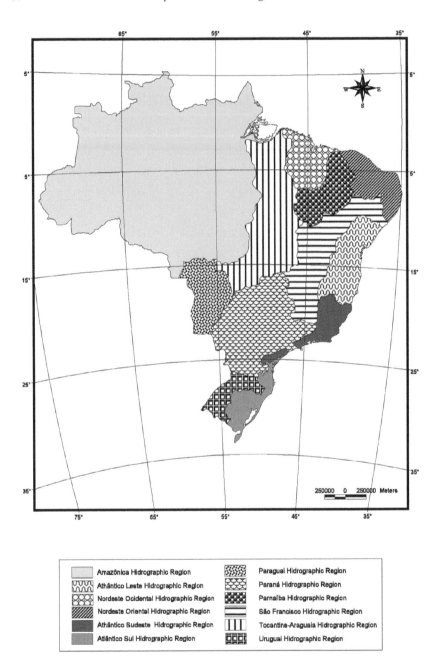

Figure 4.1   Map of Brazilian Hydrographical Regions

Source: ANA (2010)

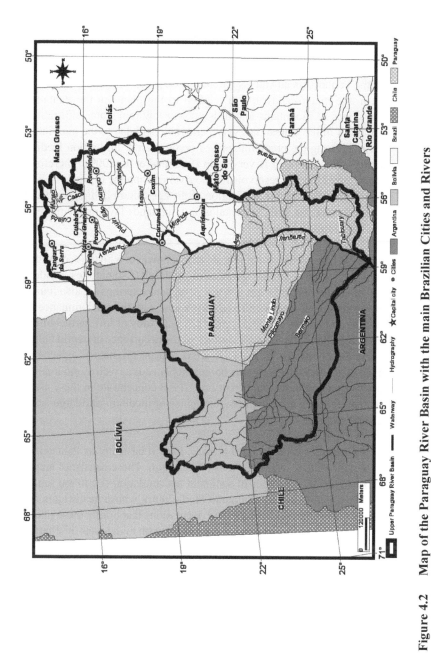

Figure 4.2   Map of the Paraguay River Basin with the main Brazilian Cities and Rivers

i) *Plateaus and Mountain Ridges*, which present mainly tabular reliefs varying from dissected to slightly dissected, with sharp and gradual scarps with interfluves, or a series of parallel mountain ridges with flat tops and inter-mountain and inter-plateau plains, with heights ranging from 300 to 850 m. They are natural water divisors with the neighbouring hydrographical regions, and concentrate several springs that commonly originate when the water table reaches the surface in hydromorphic or sandy soils. They present different physiognomies of savannah (*cerrado*) and transition vegetation with the Amazon Forest. This region is known as the "arc of the Pantanal headwaters" (Siebert 2009), due to the disposition of the elevations in the macro-landscape in relation to the plains in the Pantanal. In general, the slopes of mountains and hills present low rain water infiltration, while the main relief in the plateau has very permeable soil due to the high inter-grain porosity, suggesting the predominance of vertical water infiltration feeding continuously the water table that is very deep (Salomão et al. 2009). Therefore, the plateaus function as a water reservoir for the rivers during the dry season. The flat relief in the plateau favours mechanization of the agriculture in high scale and the different rock formations in the mountains and highlands favours the mineral exploitation of limestone and iron.

ii) *Depressions* are relatively flat and of low elevation, however with the occurrence of ramps sloping upwards, tabular formations and hills, sometimes residual elevations, with altitudes ranging from 100 to 450 m, where springs are also found. Usually, depressions are old and weathered formations with the predominance of *cerrado* vegetation in its different physiognomies, with soils of low permeability, which favours flooding and adequate conditions for the occurrence of humid fields. There is an absence of springs in the high areas and occurrence of ephemeral creeks (Salomão et al. 2009). Depressions border on plateaus as well as mountains. The higher population and industrial density of UPRB concentrate in these areas presenting diversified water uses with predominance of medium- and large-scale cattle raising and breeding.

iii) *Floodplains of the Pantanal of Mato Grosso*, known basically as Pantanal, a typical wetland, with altitudes varying from 80 to 150 m. These consist of ample flat areas, characterized by processes of sediment accumulation due to the runoff and overflow of the water streams (Salomão et al. 2009), whose headwaters are localized in the plateaus, mountain ridges and depressions that border these plains in an arc form. The rivers are commonly anastomosed, with high levels of siltation and connection with the flooding plains and marginal lagoons. Savannah vegetation is dominant, presenting different physiognomies depending on the region, rivers and soils as wells as on the rain season, wet or dry. The main economic activities carried out on the Pantanal are extensive cattle raising and breeding using natural or cultivated pasture, tourism and professional and amateur fishing. It is useful to emphasize that the watershed concept does not apply completely in the UPRB region, since the drainage areas of each river merge during the rainy season, rendering the delimitation of sub-basins difficult.

Multiple Water Uses and Impacts on the Rivers

The main economic activity in the UPRB is agribusiness, which implies the use of water for irrigation, animal consumption, processes and products in agroindustry and dilution of agro-industrial wastewaters. Animal consumption is the main water use related to cattle-raising, which is the predominant agribusiness activity in the UPRB, occupying large areas and frequently stressing fragile areas such as slopes, river margins and the surroundings of river springs, due to pasture plantation and/ or cattle trampling in search for water. There are no precise data on the amplitude and magnitude of the impacts caused by this activity on the water environment.

In the northern region of the UPRB, water is used for grain crop irrigation, especially soya bean and rice, in large and medium properties localized mainly in the high areas of the basin and plantation in small properties (family agriculture). There is no data on the water quantity impounded for irrigation in this area. However, in the southern region of the basin, c.a. $0.8 \text{ m}^3.\text{s}^{-1}$ is known to be used for irrigation from the Miranda and Correntes Rivers (Mato Grosso do Sul 2010). Considering the low amount used for irrigation, it is estimated that this water use is not significant in the UPRB due to the regular rain distribution and the well-defined seeding schedule especially in grain monoculture.

About 1.9 million people live in the UPRB, corresponding to 1 per cent of the Brazilian population with 84.7 per cent in urban areas (ANA 2010). This demonstrates that the use for domestic consumption is a small quantity compared to the availability. It is estimated that 77 per cent of the population of this area receives treated water, below the national average of 81.5 per cent (ANA 2010). This water use represents an important environmental service of some rivers of this basin, such as the Cuiabá, Paraguay and Vermelho, in whose margins the majority of this population lives.

This population concentration, together with inadequate sanitation services, emphasises the importance of water for dilution and transport of domestic effluent, urban runoff and solid residues from adjacent areas (Rondon-Lima 2009, Figueiredo 2009). In other urban areas, this water use is also common, as mentioned by SEPLAN/SEMA (2011), which pointed out 41 sites where this use occurs, mostly in the medium and higher basin. This water use is causing increases in coliform indices and impact on important environmental uses of these rivers, such as human consumption by riverside communities situated downstream of the cities as well as reduction of recreation and swimming areas (Figueiredo 2009), and increases in the cost of water treatment for public supply, leading to conflicts. The percentage of population in this hydrographical region with sewage collection system is 20 per cent, below the Brazilian average of 47.2 per cent. Regarding sewage treatment, this region presents a rate of 17.2 per cent, near the Brazilian average (i.e. 17.8%) (ANA 2010).

The inspection by environmental agencies of industrial water uses in processing end-products or for effluent dilution is more efficient, and at present, almost all industries have a water treatment system. Since the majority of the industries in this basin generate liquid effluent of organic nature, the treatment systems are biological using stabilization lagoons or activated sludge systems. The dilution and transport behaviour of this kind of effluent in the rivers is similar to domestic effluents and depends on the auto-depuration conditions of the river. It is estimated that 20 per cent of the effluents disposed of in the Cuiabá River is of industrial origin (Figueiredo 2009). The major industries in the UPRB are slaughterhouses, meat-processing, dairy, soybean processing and sugar and alcohol production. In the main, these industries are not concentrated in a specific area of the basin, with point discharge of effluents causing small changes in the natural conditions of medium and large rivers and significant alterations in small creeks, mainly those situated in the vicinities of industrial districts.

Brazilian legislation (MMA 2010) establishes the physical and chemical conditions and the concentration of contaminants allowed in the effluent that will be disposed of directly or indirectly to streams and states that the water use for effluent dilution cannot alter the class of the river. Regarding the non-consumptive uses of water, electrical energy generation is increasing significantly in Brazil in recent years. Hydroelectric power plants in this country supply more than 90 per cent of the demand.

In the UPRB, there are 23 hydroelectric plants in operation, whose reservoirs vary from the canal type to large flooded areas with long residence time (Figueiredo 2007, Mato Grosso do Sul 2010). There are another 115 projects expected to be installed in the UPRB (ANEEL 2008), with 75 per cent in Mato Grosso State (north of UPRB) where the majority of water is generated (Calheiros et al. 2009) and where larger differences in level are localised, favouring the energetic exploitation of the rivers. The majority of these plants (73%) consist of Small Hydroelectric Power Plants (PCHs), that is, schemes that generate less than 30 MW. However, many are localised or planned to be installed along the same river, resulting in a significant joint impact (Calheiros et al. 2009). The damming system in "cascade" presents additional complexity in relation to the individual systems, due to its multivariate and non-linear characteristics (Tundisi 2005).

Several PCHs, even when of the canal type, without a reservoir, cause alterations in the nutrient and suspended matter discharge and, therefore, in the nutrient cycling (Calheiros et al. 2009), due to the retention of part of this material with the river damming. In addition, these PCHs modify the hydrological dynamics of the rivers, a fundamental condition for preserving the Pantanal function that depends on the annual water level variation. In this sense it is vital that the Brazilian electricity sector and the environmental and water resources managers of both states (Mato Grosso and Mato Grosso do Sul) and of the Federal government be aware of the fragility of the Pantanal ecosystem regarding hydrological alterations. Therefore, it is important to develop an Integrated Environmental Evaluation of the impacts of utilization of the electrical potential of the UPRB (Calheiros et al. 2009). These authors also recommend that some sub-basins should be kept free of any damming and that the operation regimen of the reservoirs should not interfere in the dry and flooding pulses, in order to guarantee that the hydro-ecological functions of the Pantanal be maintained.

Regarding the modification of the sediment dynamics in the UPRB rivers, the dams of hydroelectric plants reduce sediment entry in the Pantanal. Sediment transport in the Pantanal has increased greatly in recent years due to the increase in erosive processes in fragile areas caused by the expansion of agriculture and cattle breeding (Serigatto 2006, Siebert 2009). Conversely, these power plants interfere in the natural dynamics of sediment, which is vital to flooded areas of the Pantanal, which are natural sites of deposition of the particulate generated in higher areas of the watershed. On the other hand, some hydroelectric reservoirs in the UPRB are being dredged, since the excessive retention of sediments is causing damage to the energy generation, reducing the usable reservoir volume, indicating a conflict between the agricultural and cattle breeding activities and the energy generation enterprises (personal information). There are records that in the Taquari river sub-basin some PCHs possess a device installed in the bottom of the dam structure, which is opened periodically, liberating water and sediment downstream in large quantities in short periods of time. This information is not well publicized and recent studies on sediment production, transport and deposition in the URPB, that include the existing hydroelectric plants and that estimate future scenarios under project are still lacking.

In 1997, an ample study developed by the Brazilian government (Brasil 1997) indicated the UPRB areas more vulnerable to the erosive action of rain and highly fragile, which are the ones where the main rivers and/or their tributaries are being impounded for energy generation. As examples, we can cite the Jauru river where five hydroelectric dams of medium and small size are in operation; the Correntes river, with two plants in operation and two at the project stage; the Taquari River with two in planning stage and the Manso River, an important tributary of the Cuiabá River where the largest plant of the UPRB considering the reservoir area (450 km^2) – the APM Manso is in operation, with another two hydroschemes at the project stage.

Regarding the Manso reservoir, runoff alteration in relation to historical flow records was observed, with an increase in the duration of the dry period and is believed to have negative impacts on Pantanal floodplain ecology, such as alterations in biogeochemical fluxes, biomass production, and vertical and horizontal movements of aquatic animals and their reproduction, feeding and migration habits (Da Silva and Girard 2004). Moreover, Shirashi (2003) states that besides increasing the water volume and regularizing the flow rate in the dry season, the Manso reservoir is an efficient hydraulic construction for flood control. In fact, the main purpose of this plant was to avoid flooding of downstream urban areas, mainly of the cities of Cuiabá and Várzea Grande with 800,000 inhabitants, where widespread flooding caused high social and economic damage before the river was dammed.

The use of the UPRB rivers for commercial navigation occurs basically in the Paraguay River, through the Paraguay-Paraná Hidrovia (i.e. waterway), which connects this basin to the Hydrographical Region of Paraná and serves as an important route for the economic integration of Brazil, Bolivia, Paraguay, Argentina and Uruguay. It totals 3,442 km, with 1,280 km in the Brazilian territory, specifically in the UPRB, connecting the city of Cáceres in the Northwest of the Pantanal to the confluence of Paraguay and Apa Rivers in the Brazil-Paraguay frontier in the Southwest of Pantanal (AHIPAR 2010, Mato Grosso do Sul 2010). The boats in this region are of low draught, about 6 feet, for 70 per cent of the year, since, although wide and long, this river has a low average depth causing the reduction or even interruption of navigation during the highly dry months, when the level of the Paraguay river is too low (August to November). From North to South, the direction of river flow, grains, cattle, sugar and minerals are transported, while from South to North the main products transported are wheat flour and malt. In the last ten years, on average, 8,000 tons of commodities circulated per year on this waterway between Corumbá and Asunción, where the capacity is for 25,000 tons/year (AHIPAR 2010).

The development of the Paraguay-Paraná Hidrovia is by no means a new idea and the river system has been navigated for many decades. More than 100 years ago, politicians and entrepreneurs already dreamed about a channelized waterway into the heart of South America. In the late-1980s, the La Plata Basin countries (Argentina, Bolivia, Brazil, Paraguay, and Uruguay) decided to initiate this huge project as a step towards the integration of the Basin countries (Bucher et al. 1993, cited by Gottgens 2000). They created an Intergovernmental Committee on the Hidrovia (CIH) to promote and oversee the development of this commercial waterway (Gottgens 2000). The CIH Project, however, was intensively criticized worldwide due to the major impacts expected on the riverbed of the Paraguay River due to the planned expansion of the waterways. The related engineering interventions are likely to cause irreversible impacts, mainly related to changes in the hydrological regimen of the Pantanal (Ponce 1995), as occurred in the Missouri-Mississippi Rivers system in the United States. This international reaction caused the Brazilian government to abandon most of the proposed interventions. Nevertheless, in practice, the hidrovia was gradually implemented, only with fewer interventions (when compared with the initial proposal), which are currently restricted to the control of "*camalotes*" or "*balseiros*" which are floating islands of aquatic plants that can obstruct the navigation canal, and dredging of places of high sediment deposition (AHIPAR 2010).

Wantzen et al. (1999) state that the continuous dredging of several sections of a river which are sediment deposition areas generates problems related to changes in the natural riverbed and destination of the dredged material. Moreover, the reduced width and low depth of the river allied to its high sinuosity elevate the re-suspension rate of bottom sediment and increase the processes of destabilization of the river channel and destruction of marginal vegetation caused by wave movement and by the boats during manoeuvres in the sinuous curves of the Paraguay river as it goes through the flood plain.

The AHIPAR (Paraguay-Paraná Hidrovia Administrator) mentions that an environmental monitoring of the river, of water quality and riverside vegetation, is continuously carried out during the dredging period and that part of the dredged material is deposited in the river channel in previously evaluated sites (AHIPAR 2010). The increase of urban and industrial areas in the Pantanal section of the Hidrovia, as happened in the Missouri-Mississippi rivers, is unlikely to occur considering the restrictions in the environmental law on occupation of this region, as well as the inhospitable conditions that prevail in the floodplain.

The UPRB rivers also offer touristic attractions as an environmental service, both in the upper areas of the watersheds and in the Pantanal floodplain itself, allowing its use for bathing, amateur aquatic sports and recreation. In the plateaus, the unevenness of the terrain brings on the formation of waterfalls, with good water quality, particularly those in conservation areas. However, the most attractive region for tourists, especially from other countries, is the Pantanal floodplain, where the diversity of terrestrial landscapes and aquatic ecosystems encourage rich fauna and flora form and create an environment unique in the world. The most common touristic sectors related to environmental services in the sub-basin of the Cuiabá River are ecotourism, rural tourism, fishing and adventure tourism (Garcia 2009) which also occur throughout the UPRB.

Flow Regime

In this section, the Brazilian Pantanal streamflow regime is placed within its hydrological and topographical context.[1] An analysis of flow regime seasonal and inter-annual predictability and variability is presented, followed by a local comparison of long-term aspects of the flow regime. In order to present those analyses, Brazilian Pantanal hydrography and flow data availability are described, after which data selection criteria are applied.

Streams and Flow Data

The Brazilian Pantanal is drained by the Paraguay River in a North-South direction, from Cáceres to near Porto Murtinho (c. 1,300 km). Along its left margins, the Paraguay River takes major contributions from the Cuiabá, São Lourenço, Piquiri, Taquari, Negro, Miranda and Aquidauana rivers (Figure 4.3). At its right margins, the most important contribution comes from the Jauru River discharges. The Pantanal's drainage network is also formed by several secondary channels

---

1   Source of streamflow gauging station data: Brazilian National Water Agency (ANA), available at http://hidroweb.ana.gov.br

cutting throughout the floodplains and connecting the largest rivers during floods, and in some instances during dry periods too. The drainage area of the Paraguay River increases from 32,000 km² at Cáceres to 582,000 km² at Porto Murtinho, where 27 per cent of the Porto Murtinho contributing area (158,000 km²) is in the Highlands, in the north and east portions of the UPRB. The west portion of the UPRB is formed by the Chaco and Highland areas, where flow contributions are insignificant due to low rainfall rates and to an undefined and endorheic drainage network (i.e. inland basins that do not drain to an ocean) (Tucci et al. 2005), resulting in insignificant contributions to the Paraguay River.

As it is a great conservation area, it hasn't historically received the same concerns as other regions in the country related to hydrological data acquisition. There exist only 28 streamflow gauging stations along the Pantanal's drainage network (Figure 4.3), which corresponds to a 1:5,000 km² density. Such density diminishes to almost 1:7,400 km², if the nine stations (indicated by the letters 'a' to 'i' in Figure 4.3) are discounted. Those figures are too far from the World Meteorological Organization's recommended ones for regions characterized by difficult data acquisition (c. 1,875 km² for interior plains). That situation becomes even worse considering that five (P. Bocaína, S. Gonçalo, P. Rolom, F. Rio Negro and T. Fogo) of those 28 stations are no longer operational. Regarding historical data availability, there are records for the 1939–2007 period, but only twenty – out of the 28 stations – have more than 30 per cent of the years without missing data and just three stations (Cáceres, Cuiabá and P. Murtinho) have more than 50 per cent without missing data. Most of those 28 stations (61% of total) have between 15 and 34 years without missing data.

Considering Kennard et al.'s (2010) findings on the uncertainty in hydrologic metric estimation due to length and period overlap of time series, 15 water-years of complete daily time series was adopted as a criterion for discarding gauging stations. Water-years were set as September to September periods, as most stations presented dry periods in that month. Data continuity demands were relaxed once there is rather an understanding that non-continuous information would be compensating intra-annual dependence, which is typical of the Pantanal region. As a result, a subset of 22 stations was selected, discarding four non-operational (P. Bocaína, S. Gonçalo, F.R. Negro and T. Fogo) and two still-operational stations (I. Camargo and P. Taiamã).

Flood Pulse Routing and Flow Regime Seasonality

When the Paraguay River and its major tributaries reach the Pantanal's Lowlands, they find subtle slopes and low margins, due to the particularly flat terrain – the slope in the North-South direction along Pantanal is estimated as 1 cm.km⁻¹ (Tucci, Clarke 1998). For instance, the slope along the Aquidauana River falls from more than 40 cm.km⁻¹ in the Highlands to less than 15 cm.km⁻¹ in the Pantanal, and there is a reduction of 46 per cent in the main channel cross-section area between the gauging stations at Aquidauana and Porto Ciríaco (Souza et al. in

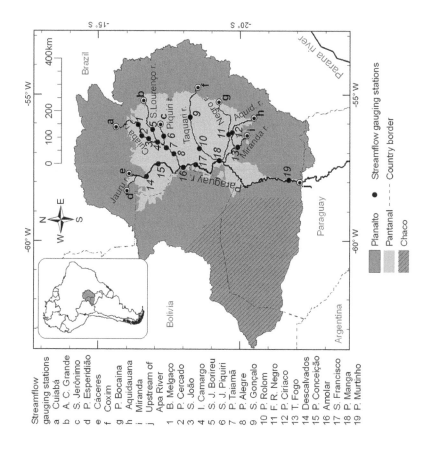

Figure 4.3   Upper Paraguay River Basin Location, Drainage Network and Gauging Stations

press, DNOS 1974). As a result of the Pantanal's very low and flat relief, rivers flowing from the Highlands seasonally inundate the floodplains and flood waters create an intricate drainage system, including vast lakes, divergent and endorheic drainage networks (Tucci et al. 1999, Bordas 1996, Brasil 1997). Indeed, there are several large alluvial fans, the largest one being the Taquari megafan, which is the most remarkable geomorphic feature of the Pantanal (Assine 2005). Another remarkable characteristic of the Pantanal is that annual rainfall is less than the potential evaporation (Bordas 1996, Tucci et al. 1999).

As a consequence of the low capacity of the river channels along the Pantanal for conducting waters flowing from the Highlands, there is a remarkable difference between flow regimes observed at consecutive stations just upstream and downstream on the Pantanal's Highland-Lowland borders. For the rivers Cuiabá, São Lourenço and Aquidauana. Along S. Lourenço River, the reduction goes from 1,000 $m^3.s^{-1}$ peak flow at A.C. Grande to less than 400 $m^3.s^{-1}$ at S.J. Borireu (250 km downstream). In the latter, peak flows reach up to 700 $m^3.s^{-1}$ at the Aquidauana station, while a marked maximum value of 150 $m^3.s^{-1}$ is observed at P. Ciríaco (Figure 4.4), located 230 km downstream.

Estimates of inundated areas provided by both satellite-based and modelling studies indicate a marked seasonal flood pulse, ranging from 10,000 km^2 in the dry period to more than 100,000 km^2 during floods (Hamilton et al. 1996, Paz et al. in press). Some areas were identified as permanently inundated, most of them located along the north and central portions of the Paraguay River, in the reach between the Descalvados and P. Manga stations, along the floodplains of the lower reaches of the Cuiabá River and along the margins of the Taquari River (Paz et al. in press).

The clearest effect of Pantanal flooding over the hydrologic regime of the Paraguay River is the delay in peak flows: from Cáceres to P. Murtinho differences in the period of peak flow occurrence is about three to four months (Collischonn et al. 1998), as illustrated in Figure 4.5. Such a delay can also be observed to a lesser extent in the mean and minimum daily discharges. Distinct discharge magnitudes, shown in Figure 4.5, result from tributaries' contributions and lateral water exchanges between the main channel and its floodplains. In the Cáceres-P. Conceição reach, the Paraguay River loses a huge volume of water to its floodplains. In a downstream reach, from P. Conceição to Amolar, there is a significant lateral gain of water from its upstream floodplains, which is also drained by Cuiabá River floodplains, besides the Cuiabá river main channel discharges. These floodplain contributions are even larger in the reach between the Amolar and S. Francisco stations, which also receive water flowing along the Taquari fan (Paz et al. in press). In the S. Francisco-P. Manga reach, channel–floodplain flow exchanges are almost negligible. At UPRB downstream reaches, downstream of P. Manga station, the Paraguay River channel receives contributions from the Negro, Aquidauana and Miranda rivers floodplains.

In general, flow seasonality is typically what was just presented for those three water-years, where higher flow events occur in the southern hemisphere fall season, except for downstream reaches where it is delayed for about three to

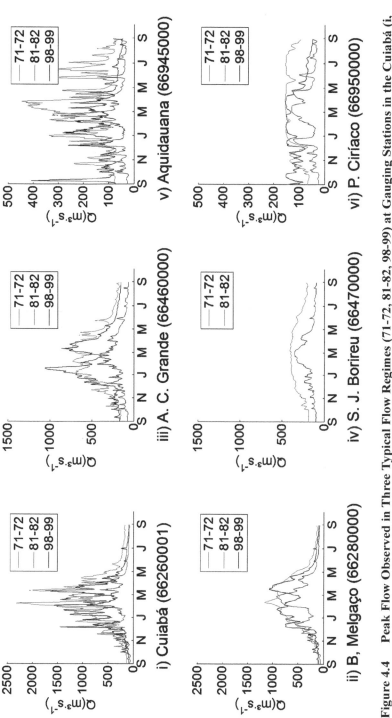

Figure 4.4 Peak Flow Observed in Three Typical Flow Regimes (71-72, 81-82, 98-99) at Gauging Stations in the Cuiabá (i, ii), São Lourenço (iii, iv) and Aquidauana (v, vi) Rivers

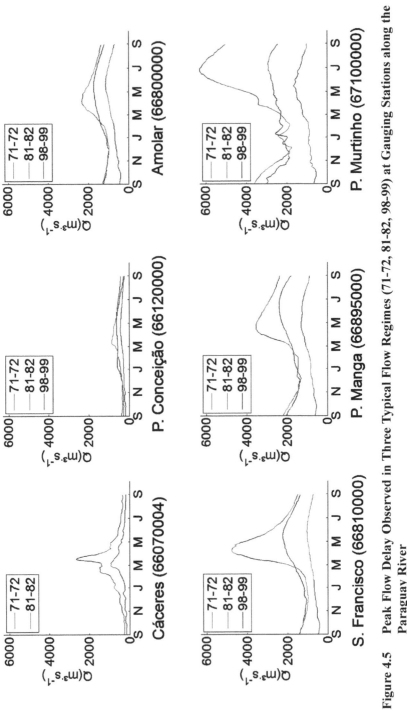

Figure 4.5 Peak Flow Delay Observed in Three Typical Flow Regimes (71-72, 81-82, 98-99) at Gauging Stations along the Paraguay River

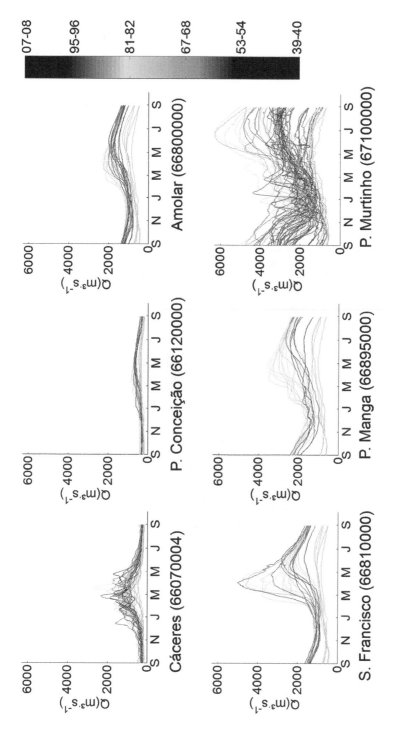

Figure 4.6   Flow Regime Hydrographs along the Pantanal Rivers

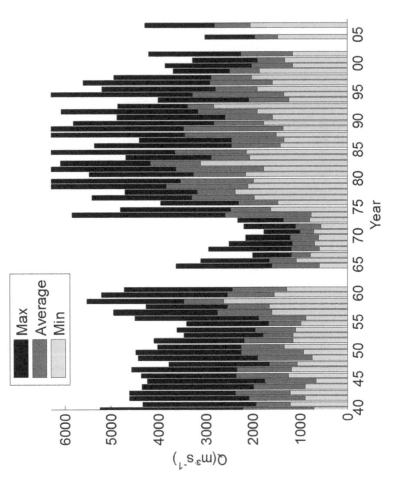

Figure 4.7   Streamflow Inter-annual Variability in the Paraguay River at the P. Murtinho Station

five months. Figure 4.6 presents flow regimes for every water-year with complete records. Some greater high flow magnitude and period of occurrence variability is observed in lower reaches, probably as a consequence of different flood pulses coming from different tributaries.

As may be observed in Figure 4.7 (sensu Collischonn et al. 1998), the period from the late 1960s and early 1970s was very dry with respect to the rest of the time series. In Figure 4.7, we present annual stats for P. Murtinho stations, where such a dry period is clearly evident. We also may point out that highest flows were observed in the 1980s and 1990s.

Long-term Flow Regime Aspects

Six long-term hydrological metrics were evaluated for characterizing streamflow regimes (Table 4.1): mean, minimum (min) and maximum (max) flows; Q90 (90% duration flow); Bs, base flow contributions, given by the ratio of Q90 and median values. ARI10 (the annual maximum flow with 10 years of average recurrence interval) was estimated by Generalized Extreme Value (GEV) distribution fitting. Every metric is presented in mm day^{-1} to ease comparison between stations, except for Bs index, which is dimensionless.

Along the Paraguay River from Cáceres to P. Murtinho, the general trend of mean, max and min metrics is to strongly decrease: from 7.05 to 0.93 mm.day^{-1} (max), 1.44 to 0.36 mm.day^{-1} (mean) and from 0.35 to 0.08 mm.day^{-1} (min), as a result of the dampening and delaying effect of the Pantanal's topography. However, as the flood pulse reaches the Amolar station, significant discharges are drained by the floodplains, and there is an increase in max and mean metrics with respect to immediate upstream and downstream reaches. Moreover, as the floodplain contribution occurs predominantly during floods, such increase is more intense in max metrics, which rises from 0.81 mm.day^{-1} at the Amolar station to 1.73 mm.day^{-1} at the S. Francisco station. Similar variation was also obtained for the ARI10 metric: a strong reduction from Cáceres station (5.22 mm.day^{-1}) to P. Murtinho station (0.71 mm.day^{-1}), but augmenting at S. Francisco station (1.18 mm.day^{-1}) and P. Manga station (0.87 mm.day^{-1}) due to discharges along floodplains. As expected, the Q90 flow index follows the variation described for the min index, with a strong reduction from upstream (0.59 mm.day^{-1}) to downstream (0.18 mm.day^{-1}) along the Paraguay River. A distinct pattern was obtained for the Bs index, which indicated that the baseflow contribution to Paraguay River flow increases from the most upstream station (0.53 at Cáceres station) to the Amolar station (0.67) and then decreases downstream to the P. Murtinho station (0.51).

At Paraguay River tributaries, main channel spills to the floodplains during floods explain the strong reduction in max and ARI10 metrics from upstream to downstream. Slighter reduction is observed in mean and Q90 metrics, while the min presents distinct patterns among these rivers. For instance, along the Cuiabá River, the max decreases from 12.19 mm.day^{-1} at Cuiabá station to 4.56 mm.day^{-1} at B. Melgaço station, 1.69 mm.day^{-1} at P. Cercado station and 1.08 mm.day^{-1} at

Table 4.1 Brazilian Pantanal's Long-term Hydrologic Metrics

Gauging station			Drain. Area (km²)	Stats (mm.day⁻¹)**					
Code*	Name	River		Mean	Min	Max	Q90	Bs	ARI10
a) 66260001	Cuiabá	Cuiabá	24,668	1.36	0.15	12.19	0.34	0.42	7.93
b) 66460000	A. C. Grande	S. Lourenço	23,327	1.23	0.24	4.34	0.54	0.56	3.97
c) 66600000	S. Jerônimo	Piquiri	9,215	2.38	0.73	6.59	1.13	0.57	6.29
d) 66072000	P. Espiridião	Jauru	6,221	1.42	0.72	4.95	0.87	0.71	4.63
e) 66070004	Cáceres	Paraguay	32,574	1.44	0.35	7.05	0.59	0.53	5.22
f) 66870000	Coxim	Taquari	28,688	1.02	0.27	3.08	0.45	0.45	2.83
g) 66886000	P. Bocaina	Negro	2,807	***	***	***	***	***	***
h) 66945000	Aquidauana	Aquidauana	15,350	0.66	0.09	3.89	0.23	0.47	3.57
i) 66910000	Miranda	Miranda	15,502	0.49	0.06	3.29	0.13	0.45	2.52
1) 66280000	B. Melgaço	Cuiabá	27,050	1.24	0.19	4.56	0.33	0.39	4.01
2) 66340000	P. Cercado	Cuiabá	38,720	0.73	0.17	1.69	0.28	0.44	1.59
3) 66360000	S. João	Cuiabá	39,908	0.57	0.15	1.08	0.26	0.44	1.00
4) 66370000	I. Camargo	Cuiabá	40,426	***	***	***	***	***	***
5) 66470000	S. J. Borireu	S. Lourenço	24,989	0.92	0.33	1.47	0.48	0.52	1.44
6) 66650000	S. J. Piquiri	Piquiri	28,871	0.91	0.23	3.16	0.31	0.48	2.86
7) 66710000	P. Taiamã	Cuiabá	96,492	***	***	***	***	***	***
8) 66750000	P. Alegre	Cuiabá	104,408	0.58	0.19	1.11	0.34	0.60	1.03
9) 66880000	S. Gonçalo	Taquari	28,688	***	***	***	***	***	***
10) 66885000	P. Rolom	Taquari	28,688	0.72	0.29	1.10	0.42	0.54	0.92
11) 66890000	F. R. Negro	Negro	14,770	***	***	***	***	***	***

12) 66950000	P. Cirıaco	Aquidauana	19,204	0.45	0.10	0.72	0.22	0.48	0.71
13) 66920000	T. Fogo	Miranda	19,020	***	***	***	***	***	***
14) 66090000	Descalvados	Paraguay	48,360	1.04	0.28	2.05	0.54	0.55	1.90
15) 66120000	P. Conceição	Paraguay	65,221	0.57	0.22	1.14	0.35	0.66	1.05
16) 66800000	Amolar	Paraguay	246,720	0.44	0.14	0.81	0.28	0.67	0.75
17) 66810000	S. Francisco	Paraguay	251,311	0.52	0.14	1.73	0.27	0.62	1.18
18) 66895000	P. Manga	Paraguay	331,114	0.48	0.15	0.93	0.25	0.54	0.87
19) 67100000	P. Murtinho	Paraguay	581,667	0.36	0.08	0.93	0.18	0.51	0.71

* Reference to Figure 1 and the eight-digit code of the Brazilian national gauging network

** Bs = base flow contribution, given by the ratio of Q90 and median values (dimensionless); ARI10 = annual maximum flow with 10 years of recurrence, estimated via GEV distribution fitting

*** Insufficient data available for calculating statistics.

S. João station, while ARI10 metric values are, respectively, 7.93, 4.01, 1.59 and 1.00 mm.day^{-1}.

For every station placed at the Highland's outlet sections, max values ranged between 3 and 5 mm.day^{-1}, except for Cuiabá and S. Jerônimo stations (6.59 mm.day^{-1}). The latter also presented the higher values for mean (2.38 mm.day^{-1}), min (0.73 mm.day^{-1}) and Q90 (1.13 mm.day^{-1}) metrics, but all of them strongly reduced at downstream sections (S. J. Piquiri station; mean = 0.91 mm.day^{-1}; min = 0.23 mm.day^{-1}; Q90 = 0.31 mm.day^{-1}). Regarding the baseflow contribution metric (Bs) for Cuiabá, Taquari and Aquidauana rivers, it presented a slight increase from upstream to downstream, while for São Lourenço and Piquiri rivers the trend was to decrease. Among all Pantanal's Lowland tributary stations, such metric ranged between 0.39 (Cuiabá River at B. Melgaço stations) and 0.60 (Cuiabá River at P. Alegre stations), while most values ranged between 0.40 and 0.50. That indicates that baseflow contributions at Pantanal's Lowland tributaries were about 20–40 per cent lower than at the Paraguay River.

Limnological Patterns and Water Quality

In this section, the spatio-temporal limnologic variations of the main rivers of UPRB are discussed, considering specifically the physical and chemical aspects and the comparison with the quality standards adequate for the multiple water uses established in the Brazilian legislation (Table 4.2). In the high and middle course of the UPRB rivers, pH, alkalinity, electrical conductivity and the main ions are mainly conditioned by the geo-pedological characteristics of their drainage area. Rivers with high values for these variables drain areas of the mountain ridges, where calcareous formations occur. These values tend to decrease in the rainy season due to dilution with the increase in water volume (FEMA 1997, Figueiredo 2009, Figueiredo 2007, Da Silva et al. 2009). On the other hand, rivers that flow from the plateaus or mountain ridges where there are no calcareous formations usually present neutral to slightly acid pH (<6.0) and low alkalinity, hardness and ions concentrations, since in these areas soils which are poor in electrolytes and of acid character predominate (Moreira and Vasconcelos 2007). However, at the beginning of the rainy season, when there is an increase in runoff, these variables tend to increase in relation to the dry season (Paula 1997, Loverde-Oliveira et al. 1999, Figueiredo 2007, 2009).

In the Pantanal, specifically in the Cuiabá River, there is a reduction in ionic concentrations and therefore also in electrical conductivity, alkalinity and hardness in relation to the high and middle course of this basin, mainly in the rainy season. This is probably due to the reduced water velocity and expansion of the river to adjacent areas, favouring sedimentation of part of the ions and/or assimilation by algae and aquatic macrophytes (Da Silva et al. 2009), in addition to the dilution effect. The electrical conductivity along Cuiabá River varies from 24 to 152 µS.cm^{-1} and from 60 to 258 µS.cm^{-1} while the alkalinity range from 10 to 97 mg $CaCO_3.l^{-1}$ and from 16 to 117 mg $CaCO_3.l^{-1}$, in the rainy and dry seasons, respectively (SEMA 2005). In the Paraguay, from its headwaters to the beginning of Pantanal, the conductivity varies from 29 to 51 $\mu S.cm^{-1}$ in the rainy and from 17 and 95 $\mu S.cm^{-1}$ in the dry season (FEMA 1997) and in the Pantanal range from 47 to 60 $\mu S.cm^{-1}$ (Calheiros and Hamilton 1998). In the Vermelho, São Lourenço, Jauru, Piquiri, Negro, Taquari and Correntes Rivers, conductivity is relatively low, varying from 3 to 60 $\mu S.cm^{-1}$ (FEMA 1997, Hamilton et al. 1997, no published data), with trends of reaching higher values in the dry season due to concentration with the reduced water volume and when the first rain occurs.

In the high and middle courses of all UPRB rivers, throughout the year, dissolved oxygen (DO) concentrations are higher than the limit established in Brazilian environmental legislation of 5.0 $mg.l^{-1}$ (Table 4.2), with the exception of some creeks that drain urban areas and are used for transport and dilution of domestic effluents. In the Pantanal plain, on the other side, there is a trend of lower average DO concentrations, including values below 5.0 $mg.l^{-1}$, mainly during the rainy season, due to the reduction in water velocity as a consequence of the low declivity and to the increase in oxygen demand by the runoff from adjacent flooded areas (Figueiredo 1997, FEMA 1997, Hamilton et al. 1997, Da Silva 2009).

In the Paraguay River, Hamilton et al. (1997) observed that the oxygen depletion associated with chemical changes is more noticeable when the river water begins its contact with the flooded areas. These authors also observed complete anoxia in this river near Corumbá during up to six weeks, causing massive death of fish. This is a natural phenomenon but one that usually occurs in the lagoons marginal to the main rivers in the floodplain. Calheiros and Ferreira (1997) state that this phenomenon comes together with water quality deterioration, with increases in CO_2 and CH_4 concentrations and is related to the rapid decomposition of organic matter at the beginning of the flood process. Its magnitude is dependent on the flooding pulse characteristics, i.e., on the conditions of the earlier dry season and on the volume and the following flooding velocity. Regarding the variables that interfere in the optical aspects of the water, the majority of the UPRB Rivers present a typically seasonal behaviour with higher turbidity, colour and suspended solid concentrations in the rainy season due to the runoff.

In the higher course of most rivers, the seasonal variation is not observed throughout the rainy season, but only during the rain event, as recorded in the Jauru River, a tributary of the Paraguay River (Figueiredo 2007), and in the Coxipó River, tributary of Cuiabá River (Paula 1997). In a few hours after rain events, turbidity, colour and suspended solid concentrations return to conditions similar to those of the dry period. It is likely that the same behaviour would be observed in the other UPRB rivers, but there are few studies in the headwaters of this basin.

In the high and middle courses of the Manso, São Lourenço and Taquari Rivers, turbidity and colour during the rainy season, especially in its beginning (October–November), present values above 100 NTU and 75 uH, respectively, which are the limits established in the Brazilian environmental legislation (Figueiredo 2007, FEMA 1997, Brasil 1997; see Table 4.2). These sub-basins are the ones in the

UPRB with higher sediment production, due to unstable riverbeds and sandy soils in the drainage area, added to inadequate occupation of permanent preservation areas at the riverside and headwaters region by crops of grains, cotton, pasture and by gold and diamond mining that occurred more intensively in the 1970s and 1980s. These natural and anthropic conditions promote the acceleration of the natural erosion processes in these sub-basins and siltation of the floodplain, as well as the increase in turbidity and colour intensity.

When compared to upstream values, an increase in suspended solids concentrations and in turbidity and colour values in the rainy season has been observed in portions of rivers and creeks which drain urban areas due to the input of domestic effluent that is discharged for dilution and transport in this environment (FEMA 1997, Figueiredo 2009, Paula 1997, Rondon-Lima, Lima 2009). Concerning the nitrogen compounds, most UPRB rivers present oligo-mesotrophic conditions, with exceptions in some periods of the year and in some areas of the rivers. As an example, the Cuiabá River presents nitrate concentrations of up to 3.4 mg.l^{-1} in the extreme dry period and at the beginning of the rainy season in the floodplain, while the annual average of the whole river, including urban area, is about 0.6 mg.l^{-1} (SEMA 2005). In the Northern UPRB, Correntes River has a total nitrogen concentration ranging from 0.06 to 0.16 mg.l^{-1}; in the Jauru River, downstream to five hydroelectric dams, this concentration varies from 0.09 to 1.5 mg.l^{-1} (unpublished personal data) and in the Miranda River from 0.05 to 0.73 mg.l^{-1} (ANEEL 2001).

In the Negro and Miranda Rivers, south UPRB, phosphorus concentrations are usually higher than 0.5 mg.l^{-1}, while in the Nabileque, Apa, Taquari and Correntes Rivers, as well as in their main tributaries, phosphorus concentrations range from < 0.2 to 0.5 mg.l^{-1} (Mato Grosso do Sul 2010). In the Cuiabá River, localized at the north of the basin, from the headwaters to the Pantanal, average phosphorus concentration is about 0.16 mg.l^{-1} (SEMA 2005), without a clear spatio-temporal variation. In many others rivers of the UPRM, from the headwaters to the Pantanal, including the Paraguay River (FEMA 1997), phosphorus concentrations are usually above 0.1 mg.l^{-1}, which is the limit established in the Brazilian environmental legislation for lotic environment.

In almost all the sub-basins at the UPRB north portion, in Mato Grosso state, phosphorus ranges from 0.1 to 0.3 mg.l^{-1} (Mato Grosso 2009). On the other hand, in the higher portions of many rivers which emerge in sandy formations, phosphorus concentrations are reduced throughout the year (Paula 1997, Figueiredo 2007), especially in those where marginal vegetation is preserved and functional. The relatively high phosphorus concentrations in the UPRB are probably related not only to localized sources such as domestic and industrial effluents discharge or to diffuse agricultural sources, such as fertilizer runoff, but also to hydrogeological characteristics of the basin. However, studies characterizing the main natural and anthropic sources of this nutrient, the autodepuration capacity of the rivers and the accumulation and/or absorption areas are still scarce.

It is worth mentioning that using the Water Quality Index, developed by the National Sanitation Foundation (NSF) and modified by Paiva and Paiva (2001), all the UPRB sub-basins, with the exception of Negro, Nabileque and Apa, which have no data, and the Cuiabá and São Lourenço Rivers and the higher reaches of the Paraguay River which were classified as MEDIUM, the waters were classified as GOOD (Mato Grosso do Sul 2010, Mato Grosso 2009, SEMA 2005). The MEDIUM results are due mainly to high dissolved solids concentration and turbidity in the Paraguay and São Lourenço Rivers and coliform indices in the Cuiabá River at the urban section.

In summary, the spatio-temporal limnologic variations in the UPRB basin are controlled mainly by the geological conditions of the drainage areas, by the rain seasonality and by the localization in the river *continuum* (high, middle and low-Pantanal sections). The anthropic activities have been altering the natural variation due to increased concentrations of organic and inorganic substances (pollution) from diffuse or point sources. These substances occur naturally in the aquatic environment and the increase in their concentrations is a consequence of domestic and industrial effluents discharge for transport and dilution by the rivers and of the increase in sediment transport from the intense erosive processes which are occurring in fragile areas.

Contamination

Regarding the contamination of river waters in the UPRB, reports of persistent chemical substances from industrial activities are sparse, since there are few industries in this region which generate effluents contaminated with these substances, except for some tanneries which utilize chromium (Migliorini et al. 2006) in the industrial process and other small industries which may discharge heavy metals and toxic chemicals, but no information is available on the occurrence of these substances in the UPRB rivers. On the other hand, there are detailed studies on mercury contamination, once it was intensively used in gold mining developed in the drainage area of Bento Gomes River, a tributary of the Paraguay River in the 1970s and 1980s (e.g. von Tümpling et al. 1995, Vieira et al. 1995, Nogueira et al. 1997, Hylander et al. 2000, Tuomolo et al. 2008). SEPLAN/SEMA (2011) pointed out ten sites with water contamination by gold mining activities, but these data are ancient and there is no recent assessment about mercury contamination

Although agribusiness has a major importance on the Brazilian economy and on the Centre-western region of Brazil, which is the region responsible for a great fraction of this growth and comprises, among others, the States of Mato Grosso and Mato Grosso do Sul where the UPRB is localized, studies on the contamination of aquatic ecosystems by agrochemicals are also limited. Presently, Mato Grosso is showing a large increase in the area under cultivation for soya bean, cotton, maize, bean and sugar cane among others, in the highlands surrounding the Pantanal mainly in the drainage areas of the Casca, Manso, Sepotuba, Paraguay, Vermelho and São Lourenço rivers. In Mato Grosso do Sul this growth is also observed in the drainage areas of the Alto Taquari, Miranda, Aquidauana and Negro rivers, where the scenario is practically the same, except that cotton is not grown in this region and for sugar cane whose increase forecast in this state is more significant than in Mato Grosso. Productivity of these crops has been increasing greatly, mainly due to the use of modern technologies. In general, these crops are grown in extensive areas (monoculture systems) demanding mechanization and intensive inputs of agrochemicals, particularly pesticides and fertilizers (Vieira and Galdino 2004).

After application in the plantations, the major destination of pesticides is the soil, where it can undergo several processes such as degradation, runoff, leaching and volatilization, suffering off-site transportation and reaching non-target areas. These processes depend largely upon the physical–chemical properties of the molecules, physical characteristics of the environment (relief, soil type, climate), soil chemical properties as well as the management practices of the crops.

The rising demand for pesticides can present a risk to the Pantanal sustainability due to their potential to alter the structure and function of aquatic ecosystems, e.g. the maintenance of the ecological processes that support the so-called environmental services of these water resources, and consequently its biodiversity (Vieira and Galdino 2004). There is an increasing number of international publications concerning environmental contamination of aquatic systems with pesticides in the tropics and subtropics (e.g. Castilho et al. 2000, Botello et al. 2000, Kammerbauer and Moncada 1998, Miles and Pfeuffer 1997, Thurman et al. 2000, Zimmerman et al. 2000, Hernandez-Romero et al. 2004, Bocquené and Franco, 2005, Silva et al. 2009, Azevedo et al. 2010).

From Brazil some studies evaluated pesticide concentrations within the environment (e.g. Caldas et al. 1999, Nascimento et al. 2004, Brito et al. 2005, Raposo, Re-Poppi 2007, Silva et al. 2008), most of these focusing on organochlorine pesticides which are already banned from use in Brazil. Recently, some studies have turned their attention to currently-used pesticides (Lanchote et al. 2000, Filizola et al. 2002, Laabs et al. 2002, Cerdeira et al. 2005, Dores et al. 2006, Bortoluzzi et al. 2007, Carbo et al. 2008, Milhome et al. 2009, Azevedo et al. 2010), but all of them in restricted areas so that a broader comprehension of contamination by pesticides in Brazil is still lacking. Few studies assessed environmental behaviour and distribution of currently-used pesticides (Laabs et al. 2000, 2002a, 2002b, Correia et al. 2006, 2007, Martins et al. 2007, Dores et al. 2009, Queiroz et al. 2009). Thus, although the interest in this area of research is increasing in Brazil, little is known about the contamination situation with currently-used agrochemicals (e.g. triazines, acetanilides, pyrethroids, organophosphates) in the environments of central and northern Brazilian territories (Cerrado, Pantanal basin, Amazonas).

Van Dijk and Guicherit (1999) noted that pesticides may be deposited tens to hundreds of kilometres from their application area and that their contribution to pesticide pollution might be substantial in remote regions. The vulnerability of remote ecosystems to long-range aerial pesticide input is stressed by van Straalen and van Gestel (1999). More recently, distribution of currently-used pesticides has been studied in ambient air collected in both rural and urban areas from around the world as reviewed by Yusà et al. (2009). Although for tropical climates an increased volatilization of pesticides, due to higher air and soil temperatures, has to be expected, no research on pesticide pollution of rainwater or gas phase atmosphere in these areas has been carried out so far (Santos et al. 2009).

Although it has been shown that pesticide dissipation in soils of the Brazilian Cerrado occurs substantially faster than is known from temperate regions (Laabs et al. 2002b), the transport of a number of pesticides (e.g. alachlor, atrazine, chlorpyrifos, endosulfan, metolachlor, profenofos, simazine, and trifluralin) from agricultural areas in the Cerrado to the edge of the Pantanal via river and rain water was proven in a pilot monitoring study (Laabs et al. 2002c). Results of this pilot monitoring showed that on the Planalto, in 70 per cent of river water (up to 128 ng.L^{-1} of individual substances) and 100 per cent of rain water samples (up to 3000 ng.L^{-1}), pesticides were detected. In many cases concentrations lie well above the established EU limits for water supplies of 100 ng.L^{-1} for individual pesticides and 500 ng.L^{-1} for the sum of pesticides and a few of them were higher than the Brazilian limits established by the Ministry of Health (Brasil 2010) and by the National Council for the Environment (Brasil 2005). It is important to point out that the Brazilian legislation establishes limits for only about 20 active ingredients and pesticides metabolites. As this pilot monitoring study did follow a regular sampling schedule (weekly/biweekly for streams/rivers) it did not cover maximum pesticide concentrations. Peak concentrations right after high intensity rainfall events have to be expected in river water at a much higher concentration range (Watts et al. 2000). Traces of pesticides were also detected in surface water (rivers, lagoons) and rainwater in Cuiabá and near Barão de Melgaço in the lowlands (67 and 74% of samples, respectively). This is a clear indication that pesticides may be transported at least 50 km from the Planalto to the Pantanal lowlands. The pesticides or derivates most frequently detected in environmental samples were: endosulfan α, β, and –sulphate, metolachlor, ametryne, trifluralin, metribuzine, malathion, alachlor, simazine and atrazine.

In another pilot study (Miranda et al. 2008) sediment samples from 25 sites in 17 rivers of the Pantanal were analyzed with the objective of evaluating pesticide contamination in sediments. A multi-residue gas chromatography-mass spectrometry method was applied to monitor 23 pesticides of different chemical classes (organochlorine, organophosphorus, triazines, anilides and pyrethroids) with some of their degradation products. Compounds identified in sediment samples included λ-cyhalothrin (1.0 to 5.0 μg.kg^{-1}), p,p'-DDT (3.6 μg.kg^{-1}), deltamethrin (20.0 μg.kg^{-1}) and permethrin (1.0 to 7.0 μg.kg^{-1}). However, Brazilian legislation established limits only for DDT in sediment samples (1.1 mg.kg^{-1}). The persistence of most pesticides in a tropical semi-field experiment was similar to the one reported in (semi-)field studies from temperate regions (summer season) and laboratory experiments (20–25°C). In contrast to observed differences between tropical and temperate soils for most of these pesticides (Laabs et al. 2002b), no distinctively faster dissipation of the studied pesticides was observed in the tropical aqueous environment in comparison to temperate (semi-)field or laboratory studies. Whether this finding is due to greater temperature differences in soils than in aquatic systems from different climates, or if it is also caused by a more enhanced volatilization of pesticides from soils than from water surfaces, merits further investigation.

The fate of pesticides in tropical soils – especially under field conditions – still requires further research (Racke et al. 1997). Pesticide dynamics in agricultural soils in Brazil were the subject of several studies (Reichenberger et al. 2002, Laabs et al. 2005, Rissato et al. 2006, Villa et al. 2006, Correia et al. 2006, 2007, Martins et al. 2007, Regitano and Koskinen 2008, Queiroz et al. 2009). However, most of these studies were done in the laboratory and did not take account of the tropical field situation with its high intensity rainfall and large temperature amplitude. Therefore knowledge of pesticide fate under tropical field conditions in Brazil is limited to few studies (Laabs et al. 2007, Dores et al. 2009). Their results indicated fast dissipation of pesticides from the topsoil ($t_{1/2} < 20$ d); but nevertheless a medium-term leaching of several substances was observed (Laabs et al. 2000, 2002b).

Ecotoxicological research on pesticides from tropical regions is so-far very sparse. Castillo et al. (1997) reviewed Central American ecotoxicological studies and concluded that for tropical ecosystems the identification of key native species (in the ecological and economic sense) for ecotoxicological tests is of utmost importance. Although a battery of test methods exists (Steinberg et al. 1995), they are so-far standardized only for temperate species. In Europe such tests are demanded by legislation for pesticide registration and are executed according to fixed guidelines (OECD 1993). Therefore detailed information about pesticide toxicity is available for selected organisms, for example water-flea (*Daphnia magna*), trout (*Salmo gairdneri*) and carp (*Cyprinus carpio*). However, results of ecotoxicological tests generally cannot be transferred to other species. Publications about toxicity of pesticides concerning native species from Brazil are also very sparse (Oliveira-Filho et al. 2004, Trevis et al. 2010). Especially for the Pantanal area with its unique fauna/flora and the economic importance of fishery, a screening of pesticide toxicity on indigenous aquatic species is needed urgently.

The microbial food web has a major role in the cycling and transference of nutrients to higher trophic levels. Pesticides used in agriculture can directly affect the microbiota, either reducing the number of individuals, inhibiting productivity, reducing phototrophic variables (ex. chlorophyll a, carbon assimilation, dissolved oxygen) or by bioconcentration (De Lorenzo et al. 1999, Lal et al. 1987, Pratt et al. 1988), causing functional and structural alterations in the microbial food web, which can eventually affect higher trophic levels. The literature review underlined that information about pesticide fate in tropical environments is very sparse, especially in wetland ecosystems. Therefore there is a need to implement further studies to move on to a detailed assessment of pesticide distribution, dynamics and ecotoxicological effects in the Pantanal wetland environment.

Water Resources Policy: Rivers Management

Water has become a main issue in the international political agenda. Since water is critical for human survival, public authorities from many countries have assumed the central responsibility for managing this resource, demonstrating that a dependency on market forces alone will not promote a satisfactory income and that some form of government intervention will always be necessary (Rodrigues 1998).

In Brazil, in the National Policy of Water Resources (Law 9433/1997), water is said to be a public asset and for this reason should be managed by public institutions. For the UPRB, SEMA (State Secretary of Environment of Mato Grosso) and IMASUL (Environment Institute of Mato Grosso do Sul) are the institutions responsible for the coordination and management of this policy. Each state of the UPRB has its own policy for water resources, with some regional adaptations, but very similar to the National Policy. All these legislations are based on the same main principles: i) water is a limited natural resource, having significant economic value; ii) water resources management must allow for multiple water uses; iii) the hydrographical basin is the territorial unit for implementation of the policy; iv) water resources management must be decentralized and count on the participation of Public Authorities, users and communities.

Regarding social participation, in both states of the UPRB, there are State Councils of Water Resources which are collegiate institutes with representatives from public institutions, organized civil societies and water users as well as representatives from the Hydrographic Basins Committees (CBHs) which is also a collegiate institute with social participation. Throughout UPRB, only two committees have been created so far, CBH Miranda in Mato Grosso do Sul and CBH Sepotuba in Mato Grosso. There are serious difficulties in mobilizing the UPRB population to participate in water resources management, due to the legislation. These difficulties are mainly related to the availability of water in quality and quantity sufficient for the demands in the UPRB, shortage of financial resources to join up the different agents in each basin and for the maintenance of the committee and lack of awareness within the populations regarding the legislation especially as far as their right to participate in basin decisions and the importance of this participation goes. The institutions which coordinate this policy in the UPRB offer supporting programmes for the creation of committees and to stimulate social participation in water resources management, but these are still isolated and also lack financial and human resources.

It is worth mentioning that, throughout UPRB, the consumptive water use (with derivation) is still smaller than the supply. In most sub-basins, the use is lower than 5 per cent of the maximum water availability for each water body aiming to guarantee the maintenance of the biotic environment, even in the dry months (Mato Grosso 2009, Mato Grosso do Sul 2010). This generates a sensation of comfort among the water users allied to the fact that water use is not yet charged in the basin. At present, in the Southern and Southeastern regions of Brazil, where there is higher population and industrial density and qualitative shortage of water, 106

CBHs were installed; in the Northeast, where the shortage is mainly quantitative due to the semi-arid climate, there are 37 CBHs, while in the Northern region where water availability is large with a small demand (Hydrographic Region of Amazon) there is only one CBH installed (SIAPREH 2010).

The main instruments of water resources management, cited in the National and State Policies and the actual phasing of their implementation in the UPRB are described below:

i) *Water Resources Plan*: this is a directive plan that aims to found and orientate the implementation of the policy on management of water resources. The National Water Resources Plan was elaborated in order to support the Water Resources Policy and the Brazilian commitments for the Millennium Development Goals and the Johannesburg Plan of Implementation of the World Summit on Sustainable Development.

Both states of UPRB possess a State Plan of Water Resources (Mato Grosso do Sul 2010, Mato Grosso 2009), in accordance with the National Plan, including diagnostics, short, medium and long-term prognostics, and directives and programmes. Both Water Resources Plans state the need to integrate the actions proposed for the UPRB in each state, such as water quality, hydrological and sedimentological monitoring, standardization of measuring methods, integrated database of water users as well as the elaboration of special plans that involve all the Paraguay Hydrographic Region. This integration, although still embryonic, is fundamental for the UPRB management, since the watershed limits do not recognize political limits and the conservation of the Pantanal depends mainly upon the natural conditions of the highlands, which have been affected by several human activities related to water and soil use.

In November 2010, with the event of the first evaluation of the National Water Resources Plan, when actions and goals for the following five years were defined, an important step for the integration of the whole UPRB was taken with the proposal for the creation of a Committee of the Pantanal States. This committee aims to promote a better articulation and exchange of experiences between the states of the Paraguay Hydrographic Region to strengthen the decisions made about strategic matters for this region (personal information).

It is worth mentioning that the processes that occur in the Paraguay Hydrographic Region are different from the other Brazilian hydrographic regions since the environmental services of its rivers have implications for integrating effects to several ecosystems of vital importance to the international, inter-state and regional communities (Adámoli 1995). The international integration of basins for water resources management is still in its infancy. Several countries, supported through entities such as the United Nations or international financial support organizations, have been organizing themselves in several ways, such as the Mekong River Commission and the network called Nile Basin Initiative (MMA 2008)

Since Brazil has most of its frontiers defined by rivers, this subject is of extreme importance. According to the National Water Agency (ANA), there are 83 frontier and trans-frontiers rivers in areas of great geographical and institutional diversity. Thus, the country has been looking forward to accomplishing such coordination efforts in order to favour dialogue and approximation with neighbouring countries aiming at the sustainable management of water resources. As such, the La Plata Basin Treaty, which includes UPRB is worth mentioning (MMA 2008).

ii) Classification of water bodies: this managing instrument requires the categorization of main uses in five classes (MMA 2010). According to the class, there may be restrictions on or increases in possible uses, which must have acceptable water quality. The main objective for the definition of these classes is to guarantee the suitability of water for the designated use of each specific class considering their most exigent uses; reduce the costs of pollution control by means of preventive permanent actions and provide elements for the establishment of the values to be charged for water use licensing and for its use (Mato Grosso do Sul 2010).

At present almost all the UPRB rivers in Mato Grosso do Sul have been classified, with restricted uses and quality requirements for the rivers at their headwaters (Class Especial and 1), followed by Classes 2 and 3 from the middle course to the mouth (Deliberation CECA MS n° 003 from 1997). On the other hand no river has been classified in Mato Grosso state, which implies that all belong to Class 2, an intermediate class regarding restrictions of use and quality requirements but which must guarantee safe human consumption after conventional treatment, protection of aquatic communities, recreation activities of primary contact (swimming, skiing, dive), irrigation of vegetables, fruit trees, parks, gardens and sports fields that people may have contact with and aquiculture and fishing. Some limits established for water quality parameters ensuring that it is adequate for these uses are shown in Table 4.2.

Table 4.2   Maximum or Minimum Limits of Some Water Quality Parameters for Class 2 Rivers (Regulation CONAMA n° 357/2005)

Parameter	Limit
pH	6,0 - 9,0
Oil and grease	virtually absent
BOD (Biochemical oxygen demand)	$< 5,0$ mg.l^{-1}
DO (dissolved oxygen)	$> 5,0$ mg.l^{-1}
Turbidity	< 100 NTU
Color	< 75 uH
Total phosphorus	$< 0,1$ mg.l^{-1} for rivers
Thermotolerant Coliform (fecal) ou *Escherichia coli*	< 1.000 NMP/100 mL

Parameter	Limit
Total dissolved solids	< 500 mg.l^{-1}
Dissolved iron	< 0,3 mg.l^{-1}
Ammoniacal nitrogen	< 3,7 mg.l^{-1} pH ≤ 7,5 < 2,0 mg.l^{-1} em 7,5 < pH ≤ 8,0 < 1,0 mg.l^{-1} em 8,0 < pH ≤ 8,5 < 0,5 mg.l^{-1} pH > 8,0

Several other variables are listed in this legislation, with the respective limits for Class 2, such as pesticides, heavy metals, organic and inorganic compounds. On the other hand, some variables mentioned in this chapter do not have limits established in this legislation but are also important indicators of natural conditions or of anthropic interference in the aquatic ecosystem:

iii) Water Use Licensing: water use licensing (system for granting a license to a given user or number of users for abstracting or applying water from a given source to a given use) is an instrument that aims to assure the qualitative and quantitative control of surface and groundwater uses, and the effective exercise of the people's right to access to water. This license is granted by the institution responsible for coordinating and managing water resources for a determined period of time. In both UPRB states, this instrument was implemented for the most significant uses, especially for economic activities of medium to large sizes or which may have high polluting potential.

iv) Water use charges: the charge for water usage is a managing instrument, integrated to the others, that aims to recognize the economic value of water and to give the users an indication of its real value, determined as a function of the actual water quality and quantity and of the required use (Mato Grosso do Sul 2010). The water use charge has not yet been implemented in any of the UPRB states due principally to the relatively low demands compared to the hydric availability, the need of effective implementation of the other managing instruments and the lack of detailed information on water quality and quantity to allow the application of the polluter-pays principle.

Conclusions

The Upper Paraguay River Basin, also known as the Paraguay hydrographic region, presents significant landscape diversity, which includes four major geomorphological units: the Plateaus, the Mountains Ridges, the Depressions and the Floodplain. From an administrative perspective, the Brazilian section of the UPRB is situated in two States, Mato Grosso and Mato Grosso do Sul. The former concentrates most of the headwaters and the majority of the catchment population, whilst the latter contains most of the floodplain wetland. The main economic activity in the UPRB is agribusiness, which implies the use of water for irrigation, animal consumption, processes and products in agroindustry and dilution of agroindustrial wastewaters, apart from public water supply, hydropower generation and fluvial navigation along the Paraguay River. Water availability is much higher than demand and less than 5 per cent of maximum river flow in most sub-basins. It suggests that the biotic functions are likely to be maintained, even in the dry months. However, in spite of the general availability, water conflicts are increasing in the UPRB, such as farm production vs. hydropower (because of sedimentation of rivers and reservoirs caused by crop production, the cultivation of riparian zones and the occupation of permanent conservation areas) and effluent dilution vs. recreation and public water supply (i.e. conflicts between upstream and downstream uses of water and discharge of urban effluents).

In this chapter the Brazilian Pantanal streamflow regime is characterized, especially the flow seasonality and flood routing within the Pantanal's hydrological and topographical context. The analysis is carried out based on daily streamflow data from 28 gauging stations, with available records varying within the 1939–2007 period. For instance, it presents how the main channel conveying capacity reduction and lateral exchanges alters high and low flow events and basal flow contributions along the Paraguay River and its tributaries.

The spatio-temporal limnologic variations in the UPRB basin are controlled mainly by the geological conditions of the drainage areas, by rain seasonality and by localization in the river *continuum* (high, middle and low-Pantanal sections). Anthropic activities have been altering the natural variation due to increased concentrations of organic and inorganic substances (pollution) from diffuse or point sources. These substances occur naturally in the aquatic environment and the increase in their concentrations is a consequence of domestic and industrial effluent discharged for transport and dilution by the rivers and of the increase in sediment transport from the intense erosive processes which are occurring in fragile areas.

Very little is known about the present impacts of water contamination by metals and pesticides on the Pantanal as well as on future possible impacts of the continuous use of these substances. Considering the intensive use of agrochemicals in the highlands neighbouring the floodplain and the influence of this region on the Pantanal, the development of thorough studies on the dynamics of these substances on this environment as well as ecoxicological ones are urgently necessary.

Thirteen years after the publication of the National Water Resources Policy, significant progress in the UPRB has been achieved in some aspects, such as the elaboration of the Water Resource Plan of each state, classification of Mato Grosso do Sul rivers, implementation of the system for granting water use licensing and strengthening of managing institutions. However, there is still a long way to go, especially concerning social participation in the managing decisions and water resources management of this basin, the effective implementation of actions and goals described in the plans, water bodies classification and integrated management of the whole basin through the articulation between the two states. The effective implementation of all the instruments described in the National Water Resources

Policy is the only way to assure water availability to present and future generations, with quality and quantity adequate to their respective uses and the maintenance of the aquatic ecosystems, especially of the Pantanal.

Acronyms

AHIPAR - Administração da Hidrovia do Paraguai (Paraguay Waterway Administration)

ANA - Agência Nacional de Águas (Brazilian National Water Agency)

ANEEL - Agência Nacional de Energia Elétrica (National Electrical Agency of Brazil)

APM - Aproveitamento Múltiplo (Multiple Uses)

ARI10 - hydrologic metric defined as the annual maximum flow with 10 years of average recurrence interval

Bs - hydrologic metric representing the base flow contribution, given by the ration of Q90 and median values

CBH - Comitê de Bacia Hidrográfica (Hydrographic Basins Committees)

CIH - Comitê Intergovernamental da Hidrovia (Intergovernmental Committee on the Waterway)

CONAMA - Conselho Nacional de Meio Ambiente (National Council of Environment of Brazil)

DNOS - Departamento Nacional de Obras de Saneamento (National Deparment of Sanitation Works)

FEMA - Fundação Estadual do Meio Ambiente (State Environment Foundation, Mato Grosso)

GEV - Generalized Extreme Value (probability distribution applied to model extreme events)

IMASUL - Instituto Ambiental de Mato Grosso do Sul (Environment Institute of Mato Grosso do Sul)

MMA - Ministério do Meio Ambiente (Brazilian Ministry of Environment)

MT - Mato Grosso State (Brazil)

MS - Mato Grosso do Sul State (Brazil)

MW - Megawatt

NTU - Nephelometric Turbidity Units

PCH - Pequena Central Hidrelétrica (Small Hydroelectric Power Plants < 30 MW)

PERH - Plano Estadual de Recursos Hídricos (State Water Resources Plan)

p,p'-DDT - 4,4' dichlorodiphenyltrichloroethane

Q90 - hydrologic metric defined as the 90% duration flow

SEMA - Secretaria de Estado de Meio Ambiente (State Secretary of Environment of Mato Grosso)

UFMT - Universidade Federal de Mato Grosso (Federal University of Mato Grosso State, Brazil)

UFSCar - Universidade Federal de São Carlos (Federal University of São Carlos, São Paulo State, Brazil)
UPRB - Upper Paraguay River Basin
UNESCO - United Nations Educational, Scientific and Cultural Organization

Bibliography

Adámoli, J. 1995. *Diagnóstico do Pantanal: Características Ecológicas e Problemas Ambientais.* Brasília: Programa Nacional do Meio Ambiente.
AHIPAR. 2010. *Obras e Serviços, Hidrovia Paraguai-Paraná, Estatística.* Available at: www.ahipar.gov.br [accessed: 25 November 2010].
ANA. 2010. *Regiões Hidrográficas.* Available at: http://www.ana.gov.br [accessed: 11 November 2010].
Assine, M.L. 2005. River avulsions on the Taquari megafan, Pantanal wetland, Brazil. *Geomorphology*, 70, 357-371.
Aulagnier, F., Poissant, L., Brunet, D., Beauvais, C., Pilote, M., Deblois, C. and Dassylva, N. 2008. Pesticides measured in air and precipitation in the Yamaska Basin (Québec): Occurrence and concentrations in 2004. *Science of the Total Environment*, 394(2), 338-48.
Azevedo, D.A., Silva, T.R., Knoppers, B.A. and Schulz-Bull, D. 2010. Triazines in the Tropical Lagoon System of Mundaú-Manguaba, NE-Brazil. Journal of the Brazilian Chemical Society, 21, 1096-1105.
Bocquené, G. and Franco, A. 2005. Pesticide contamination of the coastline of Martinique. Marine Pollution Bulletin, 51, 612-619.
Bordas, M.P. 1996. The Pantanal: an ecosystem in need of protection. International Journal of Sediment Research, 11(3), 34-39.
Bortoluzzi, E.C., Rheinheimer, D.S., Gonçalves, C.S., Pellegrini, J.B.R., Maroneze, A.M., Kurz, M.H.S., Bacar, N.M. and Zanella, R. 2007. Investigation of the occurrence of pesticide residues in rural wells and surface water following application to tobacco. Química Nova, 30(8), 1872-1876.
Botello, A.V., Ruede-Quintana, L., Diaz-González, G. and Toledo, A. 2000. Persistent organochlorine biocides (POPs) in coastal lagoons of the subtropical Mexican Pacific. *Bulletin of Environmental Contamination and Toxicology*, 64, 390-97.
Brasil. 1997. *Plano de Conservação da Bacia do Alto Paraguai (Pantanal): PCBAP. Análise Integrada e Prognóstico da Bacia do Alto Paraguai.* Brasília: Programa Nacional do Meio Ambiente.
Brasil. 2005. Resolução 357/2005. Conselho Nacional do Meio Ambiente. Dispõe sobre a Classificação dos Corpos de Água e Diretrizes Ambientais para o seu Enquadramento, bem como Estabelece as Condições e Padrões de Lançamento de Efluentes, e Dá Outras Providências. Available at: http://www.mma.gov.br/port/conama/res/res05/res35705.pdf [accessed 25 November 2010].

Brasil. 2010. Portaria 518/2004. Ministério da Saúde. Controle e Vigilância da Qualidade da Água para Consumo Humano e Seu Padrão de Potabilidade. Available at: http://portal.saude.gov.br/portal/ arquivos/pdf/portaria_518.pdf [accessed 25 November 2010].

Brito, E.M.S., Vieira, E.D.R., Torres, J.P.M. and Malm, O. 2005. Persistent organic pollutants in two reservoirs along the Paraiba do Sul-Guandu River system, Rio de Janeiro, Brazil. *Química. Nova*, 28(6), 941-46.

Caldas, E.D., Coelho, R. and Souza, L.C.K.R. 1999. Organochlorine pesticides in water, sediment and fish of Paranoá Lake of Brasilia, Brazil. *Bulletin of Environmental Contamination and Toxicology*, 62(2), 199-206.

Calheiros, D.F. and Hamilton, S.K. 1998. Limnological conditions associated with natural fish kills in the Pantanal wetland (Baía do Castelo, Paraguay River, Brazil). *Verhandlungen des Internationalen Verein Limnologie*, 26, 2189-2193.

Calheiros, D.F., Arndt, E., Rodrigues, H.O. and Silva, M.C.A. 2009. *Influência de Usinas Hidrelétricas no Funcionamento Hidro-ecológico do Pantanal Matogrossense-Recomendações*. Documento 102. Corumbá: Embrapa,

Carbo, L., Souza, V., Dores, E.F.G.C. and Ribeiro, M.L. 2008. Determination of pesticides multiresidues in shallow groundwater in a cotton-growing region of Mato Grosso, Brazil. Journal of the Brazilian Chemical Society, 19, 1111-1117.

Cerdeira, A.L., Santos, N.A.G., Pessoa, M.C.P.Y., Gomes, M.A.F. and Lanchote, V.L. 2005. Herbicide leaching on a recharge area of the Guarany aquifer in Brazil. *Journal of Environmental Science and Health B*, 40(1), 159-65.

Collischonn, W., Tucci, C.E.M. and Clarke, R.T. 1998. Further evidence of changes in the hydrological regime of the River Paraguay: part of a wider phenomenon of climate change? *Journal of Hydrology*, 245, 218-238.

Correia, F.V. and Langenbach, T. 2006. Distribution and decomposition dynamics of atrazine in an Ultisol under wet tropical climate conditions. *Revista Brasileira de Ciência do Solo*, 30(1), 183-192.

Correia, F.V., Mercante, F.M., Fabricio, A.C., Campos, T.M.P., Vargas Júnior, E.A. and Langenbach, T. 2007. Infiltration of atrazine in an Oxisol under no-tillage and conventional tillage. *Pesquisa Agropecuária Brasileira*, 42(11), 1617-1625.

Da Silva, C., Abdo, M.S.A. and Nunes, J.R.S. 2009. O rio Cuiabá no Pantanal Matogrossente, in *Bacia do Rio Cuiabá: uma abordagem socioambiental*, edited by D.M. Figueiredo and F.X.T. Salomão. Cuiabá: EdUFMT and Entrelinhas, 126-139.

Da Silva, C.J. and Girard, P. 2004. New chalenges in the managem of the Brazilian pantanal and catchment area. *Wetlands*, 12, 553-561.

De Lorenzo, M.E., Laulh, J., Pennington, P.L., Scott, G.I. and Ross, P.E. 1999. Atrazine effects on the microbial food web in tidal creek mesocosms. *Aquatic Toxicology*, 46(3-4), 241–251.

Dores, E.F.G.C., Navickiene, S., Cunha, M.L.F., Carbo, L., Ribeiro, M.L. and De-Lamonica-Freire, E.M. 2006. Multiresidue determination of herbicides in environmental waters from Primavera do Leste region (Middle West of Brazil) by SPE-GC-NPD. Journal of the Brazilian Chemical Society, 17, 866-873.

Dores, E.F.G.C., Weber, O.L.S., Spadotto, C.A., Carbo, L., Vecchiato, A.B. and Pinto, A.A. 2009. Environmental behaviour of metolachlor and diuron in a tropical soil in the central region of Brazil. *Water Air, and Soil Pollution*, 197, 185-183.

DNOS. 1974. *Estudos Hidrológicos da Bacia do Alto Paraguai*. Relatório Técnico. Rio de Janeiro: UNESCO and PNUD.

FEMA. 1997. *Qualidade da Água dos Principais Rios da Bacia do Alto Paraguai: 1995-96*. Cuiabá: MMA/PNMA.

Figueiredo, D.M. 1997. A Influência dos Fatores Climáticos e Geológicos e da Ação Antrópica sobre as Principais Variáveis Físicas e Químicas do Rio Cuiabá, Estado de Mato Grosso. MSc Dissertation. Cuiabá: UFMT.

Figueiredo, D.M. 2007. Padrões Limnológicos e do Fitoplâncton nas Fases de Enchimento e Estabilização dos Reservatório do APM Manso e AHE Jauru (Estado de Mato Grosso). PhD Thesis. São Carlos: UFSCar.

Figueiredo, D.M. 2009. Limnologia e qualidade das águas superficiais das sub-bacias alta e média, in *Bacia do Rio Cuiabá: Uma Abordagem Socioambiental*, edited by D.M. Figueiredo and F.X.T. Salomão. Cuiabá: EdUFMT and Entrelinhas, 114-125.

Figueiredo, D.M. and Salomão, F.X.T. 2009. Bacia do Rio Cuiabá: caracterização e contextualização, in *Bacia do Rio Cuiabá: Uma Abordagem Socioambiental*, edited by D.M. Figueiredo and F.X.T. Salomão. Cuiabá: EdUFMT and Entrelinhas, 41-45.

Filizola, H.F., Ferracini, V.L., Sans, L.M.A., Gomes, M.A.F. and Ferreira, C.J.A. 2002. Monitoramento e avaliação do risco de contaminação por pesticidas em água superficial e subterrânea na região de Guairá. *Pesquisa Agropecuária Brasileira*, 37(5), 659-667.

Garcia, A.B. 2009. Políticas Públicas e Demandas do Turismo, in *Bacia do Rio Cuiabá: Uma Abordagem Socioambiental*, edited by D.M. Figueiredo and F.X.T. Salomão. Cuiabá: EdUFMT and Entrelinhas, 201-210.

Gottgens, J.F. 2000. The Paraguay-Parana Hidrovia: large-scale channelization or a "Tyranny of Small Decision", in *The Pantanal of Brazil, Bolivia and Paraguay Selected Discourses on the World's Largest Remaining Wetland System*, edited by F.A. Swarts. Pennsylvania: Hudson MacArthur Publishers, 135-144.

Hamilton, S.K., Sippel, S.J. and Melack, J.M. 1996. Inundation patterns in the Pantanal wetland of South America determined from passive microwave remote sensing. *Archive Für Hydrobiologie*, 137(1), 1-23.

Hernández-Romero, A.H., Tovilla-Hernández, C., Malo, E.A. and Bello-Mendoza, R. 2004. Water quality and presence of pesticides in a tropical coastal wetland in southern Mexico. Marine Pollution Bulletin, 48, 1130-1141.

Hylander, L.D., Pinto, F.N., Guimarães, J.R.D. Meili, M., Oliveira, L.J. and Silva, E.C. 2000. Fish mercury concentration in the Alto Pantanal, Brazil: influence of season and water parameters. *Science of the Total Environment*, 261, 9-20.

Kammerbauer, J. and Moncada, J. 1998. Biocide assessment in three selected agricultural production systems in the Choluteca River Basin of Honduras. *Environmental Pollution,* 103, 171-181.

Kennard, M.J., Mackay, S.J., Pusey, B.J., Olden, J.D. and Marsh, N. 2010. Quantifying uncertainty in estimation of hydrologic metrics for ecohydrological studies. *River Research and Applications*, 26, 137–156.

Laabs, V., Amelung, W., Pinto, A.A., Altstaedt, A. and Zech, W. 2000. Leaching and degradation of corn and soybean pesticides in an Oxisol of the Brazilian cerrados. *Chemosphere*, 41, 1441-1449.

Laabs, V., Amelung, W., Fent, G., Zech, W. and Kubiak, R., 2002a. Fate of 14C-labeled soybean and corn pesticides in tropical soils of Brazil under laboratory conditions. *Journal of Agricultural and Food Chemistry*, 50, 4619-4627.

Laabs, V., Amelung, W., Pinto, A. and Zech, W. 2002b. Fate of pesticides in tropical soils of Brazil under field conditions. *Journal of Environmental Quality*, 31, 256-268.

Laabs, V., Amelung, W., Pinto, A.A., Wantzen, M., da Silva, C.J. and Zech, W. 2002c. Pesticides in surface water, sediment, and rainfall of the northeastern Pantanal basin, Brazil. *Journal of Environmental Quality*, 31, 1636-1648.

Laabs, V. and Amelung, W. 2005. Sorption and aging of corn and soybean pesticides in tropical soils of Brazil. *Journal of Agricultural and Food Chemistry*, 53(18), 718471-92.

Laabs, V., Wehrhan, A., Pinto, A.A., Dores, E.F.G.C. and Amelung, W. 2007. Pesticide fate in tropical wetlands of Brazil: An aquatic microcosm study under semi-field conditions. *Chemosphere*, 67, 975-989.

Lal, S., Saxena, D.M. and Lal, R. 1987. Uptake metabolism and effects of DDT, fenitrothion and chlorpyrifos on *Tetrahymena pyriformis*. *Pesticide Science*, 21(3), 181-191.

Lanchote, V. L., Bonato, P. S., Cerdeira, A. L., Santos, N. A. G., Carvalho, D. and Gomes, M. A. 2000. HPLC screening and GC/MS confirmation of triazine herbicides residues in drinking water from sugar cane area in Brazil. *Water, Air, and Soil Pollutution*, 118, 329-337.

Loverde-Oliveira, S.M., Figueiredo, D.M. and Nogueira, V.A.S. 1999. Avaliação da qualidade da água do córrego Arareau (Rondonópolis, MT): subsídios à gestão ambiental. *Journal of Health and the Environment*, 2(1/2), 12-23.

Martins, E.L., Weber, O.L.S., Dores, E.F.G C. and Spadotto, C.A. 2007. Leaching of Seven Pesticides Currently Used in Cotton Crop in Mato Grosso State - Brazil. *Journal of Environmental Science and Health B*, 42, 877–882.

Mato Grosso. 2009. *Plano Estadual de Recursos Hídricos.* Cuiabá: KCM.

Mato Grosso do Sul. 2010. *Plano Estadual de Recursos Hídricos.* Campo Grande: UEMS.

Migliorini, R., Dores, E.F.G.C. and Silva, E.C. 2006. Estudo da disposição final de resíduos sólidos de curtume como fonte de alteração da qualidade das águas subterrâneas. *Águas Subterrâneas*, 20, 83-96.

Miles, C. and Pfeuffer, R. 1997. Biocides in canals of South Florida. *Archives of Environmental Contamination and Toxicology*, 32, 337-345.

Milhome, M.A.L., Sousa, D.D.B., Lima, F.D.F. and Nascimento, R.F. 2009. Assessment of surface and groundwater potential contamination by agricultural pesticides applied in the region of Baixo Jaguaribe, CE, Brazil. *Engenharia Sanitária e Ambiental*, 14(3), 363-372.

Miranda, K.A., Cunha, M.L.F., Dores, E.F.G.C. and Calheiros, D.F. 2008. Pesticide residues in river sediments from the Pantanal Wetland, Brazil. *Journal of Environmental Science and Health B*, 43(8), 717-722.

MMA. 2008. *Glossário de Termos Referentes à Gestão de Recursos Hídricos Fronteiriços e Transfronteiriços*. Brasília: Ministério do Meio Ambiente.

MMA. 2010. *Legislação Ambiental*. Available at: http://www.mma.gov.br/conama [accessed: 20 November 2010].

Moreira, L.C. and Vasconcelos, T.N.N. 2007. *Mato Grosso: Solos e Paisagens*. Cuiabá: Seplan and Entrelinhas.

Nascimento, N.R., Nicola, S.M.C., Rezende, M.O.O. and Oliveira, G.Ö. 2004. Pollution by hexachlorobenzene and pentachlorophenol in the coastal plain of São Paulo state, Brazil. *Geoderma*, 121(3-4), 221-232.

Nogueira, F., Silva, E.C. and Junk, W. 1997. Mercury from gold minings in the Pantanal of Pocone (Mato Grosso, Brazil). *International Journal of Environmental Health Research*, 7(3), 181-192.

OECD 1993. *Guidelines for Testing Chemicals*. OECD, Paris.

Oliveira-Filho, E.C., Lopes, R.M. and Paumgartten, J.R. 2004. Comparative study on the susceptibility of freshwater species to copper-based pesticides. *Chemosphere*, 56, 369-374.

Paiva, J.B.D. and Paiva, E.M.C.D. 2001. *Hidrologia Aplicada à Gestão de Pequenas Bacias Hidrográficas*. Porto Alegre: ABRH.

Paula, A. M. 1997. Utilização das Comunidades de Macroinvertebrados Bentônicos e Características Limnológicas na Avaliação da Qualidade da Água do Rio Coxipó e Coxipozinho, MT. MSc Dissertation. Cuiabá: UFMT.

Paz, A.R., Collischonn, W., Tucci, C.E.M. and Padovani, C.R. 2011. Large-scale modeling of channel flow and floodplain inundation dynamics and its application to the Pantanal (Brazil). *Hydrological Processes,* in press, Available at: http:/onlinelibrary.wiley.com/doi/10.1002/hyp.7926/abstract).

Ponce, V.M. 1995. *Impacto Hidrológico e Ambiental da Hidrovia Paraná-Paraguai no Pantanal Matogrossense, um Estudo de Referência*. San Diego: San Diego State University.

Pratt, J.R., Bowens, N.J., Niederlehner, B.R. and Cairns, J. 1988. Effects of atrazine on freshwater microbial communities. *Archives of Environmental Contamination and Toxicology*, 17(4), 449-458.

Queiroz, S.C.N., Ferracini, V.L., Gomes, M.A.F. and Rosa, M.A. 2009. The behavior of hexazinone herbicide in recharge zone of Guarani Aquifer with sugarcane cultivated area. *Química Nova*, 32(2), 378-381.

Racke, K.D., Skidmore, M.W., Hamilton, D.J., Unsworth, J.B., Miyamoto, J., and Cohen S.Z. 1997. Biocide fate in tropical soils (technical report). *Pure and Applied Chemistry*, 69(6), 1349-1371.

Raposo, J.L. and Re-Poppi, N. 2007. Determination of organochlorine pesticides in ground water samples using solid-phase microextraction by gas chromatography-electron capture detection. *Talanta*, 72(5), 1833-1841.

Rebouças, A.C. 1999. Água doce no mundo e no Brasil, in *Águas Doces no Brasil*, edited by A. C. Rebouças, B.P.F. Braga, B. and J.G. Tundisi. São Paulo: Escrituras, 1-38.

Rebouças, A.C. 2004. *Uso Inteligente da Água*. São Paulo: Escrituras.

Regitano, J.B. and Koskinen, W.C. 2008. Characterization of nicosulfuron availability in aged soils. *Journal of Agricultural and Food Chemistry*, 56(14), 5801-5805.

Reichenberger, S, Amelung, W, Laabs, V., Pinto, A.A., Totsche, K.U. and Zech, W. 2002. Pesticide displacement along preferential flow pathways in a Brazilian Oxisol. *Geoderma*, 110(1-2), 63-86.

Rissato, S.R., Galhiane, M.S., Ximenes, V.F., Andrade, R.M.B., Talamoni, J.L.B., Libânio, M., Almeida M.V., Apon, B.M. and Cavalari, A.A. 2006. Organochlorine pesticides and polychlorinated biphenyls in soil and water samples in the northeastern part of Sao Paulo State, Brazil. *Chemosphere*, 65(11), 1949-1958.

Rodrigues, F.A. 1998. *Gerenciamento de Recursos Hídricos*. Brasília: Banco Mundial/MMA/SRH.

Rondon-Lima, E.B. and Lima, J.B. 2009. Qualidade da água das principais sub-bacias urbanas do município de Cuiabá, in *Bacia do Rio Cuiabá: Uma Abordagem Socioambiental*, edited by D.M. Figueiredo and F.X.T. Salomão. Cuiabá: EdUFMT and Entrelinhas, 140-145.

Salomão, F.X.T., Barros, L.T.L.P. and Cavalheiro, E.S.S. 2009. Unidades de paisagem, in *Bacia do Rio Cuiabá: Uma Abordagem Socioambiental*, edited by D.M. Figueiredo and F.X.T. Salomão. Cuiabá: EdUFMT and Entrelinhas, 154-160.

Santos, L.G., Dores, E.F.G.C. and Lourencetti, C. 2009. Análise de resíduos de pesticidas em ar atmosférico. *Revista Uniara*, 12, 185-204.

Segigatto, E.M. 2006. *Delimitação Automática das Áreas de Preservação Permanente e Identificação dos Conflitos de Uso da Terra na Bacia Hidrográfica do Rio Sepotuba-MT*. PhD Thesis. Viçosa: UFV.

SEPLAN/SEMA. 2011. *Atlas de Mato Grosso: abordagem socioeconómico-ecológica*. Cuiabá: Entrelinhas.

SEMA. 2005. *Relatório de Monitoramento: Qualidade da Água na Bacia do rio Cuiabá: 2003-2004*. Cuiabá: SEMA.

Shirashi, F.K. 2003. Avaliação dos Efeitos da Construção do APM Manso no Controle das Cheias nas Áreas Urbanas das Cidades de Cuiabá e Várzea Grande. MSc Dissertation. Rio de Janeiro: UFRJ.

SIAPREH. 2010. Relatórios 2008/2009. Sistema de Acompanhamento e Avaliação da Implementação da Política de Recursos Hídricos no Brasil. Available at: http://siapreh.cnrh.gov.br [accessed: 10 December 2010]

Siebert, D.E. 2009. Uso e Ocupação da Cabeceira do Rio Jauru: Subsídios para a Recuperação do Processo de Degradação Ambiental. MSc Dissertation. Cuiabá: UFMT.

Silva, D.M.L., Camargo, P.B., Martinelli, L.A., Lanças, F., Pinto, J.S.S. and Avelar, W.E. 2008. Organochlorine pesticides in Piracicaba River basin (São Paulo/ Brazil): a survey of sediment, bivalve and fish. *Química Nova,* 31(2), 214-219.

Silva, D.R.O., Avila, L.A., Agostinetto, D., Magro, T.D., Oliveira, E., Zanella, R. and Noldin, J.A. 2009. Monitoramento de agrotóxicos em águas superficiais de regiões orizícolas no sul do Brasil. *Ciência Rural,* 39, 2383-2389.

Souza, C.F., Paz, A.R. and Collischonn, W. 2011. Caracterização hidrológica da bacia do rio Miranda., in *Índice de Qualidade da Bacia hidrográfica do Rio Miranda (MS): Bases Ecológicas para a Gestão Integrada dos Recursos Naturais,* edited by D.F. Calheiros et al. Corumbá: Embrapa Pantanal, in press.

Steinberg, C.E.W., Geyer, H.J. and Kettrup, A.F. 1994. Evaluation of xenobiotic effects by ecological techniques. *Chemosphere,* 28(2), 357-74.

Thurman, E.M., Bastian, K.C. and Mollhagen, T. 2000. Occurrence of cotton herbicides and insecticides in playa lakes of High Plains of West Texas. *Science of the Total Environment,* 248, 189-200.

Trevis, D., Habr, S.F., Varoli, F.M. and Bernardi, M.M. 2010. Toxicidade aguda do praguicida organofosforado diclorvos e da mistura com o piretróide deltametrina em *Danio rerio* e *Hyphessobrycon Bifasciatus. Boletin do Instituto de Pesca,* 36(1), 53-59.

Tucci, C.E.M., Genz, F. and Clarke, R.T. 1999. Hydrology of the Upper Paraguay Basin, in *Management of Latin American River Basins: Amazon, Plata and São Francisco,* edited by K. Biswas, N.V. Cordeiro and B.P.F. Braga. Tokyo: United Nations University Press, 103-122.

Tucci, C.E.M., Villanueva, A., Collischonn, W., Allasia, D.G., Bravo, J.M. and Collischonn, B. 2005. Modelo Integrado de Gerenciamento Hidrológico da Bacia do Alto Paraguai (Subprojeto 5.4) in *Projeto de Implementação de Práticas de Gerenciamento Integrado de Bacia Hidrográfica para o Pantanal e Bacia do Alto Paraguai.* Porto Alegre: ANA, GEF, PNUMA and OEA.

Tucci, C.E.M. and Clarke, R.T. 1998. Environmental issues in the La Plata basin. *Water Resources Development,* 4(2), 157-173.

Tundisi, J.G. 2005. Gerenciamento integrado de bacias hidrográficas e reservatórios-estudos de caso e perspectivas, in *Ecologia de Reservatórios: Impactos Potenciais, Ações de Manejo e Sistemas em Cascata,* edited by M.G. Nogueira, R. Henry and A. Jorcin. São Carlos, RiMa, 1-22.

Tuomolo, L., Niklasson, T., Silva, E. C. and Hylander, L.D. 2008. Fish mercury development in relation to abiotic characteristics and carbon sources in a six-year-old, Brazilian reservoir. *Science of the Total Environment,* 390, 177-187.

Van Dijk, H.F.G. and Guicherit, R. 1999. Atmospheric dispersion of current-use pesticides: review of the evidence from monitoring studies. *Water, Air, and Soil Pollutution*, 115, 21-70.

Van Straalen, N.M. and Van Gestel, C.A. 1999. Ecotoxicological risk assessment of biocides subject to long-range transport. *Water, Air, and Soil Pollutution*, 115, 71-81.

Vieira, L. and Galdino, S. 2004. *Pantanal: Risco de Contaminação por Biocidas.* Available at: http://www.agronline.com.br/artigos/artigo. php?id=168&pg=1&n=2 [accessed: 1 March 2006].

Vieira, L.M., Alho, C.J.R. and Ferreira, G.A.L. 1995. Contaminação por mercúrio em sedimento e em moluscos do Pantanal, Mato Grosso, Brasil. *Revista Brasileira de Zoologia*, 12(3), 663-670.

Villa, R.D., Dores, E.F.G.D., Carbo, L. and Cunha, M.L.F. 2006. Dissipation of DDT in a heavily contaminated soil in Mato Grosso, Brazil. *Chemosphere*, 64(4), 549-554.

von Tümpling Jr., W., Wilken, R.D. and Einax, J. 1995. Mercury contamination in the northern Pantanal region Mato Grosso, Brazil. *Journal of Geochemical Exploration*, 52(1/2), 127-134.

Watts, D.W., Novak, J.M., Johnson, M.H., and Stone, K.C. 2000. Storm flow export of metolachlor from a coastal plain watershed. *Journal of Environmental Health B*, 35(2), 175-186.

Wantzen, K.M.; Da Silva, C.J., Figueiredo, D.M. and Migliácio, M.C. 1999. Recent impacts of navigation on the upper Paraguay River. *Boliviana de Ecología y Conservación Ambiental*, Memorias del Congreso Boliviano de Limnología y Recursos Acuáticos, 6, 173-182.

Yusà, V., Coscolla, C., Mellouki, W., Pastor, A. and De La Guardia, M. 2009. Sampling and analysis of pesticides in ambient air. *Journal of Chromatography A*, 1216, 2972-2983.

Zimmerman, L.R.; Thurman, E.M.; Bastian, K.C. 2000. Detection of persistent organic pollutants in the Mississippi Delta using semipermeable membrane devices. *Science of the Total Environment*, 248, 169-179.

Chapter 5

Soil and Water Conservation in the Upper Paraguay River Basin: Examples from Mato Grosso do Sul, Brazil

Carlos N Ide, Fábio V Gonçalves, Giancarlo Lastoria, Humberto C Val,
Jhonatan B Silva, Jorge L Steffen, Luiz A A Val, Maria L Ribeiro,
Sandra G Gabas, Synara A O Broch

Introduction

Soil and water are some of the most precious natural resources to be found in nature, essential for agriculture, industry, wildlife and human lives. As regards the Upper Paraguay River Basin (UPRB), the intensification of urbanisation and agricultural production has led to higher discharges of toxic substances, the release of atmospheric polluting materials, the diffuse discharges of agrochemicals and the contamination of water bodies (for example, with elements that accelerate the growth of algae and cause eutrophication). Inappropriate agricultural activity and the excessive use of agrochemicals can lead to a progressive deterioration of terrestrial and aquatic ecosystems. In addition, the expansion of cultivated areas and pastures in the uplands and depressions has relied on the removal of native vegetation, a phenomenon that started in the last century, around the 1970s, which increased superficial outflow and the production of sediments. The impacts of deforestation has been aggravated by increases in the average rate of rainfall since the 1980s – a phenomenon that is probably related to the long-term cycles of wet and dry periods in the UPRB, but may be aggravated by anthropogenic climate change – affecting the stability and morphological structure of water bodies in the surrounding plateaus and in the Pantanal floodplain (Brasil 1997a). As a result, the rate of soil erosion has increased the concentration of solids in the water bodies, causing the deposition of sediments in the rivers, lakes and water reservoirs in many parts of the river basin. Soil classification (of the Brazilian part of the UPRB) can be found in Embrapa (2006).

The following sections discuss the various aspects of soil-water interactions, especially related to the Pantanal floodplain, and will also include a specific case study of the Taquari river basin, which is certainly the most degraded area of the URPB. These discussions will be framed by reference to Brazilian environmental standards for soil and water, as stipulated by the resolution CONAMA 357/2005 (Brasil 2005) of the National Environmental Council (described below).

Geology and Hydrogeology

Geology Overview

The Pantanal Sedimentary Basin is inserted in the centre of the UPRB, occupying, in the Brazilian Territory, 38.2 per cent of its area, corresponding to 138,183 km², distributed in the States of Mato Grosso and Mato Grosso do Sul (Silva and Abdon 1998). Both its origin and the geological events related to its evolution are still not entirely well described. Some works point out the environmental diversity of the river basin (Silva and Abdon 1998, Assine and Silva, 2009); others have studied in greater details one compartment (Soares et al. 2003, Silva et al. 2007, Assine and Silva 2009, Almeida et al. 2009). The basin's geological evolution started to be discussed in the middle of the last century (Almeida 1945), but it was only in the last decades that some authors provided a more detailed description (Shiraiwa 1996, Soares et al. 1998, Ussami et al. 1999, Soares et al. 2003, Assine and Soares 2004), which, nonetheless, according to the same authors, is still only a preliminary approximation of the Pantanal's geology.

It is relevant to observe that the Pantanal is an active sedimentary floodplain, filled with quaternary sediments, deposits of alluvium of the Paraguay River and of alluvial fans in some of its tributaries, the origin of which is associated to the tectonic evolution of the Andean mountains (Assine and Silva 2009). The sediments that fill up the floodplain make up the Pantanal Formation, which is a geological structure with a maximum thickness of around 550 m, as verified through the analysis of seismic data (Assine and Soares 2004). Among the alluvial fans that form the basin, the Taquari River's fan stands out, having a nearly circular format, with a 250 km diameter, which will be the object of a case study later in this chapter. For more information on the hydrographic basins, refer to Assine and Soares (2004).

The sediments are characterized as being siliciclastic, coarse with textural fining towards the top. In its basis, coarse sandstones, as well as conglomerates predominate, sometimes reddish and with lateritic levels, possibly deposited in the late Pleistocene Age and, at the top, fine, medium, and white quartz sands predominate, having been deposited by the alluvial fans, of the Holocene Age (Assine and Soares 2004). Figure 5.1 shows a geological section of the basin at 18.5 °S, based on data from drilled wells, seismic lines and gravimetric analysis (i.e. the Bouger Anomaly), available in Ussami et al. (1999).

Hydrogeology

In hydrological terms, the headwaters of the river basin, which includes uplands and depressions (above 200 m of altitude), present natural processes that are similar to those of the traditional hydrographic basins, with permanent longitudinal-bound flux and reply time of hours or a few days in the rainfall-outflow relation. In the lower section of the UPRB (below 200 m), which includes the floodplain wetlands, the hydrological system has a more particular behaviour, due to the long time periods of drainage displacement and large flooding areas. As discussed below, changes in the UPRB's

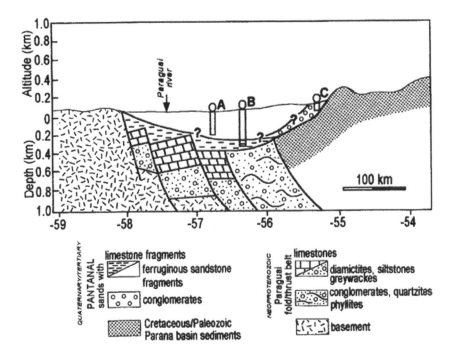

Figure 5.1   Geological Section (E-W) of the Pantanal Basin, at 18.5° S (based on drilled wells, seismic lines and gravimetric data)

Source: Ussami et al. (1999)

hydro-sedimentological behaviour have been translated into major impacts on the environment, the life of the regional population and the stability of aquatic systems.

The hydrogeological features of the Pantanal comprise the sediments of the aforementioned Pantanal Formation. The provisional designation is Quaternary Province (Brasil 1997b), Pantanal Aquifer (Brasil 2004), Cenozoic Formation (CPRM 2007), Pantanal Province (Figueiredo and Salomão 2009) and Cenozoic Aquifer System (SEMAC, 2010). Although this is not the only aquifer system of the UPRB, it is indeed the largest in the range of the outcropping area (information about the main aquifers of the Upper Paraguay River Basin can be found in Brasil, 2004). Despite the fact that groundwater is the main source of water in the Pantanal plains, no hydr-geological studies exist with details of the entire region. Limited hydrogeological surveys do exist, but only at small scales covering the entire UPRB, such as in UNESCO (1973), Brasil (1997b) and Brasil (2004). Likewise, a hydrogeological study for the State of Mato Grosso do Sul was carried out by Sanesul and Tahal (1998), but did not take into account the Pantanal floodplain.

For the purpose of this chapter, we will use the term Pantanal Aquifer System (hereinafter PAS), which is a porous, free aquifer, which presents flux towards the

Paraguay River (Brasil 1997b, Brasil 2004, Castelo Branco Filho 2005). The PAS is the source of water for most of the floodplain farms (although farmers rarely make use of the water from '*baías*', as the typical fresh-watered lagoons of the Pantanal landscape are known). The specialised literature has limited information about the PAS's groundwater storage. SEMAC (2010) introduced these calculations for the quaternary aquifer system, which has its largest part comprised of the PAS. According to such calculations, the Aquifer System would have 15.3×10^8 m^3.y^{-1} of renewable water reserves.

Drilled wells in the PAS are normally shallow, with mean depth of 46-54 (Brasil 1997b, Rebouças and Lastoria 1989, SEMAC 2010). The data of the wells inventoried by SEMAC (2010) refer to a small number of wells, certainly underestimating the actual number. Because of the weak regulatory system, there is no register of the drilled wells in the states of Mato Grosso and Mato Grosso do Sul. The static level of the wells varies from 2.4 to 7.4 m, the dynamic level from 5.54 to 22.0 m. The outflow of the wells varies from 6.0 to 22.0 m^3.h^{-1}. A synthesis of hydraulic data of PAS's wells is given in Table 5.1.

Table 5.1   Hydraulic Data of Some Wells Drilled in the Pantanal Aquifer System

City/Town*	Number of wells	Depth (m)	Static Level (m)	Dynamic Level (m)	Outflow (m^3.h^{-1})
Aquidauana (MS)	05	54.8	2.42	8.41	19.3
Barão de Melgaço (MT)	04	52.5	7.40	20.6	15.0
Corumbá (MS)	41	43.2	3.62	10.8	15.0
Poconé (MT)	01	85.0	5.00	21.0	8.3
Rio Verde de Mato Grosso (MS)	05	50.6	2.99	5.54	22.0
Santo Antonio do Leverger (MT)	01	37.0	3.00	22.0	6.0
Total/Average	57	53.9	4.07	14.7	14.3

* *MS – State* of Mato Grosso do Sul; MT – State of Mato Grosso

Source: Rebouças and Lastoria (1989)

Outflow lower (0.13 $m^3.h^{-1}$) and higher (30 $m^3.h^{-1}$) than those reported by Rebouças and Lastoria (1989) are described in Brasil (1997b), respectively for the dug wells and for some of the drilled wells in the PAS. The aquifer in the region of the Cuiabá River, in the State of Mato Grosso, presents wells with outflows between 5 to 50 $m^3.h^{-1}$, with average depth of 100 m and drawdown of 50 m (Figueiredo and Salomão 2009). Transmissivity data of $10^{-2} m^2.s^{-1}$ to $10^{-6} m^2.s^{-1}$ to the PAS in the Negro River basin has been reported by Castelo Branco Filho (2005).

In relation to the groundwater flow, it runs from East to West, with gradient towards the Southwest, in the direction of the Paraguay River, with an important water divisor, the Taquari River, the latter being an influent river, that is, it contributes to the aquifer recharging (Brasil 1997b), with a hydraulic gradient of 36 $cm.km^{-1}$ (Castelo Branco Filho 2005).

Groundwater Quality

There are no systematic hydro-chemical studies about the PAS. The physico-chemical analyses are mostly restricted to some parameters of the Brazilian standard of water potability. Despite the lack of data, the regional survey performed by Brasil (1997b) indicates the quality of PAS's water as being generally good, although it locally shows high levels of iron, characteristically having disagreeable taste and odour. Such a characteristic is also attributed to the presence of layers of organic matter in decomposition (Figueiredo and Salomão 2009). The pH increases from East to West, varying from 5 to 8, being more acid in the Northeast portion of the plain. The pH values in this interval were reported in studies in specific locations, such as Krol (1983), in the region of the Poconé, and Castelo Branco Filho (2005) in Nhecolândia. The latter author found a pH value of 8.2 in two wells in this region.

Regarding total dissolved solids, Brasil (1997b) reports values of 2 and more than 350 $mg.L^{-1}$, being these the highest located in the Nhecolândia region. Couldert (1973) quotes values of dry residue lower than 200 $mg.L^{-1}$, when the average is 65 $mg.L^{-1}$ in some wells in the region between Rivers Taquari and Negro. The author concluded that the waters present, generally speaking, low salinity, with mineralisation ranging from low to very low.

The hydrochemistry of the Pantanal's groundwater was evaluated in 17 wells on the right margin of the Rio Negro in 1971 and in 12 wells in 2004 (Castelo Branco Filho 2005). The electric conductivity varied between 46.0 to 1,000 μS. cm^{-1}, in the 1971 data, and from 67.3 to 723.8 μS.cm^{-1}, in 2004. These waters are characterized by being, predominantly, sodic bicarbonated, being some of them mixed-mixed bicarbonated and one well presenting sodic chloritic water.

Sediments, Heavy Metals and Surface Water

Discharge and Sediment Loads

According to Barbedo (2003), the transportation rates of sediments in the UPRB rivers have been increasing over time, with most sediment remaining deposited in the Pantanal (approximately 58%). Some of the watercourses (São Lourenço, Cuiabá, Upper Paraguay, Miranda, Taquari and Coxim) already have a solid load reaching undesirable values. The extinct agency DNOS (National Department of Sanitation Works) conducted measurements of discharge and sediment loads at the time of the greatest erosion in the basin, due to deforestation and the increase of agriculture in the 1970s and 1980s. The annual rates of solid load, within the period of intensification of the agricultural activity (the 1970s), increased by 66.8 per cent in the São Lourenço River, 50.8 per cent in the Cuiabá River, 48.3 per cent in the Taquari River and 41.3 per cent in the Miranda River. The permanence of the solid discharge in suspension is directly related to the permanence of the water discharge. As a means of comparing the results, permanence curves were calculated with discharges at every 30 minutes and a daily average. A total was also reached regarding the average flowing for the historical period between 1984 and 2005 available in Hidroweb (i.e. data base managed by ANA). When compared with previous assessments made by DNOS, the annual increase rates of sediment transportation in the water courses are now 1 to 2 per cent per year, much lower than previously.

The Taquarizinho River, which is an important tributary of the Taquari River with a catchment area of 494.7 km^2, has been the subject of more detailed studies that are particularly relevant for the purpose of our current discussion. The Taquarizinho River Basin is relatively vulnerable, especially if the areas next to the levees are deprived of primary or replanted vegetation. Furthermore, inside the main course of the river it was noticeable that there was sand deposited in the bed, diminishing the wet section. The river itself becomes the supplier of sediments downstream, because such material is carried even in the flux of the base outlet. Flores (2007) observed values of sediment production varying between 10 and 60 t.km^{-2}.year^{-1}. The in-suspension sediment transportation only practically occurs in the event of rainy weather. This can be stated due to the low concentrations verified in the samples of the base flux and in the permanent discharges established for some specific years. Thus, the intensity of the sampling must be concentrated in the basin's wetter period, in case there is interest of monitoring the sediment in suspension.

River Sedimentation

The rainy season in the Pantanal starts in November and continues through to March, when about 70 per cent of the annual rainfall occurs. In this period, intensive rainfall leads to a high potential for erosion (Brasil 2004). A survey

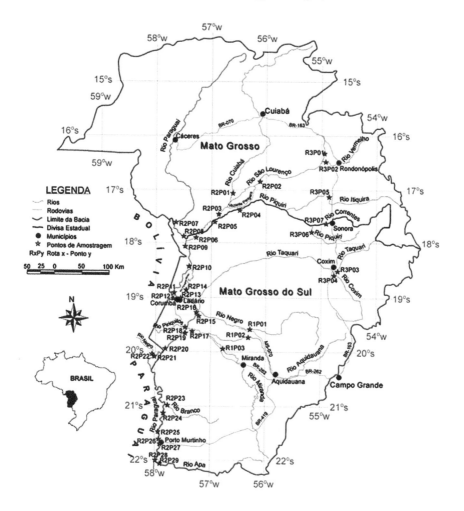

Figure 5.2   Location of Sampling Points

Source: Ide (2004), based in the Project ANA/GEF/UNEP/OAS

financed by ANA/GEF/UNEP/OAS (part of the sub-project developed by the Research Group in Environmental Technologies of UFMS that includes many authors of this chapter) was carried out to measure discharge and sediment loads and to collect water and sediment samples in the main rivers of the UPRB, both in the upland region and in the floodplains (i.e. the Pantanal itself). A map of the region under study with the location of the sampling points is depicted in Figure 5.2. Based on the regular monitoring performed between year 2002 and 2003, the current situation of UPRB's water quality could be assessed. Yet, because of the limited knowledge about the conditions of the UPRB's aquatic environments, it is impossible to identify both what has been causing their respective environmental problems and ways to solve them.

For instance, in a sampling carried out in January 2003 (in the locations included in the ANA/GEF/UNEP/OAS project), the total solid discharge was of approximately 800 tons/day at River São Lourenço in São José do Borireu (R2PO2), 29,000 tons/day in São Lourenço de Fatima, Coxim, close to the estuary (R3PO4) and 22,000 tons/day in the Taquari River (R3PO3). Thus, on the date of the sampling, only under the bridge of the Taquari River, in the town of Coxim, about 51,000 tons of sediments were recorded in the water.

The equivalent depth of the average degradation in the high basin can be between 0.092 and 0.36 mm.year^{-1}, a value that can be considered rather high. Deposition in the Pantanal over the last 25 years could have possibly reached 8.26 mm. This layer is not uniformly distributed, but its greatest part is contained in the river beds as sediment accumulates; part of it also distributed in the low-land by floods and re-suspension of the solid load. The average values resulting from the data processing, including the erosion and sediment deposition zones in the UPRB can be obtained in Brasil (2004).

Heavy Metals in the Soil

The Pantanal soils have developed from non-consolidated sediments, markedly sandy, with restricted areas of clay and organic materials, deposited throughout the Quaternary Period (Santos et al. 1997). According to Del'Arco et al. (1982), nearly the whole formation of the Pantanal happened with sediments deposited in the Cenozoic Era. The deposits along the slopes of the uplands and surrounding areas of the Pantanal occurred in the Pleistocene Era, but the present alluvia found in the low-lands of some rivers in the region was deposited in the Holocene Period (Fernandes et al. 2007).

There is limited information on the concentration of heavy metals in soils that could inform studies about the occurrence of contamination in the UPRB. Such information would also be relevant for the evaluation of the origin of the heavy metals found in the soil, whether they are natural or anthropogenic, and in the waters and sediments of the rivers. In effect, the absence of primary information has been one of the many difficulties faced by researchers and users. For instance, the region of the Taquarizinho River Basin has unique pedological characteristics, due to its geomorphology and the use and occupation of the catchment. It offers conditions for the development of environmental studies, such as the work carried out by the abovementioned Research Group in Environmental Technology of UFMS. This interdisciplinary project has the purpose of evaluating the impacts on the drainage of pluvial water, the production of sediments, the water quality, the use and occupation of the soil, among other aims. We make a special mention of this river basin, because it is an emblematic area regarding the Pantanal's problems of soil and water conservation.

Table 5.2   Average Values of Soil Fertility in the Taquarizinho River Basin

Depth	Texture	pH	pH CaCl$_2$	P	OM	K	Al	Ca	Mg	Na	H+Al	SB	CEC	CECe	V
(cm)	(%)	-		(mg.dm^{-3})	(%)					(mmol$_c$.dm^{-3})					(%)
0-10	29	5.3	4.6	7.8	2.0	2.3	7.7	12.8	16.2	0.1	41.9	31.5	73.3	39.2	42
10-20	30	4.9	4.2	4.2	1.7	1.6	8.0	10.7	14.3	0.1	33.5	26.7	60.2	34.6	42
20-30	31	4.9	4.3	3.8	1.6	1.3	9.6	6.2	11.5	0.1	40.3	19.1	59.4	28.7	32
30-40	32	4.9	4.3	2.3	1.5	1.2	11.1	5.6	11.4	0.1	53.3	18.3	71.6	29.4	26
40-50	35	5.0	4.3	1.6	1.2	1.2	12.0	4.5	11.1	0.1	55.5	16.9	72.4	28.9	26

Source: Saraiva (2007)

The Taquarizinho River drains the upland waters, and then flows into the Coxim River which subsequently flows into the Taquari River, one of the most important rivers that cross the Pantanal and a main connector of the upland to the floodplain. Saraiva (2007) evaluated the characteristics of the different types of soil of the Taquarizinho River Basin and observed that there were low levels of organic matter in all the sampled points (Table 5.2). In general, percentages of organic matter were low for all soils. If the organic matter is below 10 per cent the soil is considered inorganic or mineral and has a predominance of silica, clay and compounds of calcium, iron, manganese and others (Esteves 1988). The lack of organic matter highlights the lack of potential for adsorbing pollutants. In Brazilian soils the organic matter can contribute up to 80 per cent of the load of anions from the soil, which can explain the fact that the Cation Exchange Capacity (CEC) of such soils is mostly associated to organic matter. Thus, a variation in the quantity and quality of the organic matter can have huge effects on the proprieties and processes that occur in the soil system. The organic matter can also play important roles in nutrients recycling. Therefore, organic matter control, aiming at the conservation and improvement of its quality, is mandatory for the maintenance of the sustainability of the agro-systems in the UPRB. Clay and silt have greater capacity for retaining contaminants, thus preventing them from been leached and reaching the groundwater. The smaller-sized particles present greater potential for metal adsorbing (Förstner and Wittmann 1983).

A large variation of the CEC with depth was found in soils of the Taquarizinho River Basin, a fact attributed to the high levels of calcium (Ca) and manganese (Mg) in such soils. Agricultural inputs of these nutrients, associated with the low level of the existing organic matter, can exert a physical effect of occlusion of the oxide surfaces, which are mainly responsible for the CEC. Nevertheless, the samples with high CEC level also had high levels of clay. Probably, in this case, the clay contribution in these horizons was more relevant. The mean base saturation was low, in most depths. The concentration of copper (Cu), in the depth of 10–20cm (Table 5.3), gave a high value, relative to regulatory limits (Cetesb 2005, Brasil 2009). Saraiva (2007) verified this in two sampled points of the Taquarizinho River Basin soil, in depths of 20–30, 30–40 and 40–50 cm (Tables 5.2 and 5.3). The concentration of zinc (Zn) is above the reference values normally adopted in Brazil (Cetesb 2005, Brasil 2009); Iron (Fe), manganese (Mg) and aluminium (Al) do not have the prevention and investigation levels for soils (i.e. concentration of a certain substance in the soil or in the groundwater above which there are direct and indirect potential risks to human health, taking into consideration a standardized exposure scenario) (Brasil 2009). Metals, such as chromium (Cr) and lead (Pb) presented lower levels to those of prevention, according to Cetesb (2005) and Brasil (2009).

Table 5.3   Average Values of Metals in Soils in the Taquarizinho River Basin

Depth	Cu	Fe	Cd	Al	Cr	Zn	Ni	Pb	Mn
(cm)					(mg.kg⁻¹)				
0-10cm	8.45	3988	1.15	61.84	3.14	22.59	10.33	26.56	121.57
10-20cm	10.68	4004	1.16	63.28	3.38	27.40	11.07	27.41	118.60
20-30cm	8.58	4007	1.21	60.91	3.26	21.81	11.17	27.38	107.21
30-40cm	9.44	4026	1.23	64.98	3.38	23.54	12.16	28.16	94.70
40-50cm	9.95	4098	1.24	68.33	2.93	24.30	13.23	28.72	112.27

Source: Saraiva (2007)

The question of toxic levels of heavy metals in mineral fertilizers has given rise to many discussions. There is a broad band of variation within the tolerable limits in the levels of such metals in the fertilizers, in the rates of their application per hectare and in their maximum levels in the soil, in the legislation of several countries (Malavolta 2006). According to Dynia (2000), the higher the phosphorus (P) level observed in the soil, the greater the level of the obtained extractable cadmium (Cd). The total content detected in the soil does not allow a direct observation of its bioavailability. That is important because metals such as mercury (Hg), cadmium (Cd) and lead (Pb) present in the soil have been highly correlated with the total phosphorus in the soil, which can also provide evidence that the use of phosphorus fertilizers adds heavy metals to the soil (Jia et al. 2010). So as to lessen the transfer of the cadmium in the soil to plants, bioavailability of such an element must be reduced in the soil through the agricultural maintenance of the pH close to neutrality (Nawrot et al. 2010). Albeit just one of the points studied by Saraiva (2007), in the 20–30 cm layer of depth, presented more than 3 mg of Cd.kg⁻¹ of dry soil (Figure 5.3). Ten other points gave values greater than the regulatory standards of CONAMA 420/2009 (Brasil 2009), that is, more than 1.3 mg of Cd.kg⁻¹ of dry soil.

The above results are significantly important, because numerical parameters have been used for decision-making and to direct repairing or prevention strategies towards contaminated areas. Metal concentration allows the comparison of behaviours of elements in different systems, due to information about the

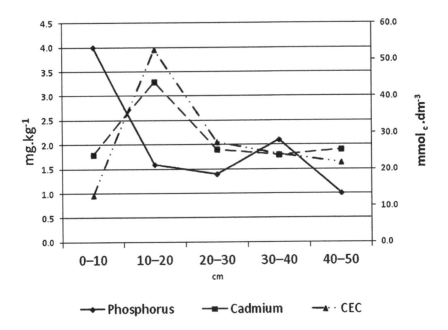

Figure 5.3   Relationship Between the Available of Phosphorus, Cadmium and the CEC

retention magnitude of potentially toxic elements. Because the adsorption of metals depends on many of the soil properties the concentration, when introduced in various chemical models, allows the performance of estimates of the quantity of metal diluted in the soil solution and the foreknowledge of its mobility, as well as the potential of loss through lixiviation and adsorption by the plants.

Monitoring Heavy Metals in Water and Sediments

The presence of heavy metals in high concentrations in the aquatic environment causes fish mortality and affects benthonic, periphytic, planktonic and nektonic communities and photosynthesising beings. Table 5.4 shows the heavy metals concentrations found in the water samples. Some rivers showed a concentration of heavy metals in the water, above the reference values for Class I of the CONAMA 357/2005 (Brazil 2005). The highest concentrations of Cu, Mn and Fe were found in the Vermelho and Coxim Rivers, in the rainy season. Lead was the element that was found throughout the UPRB at concentrations above the reference values for Class III of the CONAMA 357/2005 (Brazil 2005), which is 0.033 mg.L^{-1}.

The bed sediments perform the most important role in the pollution scheme of the river systems, because they reflect the current quantity of heavy metals in the aquatic system and can be used to detect the presence of contaminating substances

Table 5.4   Concentrations of Heavy Metals in the Water

River	Location	Cr	Cu	Mn	Fe	Ni	Cd	Pb	Zn
					(mg.L^{-1})				
Negro	Fazenda Rio Negro	<DL	0.013	0.178	1.692	0.024	0.008	0.342	0.093
Aquidauana	Fazendo Porto Ciríaco	<DL	0.023	0.090	3.120	0.021	0.008	0.376	0.130
Miranda	Fazenda Guaicurus	<DL	0.009	0.084	2.487	0.022	0.009	0.299	0.040
Cuiabá	Fazenda São João	<DL	<DL	0.066	3.892	<DL	0.002	0.085	0.079
São Lourenço	Fazenda São José do Borireu	<DL	<DL	0.082	5.884	<DL	0.002	0.148	0.123
Piquiri	Fazenda São José do Piquiri	<DL	<DL	<DL	0.951	<DL	0.001	0.107	0.094
Paraguay	Bela Vista do Norte	<DL	<DL	0.003	1.992	<DL	0.001	0.178	0.117
Paraguay	Ladário	<DL	<DL	0.009	2.564	<DL	<DL	0.185	0.125
São Lourenço	São Lourenço de Fátima	<DL	0.006	0.107	7.249	<DL	<DL	0.164	0.165
Vermelho	upstream S. Lourenço River	0.025	0.032	0.940	34.891	<DL	0.001	0.245	0.224
Taquari	upstream of the Coxim River	<DL	0.006	0.176	7.154	<DL	<DL	0.140	0.165
Coxim	upstream of the Coxim River	<DL	0.031	0.522	21.567	<DL	<DL	0.169	0.189
Itiquira	Fazenda Porto Seguro	<DL	<DL	0.045	2.355	<DL	<DL	0.113	0.136
Piquiri	Montante do Corrente	<DL	<DL	0.096	4.492	<DL	<DL	0.143	0.221
Correntes	Ponte de Pedra hydropower	<DL	<DL	0.003	0.881	<DL	<DL	0.110	0.092

Note: DL – Detection Limit concentration (mg.L^{-1}): Chromium - 0.006; Copper – 0.003; Nickel - 0.001; Cadmium – 0.001; Zinc - 0.001

Source: Ide (2004)

Table 5.5　Heavy Metals Concentrations in Sediments

Collection Points	Layer (cm)	Ca	Al	Mg	Fe	Cr	Ni	Cu	Pb	Mn	Co	Cd	Zn
							($mg.kg^{-1}$)						
R1P01-1	0 – 4	125	2978	372	1030	5.4	6.7	5.8	9.0	351	13.9	1.2	23.9
R1P02-1	0 – 4	154	7439	1303	1201	12.4	20.0	45.2	14.9	323	25.0	1.9	48.4
R1P03-1	0 – 3	156	4081	894	921	7.5	9.3	12.0	6.8	151	9.8	0.9	27.0
R2P01-1	0 – 5	220	6059	1272	4769	25.9	18.8	14.9	21.6	250	17.5	1.2	45.9
R2P02-1	0 – 4	211	4455	1015	4562	24.4	13.7	12.2	15.9	239	17.1	1.3	39.6
R2P04-1	0 – 4	157	5560	587	4289	20.2	16.1	14.3	18.2	196	18.1	1.0	36.8
R2P07-1	0 – 5	210	5073	865	4916	21.5	14.1	13.2	16.6	209	13.4	0.9	36.0
R2P12-1	0 – 4	216	1861	513	4753	9.9	5.7	5.2	8.5	236	10.3	0.6	31.9
R2P13-1	0 – 4	212	4567	834	4645	19.8	10.2	10.1	17.8	229	16.3	1.1	39.7
R2P14-1	0 – 5	204	5017	864	4953	19.8	11.8	15.2	18.4	272	19.7	1.4	39.3
R2P22-1	0 – 5	196	2655	846	4651	11.2	6.8	7.5	12.1	188	11.0	1.3	32.4
R2P27-1	0 – 5	199	3356	791	4780	13.8	8.2	6.6	12.8	245	12.7	1.2	32.5
R2P29-1	0 – 4	183	6875	1471	4847	22.6	17.0	24.0	21.5	276	20.5	1.7	43.1

R3P01-1	0 – 5	178	5511	664	807	8.7	12.7	7.9	10.5	200	10.3	1.1	30.4
R3P02-1	0 – 5	193	2336	549	804	12.8	16.5	10.8	22.5	336	20.5	2.0	45.8
R3P03-1	0 – 5	200	4743	919	677	8.3	11.2	5.9	17.4	250	12.5	1.6	29.3
R3P04-1	0 – 8	213	2951	977	601	3.6	6.0	7.6	13.8	170	11.8	1.7	22.1
R3P05-1	0 – 5	111	3300	438	549	2.9	7.0	5.3	13.6	117	11.4	1.3	19.6
R3P06-1	0 – 6	192	4857	880	546	7.7	12.8	7.3	20.2	351	22.6	1.9	40.3
R3P07-1	0 – 6	40	8607	302	538	9.1	15.7	15.2	29.6	93	16.6	1.7	31.9

Source: Sampaio (2003)

which do not remain soluble after having been launched into the surface waters. Even more than that, the sediments act as carriers and possible sources of pollution, once the heavy metals are not permanently fixed by them and can be re-distributed in the water, as a consequence of changes in the environmental conditions, such as the pH, potential redox or presence of organic chelates.

The natural mechanisms for sediment formation have been intensely changed by the behaviour of human beings in the UPRB. Table 5.5 displays the summary of the results obtained from the analysis of heavy metals in the sediments. Only the results of the superficial layer are presented, due to their being more prone to being carried by pluvial drainage. Variations have been noticed in the concentration, in the deeper layers and in the different sampling locations, which have also been analysed. The complete results can be found in Sampaio (2003).

Organic and Inorganic Pollution in Surface Water

Water pollution has as its major sources domestic effluent, the urban and rural diffuse load and industrial effluents. In the case of the UPRB the main pollutant sources are domestic and rural, except near the major urban centres where some industrial activity exists. To achieve pollution control of the water of rivers and reservoirs, quality standards are used to define concentration limits, with which each substance present in the water must comply. In Brazil such standards depend on the classification of the interior waters (freshwater), which is established according to their predominant uses. Brazilian legislation, especially the regulation CONAMA 357/2005 (Brasil 2005), established the classification of water bodies and rules for the discharge of effluents. According to CONAMA 357/2005 (Brasil 2005, the typology of water bodies goes from Special Class, the most pristine, to Class IV, the worst.

The concern about pollution as a consequence of the pluvial waters is not new. In fact, in many areas, one realizes that such waters contain substantial amounts of impurities, being comparable to the most severe pollutant sources, such as domestic sewage (Ide 1984). When the rainfall starts to run off, it can carry nutrients, fertilizers, pesticides, heavy metals, biological species, hydrocarbons derived from petrol and eroded materials. In general the pluvial runoff contains all the pollutants deposited in the soil surface. Whenever it rains the runoff carries such accumulated materials away towards the superficial water bodies, thus constituting a pollution source. Several of such pollutants can possibly come from other areas, brought by the wind. The use of the soil is presumably the main variable, since it affects all the types of pollutant generators, through the activities performed by humans and by nature. For example, high concentration of metals in the dust and litter are expected in the surroundings of industrialized areas; petrol derives from areas with traffic; nitrogen, phosphorus and potassium in agricultural areas. Especially in rural areas erosion can be the major supplier of organic matter and of nutrients to the watercourses. Soil erosion is a selective process in which the fine particles are more vulnerable than the coarser soil fractions. Erosion, besides polluting the watercourses with organic matter, impels sediment deposition in their beds and in the reservoirs existing therein. It is also worth acknowledging the impact generated by urban runoff on the watercourses.

Table 5.6 presents the arithmetic average, the standard deviation and the extreme values (minimum and maximum) of the analysed parameters in the six sampled events in the Taquarizinho River. Concerning the qualitative observation of the water, it became clear that there was variation of some parameters, such as colour and turbidity of the river water. A discrepancy between the peak times in the pollutogram compared to the hydrogram was observed in all the events, which suggests a delay in the pollutogram in relation to the comparable record in the hydrogram. Oliveira (2007) noted the impact of diffuse loads in the Taquarizinho River Basin (rural area). Among the parameters analyzed, there are those that exceeded the standards established by CONAMA Resolution 357/2005 (Brasil 2005), some even for a short time, exceeded the standards of Class IV.

Table 5.6   Water Quality Parameters of the Taquarizinho River

Parameters	Unit	Average	Standard Deviation	Extremes Minimum	Extreme Maximum
Air temperature	°C	24	4	14	33
Water temperature	°C	25	3	19	29
Dissolved oxygen	$mg.L^{-1}\ O_2$	6.7	0.3	5.8	7.1
$BOD_{5,20}$	$mg.L^{-1}\ O_2$	2.7	2.2	0.9	13.1
COD	$mg.L^{-1}\ O_2$	30.17	15.74	6.69	63.29
Total coliforms	MPN.100mL^{-1}	1.21E+05	1.11E+05	1.47E+03	2.60E+05
Escherichia coli	MPN.100mL^{-1}	5.01E+04	6.92E+04	1.80E+02	2.40E+05
TKN	$mg.L^{-1}\ N$	1.26	0.72	0.21	3.10
Ammoniacal nitrogen	$mg.L^{-1}\ N$	0.68	0.51	0.11	2.39
Total phosphorus	$mg.L^{-1}\ PO_4^{-3}$	0.278	0.200	0.001	0.809
Clorets	$mg.L^{-1}\ Cl^-$	5.0	1.9	0.9	7.9
Apparent colour	$mg.L^{-1}Pt$	89	212	0	880
Turbidity	NTU	448.2	694.6	8.7	2560.0

Parameters	Unit	Average	Standard Deviation	Extremes Minimum	Extreme Maximum
Conduciveness	µS.cm^{-1}	15.48	3.73	10.02	22.30
PH	-	6.57	0.30	5.99	7.00
TS	mg.L^{-1}	413.4	425.3	39.0	1420.0
Cadmium	mg.L^{-1}Cd	0.010	0.005	<DL	0.022
Lead	mg.L^{-1}Pb	0.149	0.062	0.033	0.265
Zinc	mg.L^{-1} Zn	0.096	0.082	<LD	0.348
Chromium	mg.L^{-1}Cr	<DL	<DL	<DL	<DL
Copper	mg.L^{-1}Cu	0.007	0.011	<DL	0.047
Manganese	mg.L^{-} Mn^{1}	0.222	0.225	0.012	0.933
Iron	mg.L^{-1} Fe	4.033	2.891	0.658	10.688
Nickel	mg.L^{-1} Ni	0.052	0.026	<DL	0.108

Note: DL – Detection limit concentration (mg.L^{-1}): Cadmium – 0.002; Chromium - 0.006; Zinc - 0.001; Copper– 0.003; Nickel - 0.001.

Source: Oliveira (2007)

In general the diffuse loads' impact in urban basins is greater than in the rural areas. It is clearly noticeable that there are nutrients loss and transportation from agricultural and grazing areas. In the Taquarizinho Basin the impact caused by the diffuse discharges in the rural area was noticed. The metals concentrations found in the runoff of the pluvial waters showed wide fluctuations, zinc being the one that showed the highest variation. The Taquarizinho Basin has an extensive cattle breeding area, which is probably the main cause of the concentration of nutrients. It is relevant to add that about 16.6 per cent of the total drainage area is covered with forest and natural prairies, which reasonably suggests the presence of wild animals. In the drainage area there are no discharges of domestic and industrial effluent upstream to the monitoring section; therefore, the main sources of *E. coli* are likely to be from cattle and undomesticated animals (Oliveira 2007).

Figure 5.4 shows the key-curves of in-suspension sediments of the Taquarizinho River. When relating the runoff data and the discharge of in-suspension sediments (Qss in Figure 5.4 left), it is noticeable that the greatest data dispersion occurred for the flow-off higher than 5 m^3.s^{-1}. There was a strong relationship between SS concentration and discharge (Css in figure 5.4 right).

Considering now the calculation of the water quality index (WQI), it can be noted that the NSFWQI most reliably represents the actual quality of the water body, whereas the application of the WQI$_{SMITH}$ reveals, in a more explicit way, the responsible parameter for the most critical quality of the water. In the event of a

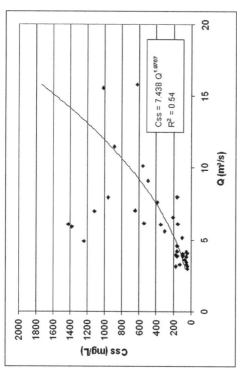

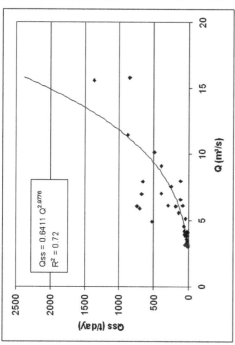

Figure 5.4   Key-Curves of In-suspension Sediments

Source: Flores (2007)

monitoring plan, the application of both the rates for the follow-up of the quality evolution is desirable. Nonetheless, the regionalisation of the aforementioned rates must be carefully studied. The application of such rates indicates that, pursuant to the NSFWQI modified by Cetesb (WQI_{CETESB}), in most of the collection points the water quality was 'good' on the sampling date except in the Vermelho and Coxim Rivers. Both had their quality influenced by the waters of the pluvial drainage, once the collections were performed in the rainy periods, which matches the conditions raised in the field work. Nevertheless, the WQI_{SMITH} showed, for the same collection points, quality variations ranging from 'very bad' to 'good', with the high levels of Turbidity and the *E. coli* responding for the classification. However, as a means of acknowledging such a characteristic as correct, a more systematic monitoring of such rivers would be required.

Comparisons with the limits of CONAMA 357/2005 (Brasil 2005) show that some parameters do not seem to correspond to the UPRB's reality. An example is the concentration of phosphorus, where the limit value is rather below the obtained results that can be considered as natural in some of the monitored points of the UPRB rivers, due to the contribution of undomesticated animals. In others there is also the contribution of agricultural and cattle breeding activities and of domestic and industrial effluents. This same observation is valid for several parameters, such as *E. coli*. The complete results and discussions regarding the quality of the water in the rivers, focusing on the large area investigated and the large number of sample points, can be found in the website of the National Water Agency (www. ana.gov.br).

Silva (2004), in a study on the distribution and transportation of mercury in the UPRB, realized that mercury (Hg) is associated to the particulate matter in suspension (7 at approximately 2,700 ng/g). He also found higher values in the active sediments, in the Barão do Melgaço region, probably resulting from the water characteristics, prospecting activities and from the geomorphology of the region which facilitated their accumulation. In some species of fish, such as *barbado* (*Pinirampus pinirampu*) and *dourado* (*Salminus maxillosus*), he found high levels of mercury. In other species mercury levels between 200 ng/g and 815 ng/g were found. Nonetheless, despite all the collected species of fish presenting mercury levels lower than the established limit in the Brazilian legislation, 37 per cent of the samples were above the limits for river-water fish, indicating the existence of mercury in the UPRB aquatic environment.

Contamination by Pesticides

Most of the pesticides normally used by farmers have a short average lifespan, although their active principles (i.e. the main ingredient) can have a longer lifespan. Table 5.7 presents examples of the daily pesticide loads estimated at several sampling points. Pesticides comprise numerous chemical substances, which have diverse form of action and different toxicity levels. Currently, there are about 1,000 active principles in the world. In Brazil, there are 400 active principles with 7,000 different formulations registered with the Sanitary Protection Division of the Ministry of Agriculture. Hormones that regulate growth and chemical and biochemical products for veterinary use pose similar risks to environment and society.

Table 5.7   Daily Charges of Pesticides

Location	Pesticide	Concentration (ppb)	Discharge ($m^3.s^{-1}$)	Charge ($kg.day^{-1}$)
Cuiabá River (São João farm)	Diazinone	1.77	246.00	37.62
	Prometryne	38.64		821.27
	O,P' DDE	6.29		133.69
	P,P' DDE	14.65		311.38
São Lourenço River (São José do Borireu farm)	Prometryne	2.98	180.72	46.53
	P,P' DDE	0.78		12.18
Coxim River (mouth of the river)	Epoxide Heptachlor	0.19	288.82	4.74
Itiquira River (Porto Seguro farm)	Epoxide Heptachlor	0.04	159.54	0.55
Piquiri River (upstream of Correntes River)	Heptachlor	0.04	38.68	0.13
Correntes River (UHE Ponte de Pedra)	Prometrina	0.15	139.82	1.81
	Epoxide Heptachlor	0.32		3.87

Source: Troli (2004)

Since 1985 some pesticides have had their sale and utilisation prohibited in Brazil. These are known worldwide as the 'dirty dozen' pesticides. Even so, in the Pantanal, the concentrations of Epoxide Heptachlor (one of the 'dirty dozen') still widely exceed the maximum levels allowed by Brazilian legislation, CONAMA 357/2005 (Brasil 2005). Another forbidden substance is DDE, a metabolite of DDT and a potentially cancerous substance, which has been found in the Cuiabá and São Lourenço Rivers at extremely high levels. What is worrying is the possibility of bioaccumulation of such active principles in the aquatic fauna, consequently, reaching the top of the food chain.

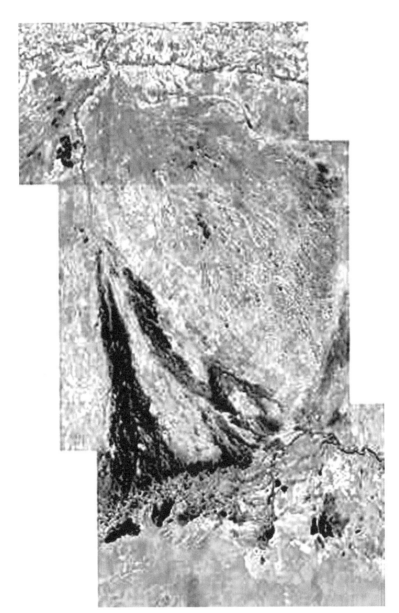

Figure 5.5   Alluvial Fan of the Taquari River

Source: Gonçalves (2005)

The Degradation of the Taquari River Basin: A Case Study of Water and Soil Degradation

The Taquari River Basin has two different regions: the upland area *comprised* of lands with elevations above 200 m, called Upper Taquari Basin (UTB), and the area called Low Taquari, which is within the Pantanal. The UTB has as its main tributaries, the Rivers Taquari, Coxim and Jauru. In the region of the Pantanal, the Taquari River does not have any other tributaries, instead the drainage in the region is divergent (Collischonn and Merten 2000). This unusual morphological situation forms the alluvial fan that characterizes the region (Figure 5.5). As the river flows to the interior of the Pantanal its margins become lower, gradually taking the water downstream to the alluvial plain along its course, with the water being partially lost to the plain during the great floods until it reaches the main course: the Paraguay River. The alluvial fan covers an area of 50,000 km^2 (36% of the Pantanal area) and is characterized as the deposition zone of the sediments that come from the high basin, in the adjacent uplands (Mato Grosso do Sul 2004).

Another phenomenon that occurs in the region, due to the influence of the loss of soil and transportation of sediments coming from the uplands, is shown in Figure 5.6.

The situation is critical where there is straightforward evidence of the significant volume of water lost by the "*arrombados*", an avulsion or rupture of the levee occurring in the situation of a permanent flood (Ide 2004). The phenomenon of "*arrombado*" represents an environmental problem with socioeconomic consequences for the Pantanal region, including the abandonment of almost 180 productive farms and the exodus of approximately 1,000 families (Figure 5.7). In the measurements performed in November 2002, we estimated that 47.5 per cent of the water discharge of the Taquari River was lost through the "*arrombados*".

Between 2002 and 2003 additional surveys were conducted in the Taquari River to study the effect of river migration and to propose a solution to such an effect in areas under permanent risk of flooding. The case study area was in the Nascente Farm at 18°12'45" S (Latitude) and 55°45'18" W (Longitude), approximately 200 km from the city of Coxim. The area is near the "Arrombados do Caronal", a well-known, permanently flooded area responsible for the loss of 1,000,000 ha of productive soil; this case study area is highly important because it is a main area for the formation of a new "*arrombado*" predicted to flood nearly 4,000 ha.

The formation of those "*arrombados*" is the result of a combination of factors. These include the lateral migration of the river with the rupture of the Taquari River levee (in the Pantanal plains), the large supply of sediments (due to deforestation) and the absence of a forested bank levee. To verify the real situation in the case study area bathymetric observations were made using a geodesic theodolite in order to portray an accurate map and the profile of the area. Based on such results a control area was determined and it could then be deduced that this area had some elements that induce lateral migration. The detailing of the control section showed that the section had the same configuration that appeared in Julien (2002), who

Figure 5.6   Voçoroca (gully erosion) in the Uplands of Taquari River Basin, February 2002

Source: Gonçalves (2005)

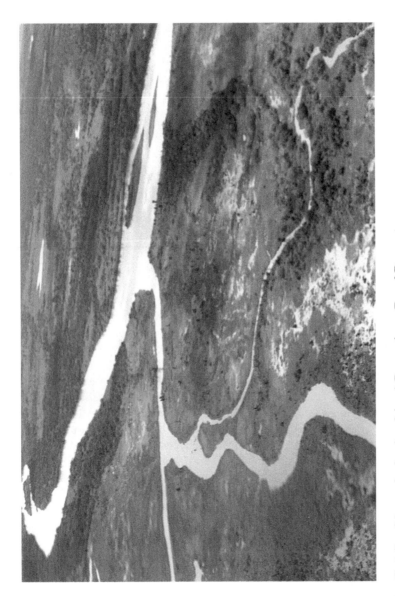

Figure 5.7   Details of Arrombados (avulsions) Damaging a Rural Property

Source: Gonçalves (2005)

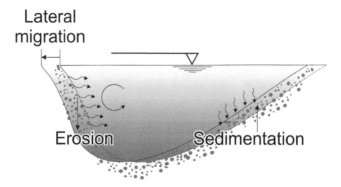

Figure 5.8   Lateral Migration Process

Source: Julien, P.Y. 2002. River Mechanics. Cambridge University Press (reproduced by authorisation)

explains that in the case of erodible bed material, secondary flows will induce scour at the toe of the outer bank, leading to bank caving and lateral migration. Stability can only be reached when the bank caving balances the deposition on the point bar on the inner bank (Figure 5.8). The control section is shown in Figure 5.9 and demonstrates the possibility of lateral migration.

In order to follow the lateral migration and its velocity a series of assessments were carried out on the right bank of the Taquari River. The average bank loss was of nearly three meters in one year of observation with some losses of six to seven meters. Those effects are visible after the flooding of the Pantanal. In the summer, with the beginning of torrential rains in the region, the river level increases, flooding some areas. In the winter, with the dry season, the level returns to the river channel, leaving a degraded bank with no support and consequent landslides, increasing the velocity of the lateral migration. The results showed that the thalweg (i.e. the deepest channel of a river bed) runs near to the levee bank increasing the erosion due to the speed of high water; it should also be noted that the section control point of the levee is short, which increases the possibility of further ruptures and, ultimately, the permanent flooding in the area. Analysis of particles and the visual appearance of the soil samples show a soil with a predominance of silty, fine sand in the surface samples (i.e. 50 cm deep) and with the presence of layers of clayey silt around 1.0 meter deep.

Two campaigns were conducted aiming at collecting samples of the water to verify the sediment transport in the wet season and in the dry season. Fieldwork data indicate that the mass of total sediment discharge easily rises up to 27,000 tonnes a day in the summer. The data evidenced that the channel is extremely movable, preventing the creation of a feasible discharge control point in the case study. Based on all the information collected in the field and out of concern to provide a solution to avoid the levee rupture and consequent productive land loss, the idea of using textile geotube was taken into consideration. The proposal is to

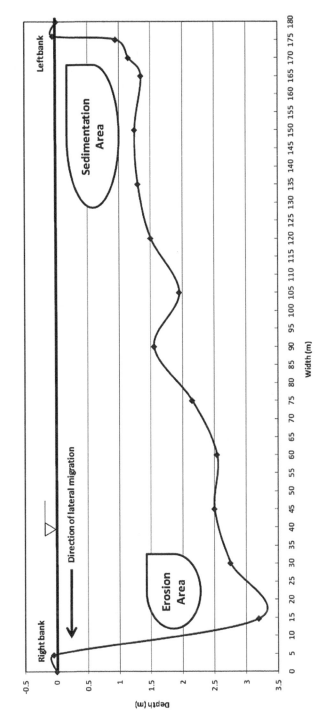

Figure 5.9   Control Section Bathymetry

Source: Gonçalves (2005)

use the geotube to deflect the thalweg to the centre of the river channel and thereby diminish the high velocity at the banks, decreasing the velocity of the erosion in problematic areas or simply stopping the erosion. On the other hand, sediment transport modelling was devised so as to determine the river characteristics and the use of the models HEC-6 and HEC-RAS (mathematical models from the Hydrologic Engineering Center (HEC) of the US Army Corps of Engineers) to determine what actually happens at different flow rates, thus realizing how the river works hydrologically with flows between 100 and 500 cubic meters per second. By analysing this behaviour, we conclude that geotubes may be sized according to their position. All models were calibrated based on data collected in such a way that put the predicted modelling the closest to reality. At the end of the mathematical modelling, a physical assessment was carried out to determine installation details of each geotube; computer-aided design (CAD) blue plants were also made for all geotubes. The conclusion is that such technology can be used in at-risk areas to create a barrier to the continued advancement of the lateral migration, consequently reducing the risk of the appearance of a larger area constantly flooded in that region.

An Overview of the Perspectives for an Integrated Management of Soil and Water

The superficial waters of the UPRB, as well as the soil used for agriculture, are currently the main foci of contamination by pesticides. Thousands of litres of pesticides are applied to the plantations, to the lands alongside UPRB's tributary rivers, not to mention the certified seeds, which are planted after having already been protected by fungicides. This happens in the rainfall season, when they are carried into the rivers, where only a minimal amount of dilution occurs and the rest is transported and deposited onto the riverbeds and floodplains, in which the finer particles remain, especially the clay. Modern pesticides have a very short average lifespan and, thus, their presence is rarely detected. The direct effect on the structure and fertility of the soil is questionable, but there are indirect effects that are much more difficult to analyse or quantify.

The unsustainable use of the soils is likely to bring about consequences that have not yet been adequately appreciated. It is true that erosion enlargement, sediment transportation in the rivers and the resulting sediment deposition will cause damage of several kinds. More frequent floods are expected, besides various other effects arising especially from sediments deposited in the rivers. Some water courses (São Lourenço, Cuiabá, Upper Paraguay, Miranda, Taquari and Coxim) already have significant solid loads; they are increasing with time and could reach undesirable values in the very near future. The "*arrombados*" (avulsions) in the Taquari River Basin are caused by vast amounts of sediment transported from the plateau to the plain area allied to low river flow speeds and fragile soil margin formation. This reinforces the conclusion that, in order to preserve soil and water in the UPRB, one needs to enlarge and implement a monitoring network to continuously assess water quality, to develop and implement reliable models, to establish an evaluation and decision support system, as well as establishing criteria for sediment quality.

Understanding the conditions of the Pantanal's aquatic environment, identifying what has been causing its environmental problems and establishing ways to solve them, necessarily involves the implementation and maintenance of permanent monitoring programmes, able to supply data showing temporal trends. The permanent data flux stemming from such monitoring programmes can use regionalized indicators for the control of the aquatic environmental conditions, which would allow immediate decision-making for its conservation and improvement. Such indicators would sum up the obtained information through the monitoring programme parameters, leading to the verification of the Pantanal's situation of evolution through time. Based upon such knowledge, the decision-makers who are concerned with the Pantanal could adopt more adequate measures.

It is vital to develop mechanisms to encourage the rational use of water, the disclosure of knowledge to the different sectors of users, the furtherance of education and the qualification of technical staff for catchment management, and also the transformation of scientific data into easily understandable information. As a means of guaranteeing equitable water access for all, sound management is required. This entails a combination of governmental structures, the identification of users and of the involved players, which can influence the decision-making related to hydrological matters, taking into account the different interests and objectives for its utilisation. In a basin that comprises trans-boundary waters, in an ever-more interdependent world, shared sovereignty among countries for integrated management, according to Calazans (2006), would be the most satisfactory way to encourage cooperation among the States towards the solution of disputes.

When the environmental issues result in trans-frontier effects, the national states cannot explore the natural resources without bearing in mind their neighbours (Preste 2000). This means that the rational use of regional water resources cannot be construed as a matter of convenience for each of the countries that share those waters. Legal harmonisation in the UPRB depends on the governmental action of each country involved, which is hindered by the different stages regarding the implantation of their respective policies for water management. There is no evidence that the sovereign position adopted by Brazil has impaired action towards the management of the trans-boundary waters at the UPRB. However, the international acts that established the basis for the integrated management of the shared resources, all supported and adopted by Brazil, Bolivia and Paraguay, continue to be elusive and have not actually been adopted in the UPRB.

Adequate action is needed for efficient, integrated management of the trans-boundary resources at the UPRB, particularly, and mainly, involving the political desire that this should be put into effect. As observed by Broch (2008) a new approach is necessary and should include a series of activities supported by common objectives and responsibilities and shared by governmental institutions, by society and by water users. This would feature a wider process than that induced only by government, since it would be supported by the participation of society in general and of the water users at the time of decision-making for planning in the UPRB. Integrated management of the UPRB would require a clear perception of the interdependent dynamics of the socio-ecological systems, which needs to assume a trans-disciplinary profile, with various aspects related to the fulfilment of the fundamental needs of human beings – local economies, means of subsistence, development – and also through participative processes of political decision-making (Vieira 2005). When analysing implemented or still-in-progress programmes and projects, we realise that, albeit their being incomplete, they have produced positive results, but the implementation of integrated management of catchment resources is still a challenge.

Planning and the management of the shallow waters in Brazil belong to the Federal and State Governments, in combination with sovereign control of the waters in Brazil (Brasil 1988), whereas the regulation of groundwater is a responsibility of state (provincial) administrations (Brasil 1997b) and the regulation of soil use and its occupation fall to the municipal governments. The managerial system, supported exclusively by the performance of inspections and by the policy of governmental control bodies, has been barely sufficient regarding groundwater and the absence of actual operation of the tools for the preservation of groundwater in the three countries, and poses a factor that brings difficulties to the integrated and shared management of the UPRB resources. In such a context the sound management of catchment resources means much more than the integration of intentions, since it requires the integration of common interests: of the involved governments, of the different users and of the society, with the objective of reducing the occurrence of conflicts arising from the use of the water shared by the aforementioned countries.

Ultimately, the success of integrated management in the UPRB depends on political, institutional, technical, professional, financial and human capacity, and also on the capacity for regulatory compliance, on the part of those who control and inspect and on the part of those who have to cope with restrictions, through the formal and informal structures. The governmental and non-governmental bodies would have to start, while performing their respective competences, "acting in an integrated way", so that the expected and desirable forms of management do not remain just "on paper".

Acronyms

ANA - Brazilian Water Agency

$BOD_{5,20}$ - Biological Oxygen Demand (oxygen consumed per litre of sample during 5 days of incubation at 20 °C; it is a surrogate of the degree of organic pollution of water)

CEC - Cation Exchange Capacity

CECe – effective Cation Exchange Capacity
Css - In-suspension Sediments concentration
COD - Chemical Oxygen Demand
CONAMA - Conselho Nacional do Meio Ambiente (National Environment Council)
CPRM - Serviço Geológico do Brasil (Brazilian Geological Survey)
DNOS - Departamento Nacional de Obras e Saneamento (National Department of Sanitation Works)
GEF - Global Environmental Facility
MPN - Most Probable Number
ng - nanogram (1/1,000,000,000 gram)
NSFWQI - Water Quality Index developed by the NSF
NTU - Nephelometric Turbidity Units
OAS - Organisation of American States
Qsa - Entrainment Sediment Discharge for Cross-section
Qss - Discharge of In-suspension Sediments
PAS - Pantanal Aquifer System
SEMAC - Secretaria de Estado do Meio Ambiente, do Planejamento, da Ciência e Tecnologia (Mato Grosso do Sul state's Secretariat of Environment, Planning, Science and Technology)
TKN - Total Kjeldahl Nitrogen
TS - Total Solids
TDS - Total Dissolved Solids
UFMS - Federal University of Mato Grosso do Sul
UNEP - United Nations Environment Programme
UNESCO - United Nations Educational, Scientific and Cultural Organization
UPRB - Upper Paraguay River Basin
UTB - Upper Taquari Basin
WQI - Water Quality Index
WQI_{CETESB} - Water Quality Index developed by CETESB
WQI_{SMITH} - Water Quality Index developed by Smith

References

Almeida, F.F.M. de. 1945. *Geologia do Sudoeste Matogrossense*. Boletim N° 116. Rio de Janeiro: Departamento Nacional da Produção Mineral, Divisão de Geologia e Mineralogia.
Almeida, T.I.R., Paranhos Filho, A.C., Rocha, M.M., Souza, G.F., Sigolo, J.B. and Bertolo, R.A. 2009. Estudo sobre as diferenças de altimetria do nível da água de lagoas salinas e hipossalinas no Pantanal da Nhecolândia: um indicativo de funcionamento do mega sistema lacustre. *Geociências*, 28(4), 401-415.

Assine, M.L. and Silva, A. 2009. Contrasting fluvial styles of the Paraguai River in the northwestern border of the Pantanal wetland, Brazil. *Journal of Geomorphology*, 113(3-4), 189-199.

Assine, M.L. and Soares, P.C. 2004. Quaternary of the Pantanal, wet-central Brazil. *Quaternary International*, 114, 23–34.

Barbedo, A.G.A. 2003. *Estudo Hidrossedimentológico na Bacia do Alto Paraguai - Pantanal*. MSc Dissertation. Campo Grande: UFMS.

Brasil. 1988. *Constituição da República Federativa do Brasil*. Available at: http://www.senado.gov.br/sf/legislacao [accessed: 19 November 2007].

Brasil. 1997a. *Lei 9433 - Institui a Política Nacional de Recursos Hídricos*. Brasília: Diário Oficial da República Federativa do Brasil.

Brasil. 1997b. *Plano de Conservação da Bacia do Alto Paraguai - Pantanal - PCBAP*, Volume 2, Tomo 2 - Hidrossedimentologia. Brasília: MMA/PNMA.

Brasil. 2004. *Implementação de Práticas de Gerenciamento Integrado de Bacia Hidrográfica para o Pantanal e Bacia do Alto Paraguai: Programa de Ações Estratégicas para o Gerenciamento Integrado do Pantanal e Bacia do Alto Paraguai, Relatório Final*. Brasília: ANA.

Brasil. 2005. *Resolução CONAMA nº 357 - Define Critérios e Procedimentos para o Uso Agrícola de Lodos de Esgoto Gerados em Estações de Tratamento de Esgoto Sanitário e seus Produtos Derivados, e Dá outras Providências*. Brasília: Diário Oficial da República Federativa do Brasil.

Brasil. 2009. *Resolução CONAMA nº 420 - Dispõe sobre Critérios e Valores Orientadores de Qualidade do Solo quanto à Presença de Substâncias Químicas e Estabelece Diretrizes para o Gerenciamento Ambiental de Áreas Contaminadas por essas Substâncias em Decorrência de Atividades Antrópicas*. Brasília: Diário Oficial da República Federativa do Brasil.

Broch, S.A.O. 2008. *Gestão Transfronteiriça de Águas: O Caso da Bacia do Apa*. PhD Thesis. Brasília: UNB.

Calazans, J.T. 2006. *Águas Transfronteiriças - Apostila do Curso de Capacitação para Jornalistas: "Água: Uma boa notícia"*. Cidade de Goiás: MMA.

Carvalho, N.O., Filizola Júnior, N.P., Santos, P.M.C. and Lima, J.E.F.W. 2000. *Guia de Avaliação de Assoreamento de Reservatórios*. Brasília: ANEEL.

Castelo Branco Filho, H. 2005. *Distribuição Espacial e Temporal das Características Hidroquímicas das Águas Subterrâneas do Pantanal do Rio Negro*. MSc Dissertation. Rio de Janeiro: UFRJ.

Cesteb. 2005. *Relatório de Estabelecimentos de Valores Orientadores para Solo e Águas Subterrâneas no Estado de São Paulo* Decisão de Diretoria Nº 195-2005- E. São Paulo: Cetesb.

Couldert, P. 1973. *Reconhecimento Hidrogeológico Preliminar no Pantanal Mato-Grossense entre os Rios Negro e Taquari*. Rio de Janeiro: Editorial DNOS.

Collischonn, W. and Merten, G. 2000. Análise de estabilidade de um rio no Pantanal utilizando um modelo matemático de transporte de sedimentos, in *Caracterização Quali-quantitativa da Produção de Sedimentos*, edited by J. B. D. Paiva. Santa Maria: ABRH-UFSM, 131-154.

CPRM. 2007. *Mapa de Domínios e Subdomínios Hidrogeológicos do Brasil.* Escala 1:2.500.000. Brasília: CPRM.

Del'Arco, J.O., Da Silva, R.H., Tarapanoff, I., Freire, F.A., Mota Pereira, L.G., Souza, S.L., Palmeiras, R.C.B., Tassinari, C.C.G. 1982. *Projeto RadamBrasil.* Folhas SE. 20/21, Levantamento de Recursos Naturais, 27. Rio de Janeiro: IBGE.

Dynia, J.F. 2000. Nitrate retention and leaching in variable charge soils of a watershed in São Paulo State, Brazil. *Communications in Soil Science and Plant Analysis,* 31, 777-791.

Embrapa. 2006. *Sistema Brasileiro de Classificação de Solos.* Rio de Janeiro: Editora Embrapa.

Esteves, F. A. 1988. *Fundamentos de Limnologia.* Rio de Janeiro: Interciência.

Fernandes, F. A., Fernandes, A.H.M., Soares, M.T.S., Pellegrin, L.A. and Lima, I.B.T. 2007. *Atualização do Mapa de Solos da Planície Pantaneira para o Sistema Brasileiro de Classificação de Solos.* Comunicado Técnico 61. Corumbá: Embrapa Pantanal.

Figueiredo, D.M. and Salomão, F.X.T. 2009. *Bacia do Rio Cuiabá: Uma Abordagem Socioambiental.* Cuiabá: Editora UFMT.

Flores, A.M.F. 2007. Análise e Estimativa do Transporte de Sedimentos em Suspensão Durante Eventos Chuvosos. MSc Dissertation. Campo Grande: UFMS.

Förstner, U. and Wittmann, G.T.W. 1983. *Metal Pollution in the Aquatic Environment.* 2nd edition. Berlin: Springer Verlag.

Gonçalves, F.V. 2005. *Controle de Migração de Margens do Rio Taquari Utilizando Geotubos como Elemento de Contenção.* MSc Dissertation. Campo Grande: UFMS.

Ide, C.N. 2004. Distribuição e transporte de agroquímicos e metais pesados na Bacia do Alto Paraguai (Subprojeto 1.5), in *Projeto Implementação de Práticas de Gerenciamento Integrado de Bacia Hidrográfica para o Pantanal e Bacia do Alto Paraguai,* edited by C.N. Ide et al. Brasília: ANA/GEF/PNUMA/OEA. Available at http://www.ana.gov.br/gefap/arquivos/Resumo%20Executivo%20Subprojeto%201.5.pdf [acessed: 10 november 2010].

Ide, C.N. 1984. *Qualidade da Drenagem Pluvial Urbana.* MSc Dissertation in Civil Engineering, Water Resources and Sanitation. Porto Alegre: UFRGS/IPH.

Jia, L., W. Wang, Y. Li, and Yang, L. 2010. Heavy metals in soil and crops of an intensively farmed area: A case study in Yucheng City, Shandong Province, China. *International Journal of Environmental Research and Public Health,* 7 (2), 395-412.

Julien, P.Y. 2002. *River Mechanics.* Cambridge: Cambridge University Press.

Krol, J.G. 1983. *Geological Studies of the Upper Paraguay River Basin (Pantanal).* Paris: UNESCO Press and Editorial DNOS.

Malavolta, E. 2006. *Manual de Nutrição Mineral de Plantas.* São Paulo: Editora Agronômica Ceres.

Mato Grosso do Sul. 2004. *Relatório de Qualidade das Águas Superficiais da Bacia do Alto Paraguai – 2002.* Campo Grande: SEMA.

132 *Tropical Wetland Management*

Meurer, E.J. 2004. *Fundamentos de Química do Solo*. 2nd edition. Porto Alegre: Gênesis.

Nawrot, T., Staessen, J., Roels, H., Munters, E., Cuypers, A., Richart, T., Ruttens, A., Smeets, K., Clijsters, H. and Vangronsveld, J. 2010. Cadmium exposure in the population: from health risks to strategies of prevention. *Biometals, Springer Netherlands,* 23(5), 769-782.

Oliveira, H.A.R. 2007. *Qualidade da Drenagem Pluvial Rural: Rio Taquarizinho – MS*. MSc Dissertation. Campo Grande: UFMS.

Preste, P. 2000. *Ecopolítica Internacional*. São Paulo: SENAC.

Rebouças, A. and Lastoria, G. 1989. *Interaction of Wetlands and Groundwater in Pantanal, Brazil. Proceedings of the 28th* International Geological Congress, 2, 680-681. Washington, 9-19 July 1989.

Sampaio, A.C.S. 2003. *Metais Pesados na Água e Sedimentos dos Rios da Bacia do Alto Paraguai*. MSc Dissertation. Campo Grande: UFMS.

Sanesul and Tahal. 1998. *Estudos Hidrogeológicos de Mato Grosso do Sul*. Relatórios v. I a V, 14; Mapas escala 1:500.000. Campo Grande: Sanesul.

Santos, R.D. dos, Carvalho Filho, A., Naime, U.J., Oliveira H., Motta P.E.F., Baruqui, A.M., Barreto, W.O., Melo, M.E.C.C.M., Paula, J.L., Santos, E.M.R. and Duarte, M.N. 1997. Pedologia, in *Plano de conservação da Bacia do Alto Paraguai – PCBAP,* Volume 2, Tomo 1 - Diagnóstico dos Meios Físico e Biótico. Brasília: MMA/PNMA.

Saraiva, M.A.M. 2007. *Metais Pesados em Amostras de Horizontes Superficiais de Solos como Indicadores Ambientais*. MSc Dissertation. Campo Grande: UFMS.

SEMAC. 2010. *Plano Estadual de Recursos Hídricos de Mato Grosso do Sul*. Campo Grande: Editora UEMS.

Shiraiwa, S. 1996. *Flexura da Litosfera continental sob os Andes Centrais e a Origem da Bacia do Pantanal*. PhD Thesis.São Paulo: USP.

Silva, E.C. 2004. Distribuição e transporte de mercúrio na Bacia do Alto Paraguai-MT (Subprojeto 1.4), in *Projeto Implementação de Práticas de Gerenciamento Integrado de Bacia Hidrográfica para o Pantanal e Bacia do Alto Paraguai,* edited by E. C. Silva et al. Brasília: ANA/GEF/PNUMA/OEA. Available at: http://www.ana.gov.br/gefap/arquivos/Resumo%20Executivo%20 Subprojeto%201.4.pdf *[accessed: 10 November 2010].*

Silva, A., Assine, M.L., Zani, H., Souza Filho, E.E. and Araújo, B.C. 2007. Compartimentação geomorfológica do Rio Paraguai na borda norte do Pantanal Mato-Grossense, região de Cáceres-MT. *Revista Brasileira de Cartografia,* 50(01). 73-81.

Silva, J.S.V. and Abdon, M.M. 1998. Delimitação do Pantanal Brasileiro e suas sub-regiões. Pesq. Aqropec. Bras., 33(Especial), 1703-1711.

Soares, P.C., Rabelo, L. and Assine, M.L. 1998. *The Pantanal Basin: Recent Tectonics, Relationship to the Transbrasiliano Lineament. Proceedings of the 9th Simpósio Brasileiro de Sensorimento Remoto,* CD-ROM, Santos, 11-18 September 1998.

Soares, A.P. Soares, P.C. and Assine, M.L. 2003. Areiais e lagoas do Pantanal, Brasil: herança paleoclimática? *Revista Brasileira de Geociências*, 33(2), 211-224.

Troli, A.C. *Praguicidas em rios da Bacia Hidrográfica do Alto Paraguai*, Campo Grande: MSc Dissertation. UFMS.

UNESCO. 1973. *Hydrological studies of the Upper Paraguay River Basin (Pantanal). Paris: UNESCO.*

Ussami, N., Shiraiwa, S. and Dominguez, J.M.L. 1999. Basement reactivation in a sub-Andean foreland flexural bulge: The Pantanal Wetland, SW Brazil. *Tectonics*, 18(1), 25-39.

Vieira, P.F. 2005. Gestão de recursos comuns para o ecodesenvolvimento, in: *Gestão Integrada e Participativa de Recursos Naturais: Conceitos, Métodos e Experiências*, edited by P.F. Vieira, F. Berkes and C.S. Seixas. Florianópolis: Secco/APED, 333-378.

Chapter 6

Systematic Zoning Applied to Biosphere Reserves: Protecting the Pantanal Wetland Heritage

Reinaldo Lourival, Matt Watts, Guilherme M Mourão,
Hugh P Possingham

Introduction

The Biosphere Reserve Concept

Biosphere Reserves (BR) were introduced into the conservation arena in the late sixties almost concomitantly with the launch of the United Nations' Man and the Biosphere Programme (MAB-UNESCO). They were conceived as a "way to a more sustainable future" or "spaces to reconcile people and nature" (UNESCO 2002). They represent the concerns of the United Nations to secure the protection of natural diversity while maintaining the cultural heritage of traditional communities. This concept is intended to be implemented through a network of reserves representative of all ecosystems, and should serve as models of "sustainable societies" and sustainable development (Batisse 1990). The current network includes 529 sites in 105 countries (UNESCO 2002, 2007), six of them located in Brazil.

The BR zonation model evolved with little formal guidance and was based on a nested three-zone scheme, popularised as the "egg model" (Fig. 6.1). The criteria/guidelines for the establishment of BRs were initially defined in the Seville strategy (Poore 1995), where the functional objectives of BRs were addressed (i.e. biological and cultural diversity, sustainable development and logistical role) through four objective-oriented tasks (UNESCO 1995):

1. Use BR to protect and conserve natural and cultural diversity.
2. Utilize BR as models for sustainable development.
3. Use them as logistic support for research, monitoring, educational and training.
4. To fully implement the concepts of BR, through the harmonization of the above functions.

The spatial configuration of Biosphere Reserves, based on "the egg" (Figure 6.1), have core areas, which are necessarily strict reserves (IUCN categories I - IV), and are nested within buffer zones which allow some non-consumptive use of biodiversity. The buffer is then surrounded by transition zones which accommodate a broader array of land uses, including agriculture, urban centres and other more intensive land use forms (Dasmann 1988, UNESCO 2002). When the BR model was conceived the key principles of systematic conservation planning were not yet developed and the goals and outcomes of each zone were laid out more as a checklist than a framework for planning (Possingham et al. 2006a).

Currently however the UNESCO-MAB is revising the goals and objectives of BRs in order to provide more objectivity for the BR model. Their aim is to review the framework of the BR in order to deal with the challenges of fragmentation and the effects of climate change. The Madrid Congress (MAB Programme 2008) aimed to substantially improve the BR model in four components: adaptive governance, zonation, science and capacity building and partnerships. However a review of the proposed Action Plan overlooks the key principles of Comprehensiveness, Adequacy, Representation and Efficiency known by the acronym CARE (Possingham et al. 2006a), which are central to systematic conservation planning and essential for accountability of BRs.

Currently the limitations of the BRs are similar to the flaws found in traditional *ad hoc* reserve design (Pressey 1994). The lack of clarity in problem definition and inexistence of an accountability framework and a mathematical formulation, prevent BR planners from optimising the selection of reserves according to their contribution to their multi-objective problem. Thus, systematic planning for BRs, need to integrate the qualitative and quantitative tradeoffs between objectives within and between zones (Batisse 1997, Rosova 2001), while providing quantitative assessment of target achievements (i.e. biodiversity, socio-cultural and economic sustainability).

The Upper Paraguay River Basin (UPRB) is a 360,000 km^2 watershed in central South America. It is part of the broader La Plata Basin, which is shared by Brazil, Paraguay, Bolivia, Argentina and Uruguay. It comprises an extensive network of rivers and wetlands of global biodiversity importance (Higgins et al. 2005). The Pantanal, the biggest of those wetlands, is located in the northern reaches of the UPRB (Assine and Soares 2004) with 70 per cent in Brazil and 30 per cent shared between Bolivia and Paraguay. Currently the UPRB is under pressure from several development projects including agricultural encroachment, expansion of primary industry and large investments in energy and transportation infrastructure (Junk and Nunes da Cunha 2005). The outcomes of such threats can be easily identified in and around the Pantanal wetland by the rates of conversion of natural habitats, accelerating erosion of headwaters, and siltation in the floodplain (Padovani et al. 2004, Veneziani et al. 1998, Wantzen 2003).

The original planning for the Pantanal BR was largely an *ad hoc* process, based on the results of the first series of priority-setting workshops of the National Biodiversity Programme (Ministério do Meio Ambiente 1999). Some of these flaws

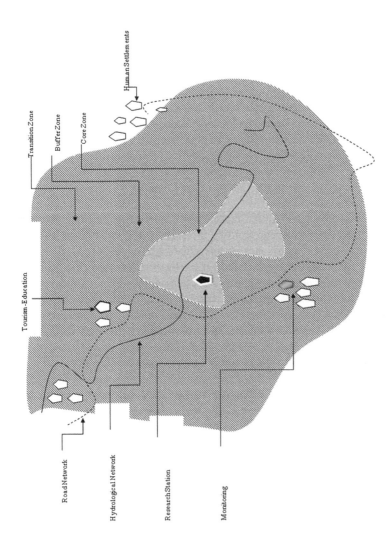

Figure 6.1 Biosphere Reserve Zonation Scheme Based on Nested Zones "The Egg Model" (The colours represent a 3 zone scheme comprised of core areas (doted), buffer zones (opaque) and transition zones (transversal)) Adapted from UNESCO-MAB at www.mab.com

were addressed for the Pantanal in a second round of priority-setting workshops (Ministério do Meio Ambiente 2007); this time around, plans were guided by some of the principles of systematic planning. Nevertheless the results of such plans have not been integrated in the Brazilian Biosphere Reserves System. For example indigenous territories and other key historical and biodiversity sites are absent inthe PBR map (e.g. enduring rockeries, archaeological sites and fortresses from the Paraguay war or even memorials of the Pantaneiro ranching history - Figure 6.2 a, b and c).

For the above reasons, Biosphere Reserves are an interesting case in systematic planning because of the opportunity they offer to shape the "reserve design problem" under an existing multiple objective/stakeholder framework, allowing researchers to explore emergent issues such as systematic zoning. This new BR model needs to be able to report on ecological, social and economic sustainability indicators, so that UNESCO-MAB programme has a solid quantitative basis, repositioning Biosphere Reserves as a true alternative for sustainable development.

Zoning in the Context of Systematic Conservation Planning

According to Possingham et al (2006a) systematic conservation planning involves finding the best set of potential areas to be protected while satisfying a number of principles (e.g. comprehensiveness, representativeness, adequacy, efficiency, flexibility, risk spreading, and irreplaceability). Systematic zoning aims to go a step further and help mainstream the principles of systematic planning to other sectors of society and their objectives, including traditional landholders, agriculturalists and even to the urban centers. These new challenges will enhance old problems such as data scarcity, bias and uncertainties associated with new disciplines (e.g. social science), therefore systematic zoning assumes a new dimension in complexity for planners (Gaston and Rodrigues 2003, Schoemaker 1991).

Although zoning has been used in urban planning and conservation for a long time (Murphy 1958, Ukeles 1964, Werner 1926) not surprisingly sharings the same *ad hoc* characteristics observed in non-systematic reserve design (Pressey 1994), it has the potential to incorporate the contribution of reserved and non-reserved areas, not only to biodiversity conservation but as part of a broader set of objectives within the planning framework, where different kinds of zones could cover a wider range of land management regimes (i.e. from strict reservation to intensive agriculture and urban development).

Systematic zoning for a BR needs to be, as generally postulated for all systematic conservation planning: data driven, goal oriented, efficient, spatially explicit, transparent, flexible and inclusive (Pressey 1999). Only recently however, systematic zoning started to be investigated (Moilanen et al. 2005) enabling planning in multi-objective/stakeholder contexts (Wilson et al. 2010). For example, forest managers used simulated annealing to harmonize site suitability and different forestry regimes (Bos 1993). Using similar tools Verdiell and Sabatini (2007, 2005) internally zoned a protected area designated as a world

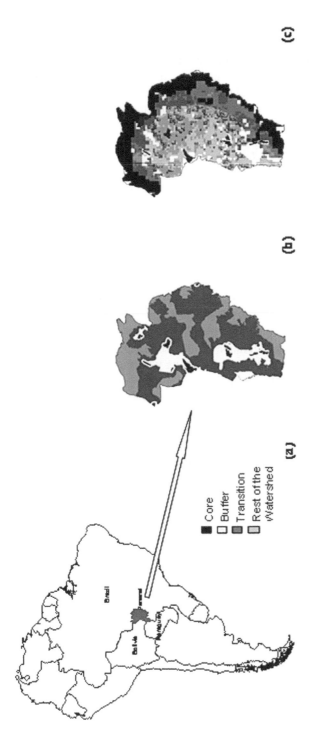

Figure 6.2   Pantanal in South America and Conservation Areas: (a) Map of the Pantanal Biosphere Reserve (ad hoc scenario), (b) Black represent core areas, light grey - buffer zones and dark grey the transition zone. Intermediary grey refers to areas within the boundary of the UPRB that were not included in the original ad hoc plan. Average acquisition cost surface for the Upper Paraguay River Basin, extracted from Lourival et al. (2008). Costs were scaled in 14 greyscale classes; (c) Existing protected areas are represented in black, indigenous territories in light grey and disturbed areas in light grey as in Padovani et al.(2004).

heritage site in Argentina. However, in both cases the claims of spatially explicit zoning were based only on spatial attributes such as compatibility and connectivity. They overlooked the explicit incorporation of other objectives in the optimization process, failing to explicitly target representation, complementarity and constraints of conflicting objectives in the actual problem definition (Possingham 2001).

In this chapter we use systematic zoning not just as a tool to spatially characterize management schemes, but also to integrate the principles of comprehensiveness, adequacy, representation and efficiency of CARE (Possingham et al. 2006a) in the BR model. We show how spatial variability of ecological, social and economic processes can be used within a spatially explicit tool for public engagement into the decision process, whereby objectives and goals are laid out and compromises are negotiated explicitly by stakeholders (Cabeza and Moilanen 2006, Dhargalkar and Untawale 1991, Higgins et al. 2004, Wilson et al. 2010).

Study Region: The Pantanal Biosphere Reserve

The Pantanal Biosphere Reserve (PBR), officially instated in the year 2000, extends through 25,156,905 ha of the UPRB. With a total of 664,245 ha, in nine core areas, and 5,392,480 ha, in buffer zones, and the other 19 million ha in transition zones. The PBR spreads through three Brazilian states; Mato Grosso, Goiás and Mato Grosso do Sul. The configuration of the PBR proposed to UNESCO was based on the recommendation of a priority setting workshop, conducted in 1998 (Ministério do Meio Ambiente 1999). A review of these strategies by Lourival et al (2009) found poor representation of important conservation features, lack of socio-cultural consideration, and low efficiency, indicating the need for a careful re-evaluation of the PBR configuration. In 2007, however, the original recommendations of the priority setting workshop were revised based on some of the principles of systematic conservation planning (Ministério do Meio Ambiente 2007). So far none of these recommendations have been used to review the PBR.

Zoning with Marxan

Marxan with zoning capabilities is a new development (Watts et al. 2009) of the widely used reserve design software Marxan (Ball and Possingham 2000 - called Marxan with Zones). This development represents a change from the traditional framework of binary decisions in systematic conservation planning tools, towards a multi-zone scheme that allows planning unit's assignment to several management regimes. Each planning unit, however, can be assigned to just one zone, and these assignments are conducted by the algorithm to maximize the efficiency across a range of land uses (Possingham et al. 2000). The software is able to optimise the compromises between site suitability and availability within a multi-objective and variable cost context.

The formulation of Marxan with Zones aims to capture the contribution of each distinct zone (management regime or land use) towards a predefined set of objectives, while still fulfilling the representation targets for each feature. These contributions may vary across sites and zones and are related to the constraints imposed on the software (e.g. acquisition costs and compatibility between zones, etc.). The achievement of such objectives, on the other hand, is pursued based on an agreed set of targets, using simulated annealing and iterative improvement algorithms according to the mathematical formulations described by Watts et al. (2009). Marxan with Zones generalizes Marxan's formulation (Ball and Possingham 2000) by increasing the number of states or zones to which a planning unit can be assigned.

Feature Representation in Biosphere Reserves

Regrettably we could not find any quantitative indicators or formally prescribed targets for any of the zones of Biosphere Reserves in the peer reviewed literature. The guidelines contained in the Madrid Action Plan (MAB Programme 2008) however, state that every zone should contribute to the achievement of the biodiversity, socio-cultural and sustainability objectives of a BR. This provides us with a crude idea on how we could zone BRs systematically.

Targeting features in a multi-objective scheme allowed control over the trade-off between feature representation and complementarity between zones, within a BR. It also provided an opportunity to regulate the spatial distribution of features only to suitable zones avoiding spatial juxtaposition of incompatible land-uses (emergent principle of permissibility). We have explored the spatial compatibility between management regimes in some detail, while assigning planning units to specific zones of the Pantanal Biosphere Reserve (PBR).

Our aim was to provide the framework and the guidelines to systematically zone Biosphere Reserves. We compared the representation, comprehensiveness and spatial configuration of *ad hoc* and systematically redesigned the Pantanal BR. We used a set of six scenarios to illustrate the capacity and flexibility of Marxan with Zones to: evaluate existing BRs, support the design of new core reserves, and help the rezoning process of existing reserves under the principles of systematic planning.

Our proposition uses the principles of the Madrid Action Plan (MAB Programme 2008), however, supported by the principles of systematic planning (Possingham et al. 2006a), which are missing in the official Action Plan. We specifically wanted to respond to two questions related to the BR model. How to optimise spatially explicit compromises of representation under a multi-zone and multi-objective context? Furthermore, what features of Marxan with Zones could be useful to handle the spatial constraints of the "egg model" (nested zonation), in the context of Biosphere Reserves in Brazil?

In order to respond to these questions we used the datasets from the Upper Paraguay Basin Conservation Programme as well as the targets and results of the priority setting exercise for the Cerrado and Pantanal (Ministério do Meio Ambiente 2007). We intend our solutions to support the biodiversity, socio-cultural and sustainability and logistic role of Biosphere Reserves (UNESCO 2002). The zoning problem as previously defined is solved by Marxan with Zones, allowing for spatially explicit trade-offs between these objectives for Biosphere Reserves. Comparing the existing map with the proposed new spatial configuration we aim to provide MAB UNESCO with some methods for using systematic conservation planning principles in Biosphere Reserve planning.

Method

This section describes the planning framework, the problem formulation and information layers used in systematic zoning, as well as the data treatments necessary to parameterize Marxan with Zones. Later we explore the sensitivity of the software to parameters such as zone compatibility, representation targets and costs. Finally we describe the scenarios we used to redesign the PBR based on a multi-objective framework, exploring the potential of Marxan with Zones for planning Biosphere Reserves in general.

Planning Framework for the Pantanal Biosphere Reserve

Considering that watershed-based management forges an overlap between several administrative and management structures, including the municipality, catchment authority, national and state parks, the RAMSAR areas and world heritage sites. We decided to expand the planning region designated to rezone the PBR to include the boundaries of the upper Paraguay River basin. This watershed was divided into a grid with 3,727 planning units, each one with an area of 10,000 hectares. We used 293 features including 117 modelled species distributions, environmental features such as soil, distance to rivers and vegetation, and socio-cultural elements such as indigenous land and traditional ranching as targetable features. They were systematically structured to provide representation and complementarity between the core, buffer and transition zones.

We searched for the most efficient spatial configuration for the PBR, under constrained and unconstrained scenarios, the Marxan with Zones algorithm was able to select and discharge planning units according to their contribution to targets.

Problem Formulation for Multiple Objective Zoning under Multiple Costs

Heuristic algorithms have been the method of choice to efficiently achieve biodiversity feature representation in systematic conservation planning (Possingham et al. 2000, ReVelle et al. 2002). Nevertheless, zoning, as assumed by the BR model, offers extra complexity to systematic planners for two reasons: (a) because each management regime/land use offers differential feature contribution to targets (MAB Programme 2008, Watts et al. 2009) and (b) because each land parcel, when allocated to a particular zone, has a variable cost structure (i.e. acquisition, management and maintenance) both affecting the way in which efficiency of BRs are measured (Pressey and Nicholls 1989).

The traditional formulation of the objective function used by Marxan (Ball and Possingham 2000) to optimise site selection under the minimum set coverage problem can be summarized by the expression:

Minimize the cost of the reserves system + Boundary cost + Feature representation shortfall penalty,

The objective function according to Watts et al. (2009) is now written as:

Minimize the configuration cost of all zones + Boundary compatibility cost + Feature and zone representation shortfall penalty, and represented by the equation:

Equation 6.1

$$\text{Min}\left\{\sum_{k=1}^{p}\sum_{i}^{m}x_{ik}c_{ik}+b\sum_{k}^{p}\sum_{i}^{m}\sum_{j}^{m}cv_{ij}z_{e_ie_j}+\sum_{l}^{p}\sum_{k}^{n}FPF_kFR_kH\left(ta_{kl}-\sum_{i}^{m}x_{il}a_{ikl}\right)\left(\frac{ta_{kl}-\sum_{i}^{m}x_{il}a_{ikl}}{ta_{kl}}\right)\right\} \quad (1)$$

where the element x_{ik} is 1 if planning unit i has been assigned to zone k and zero otherwise $\sum_{k=1}^{p}x_{ik}=1$ and where p is the number of zones, and m is the number of planning units. Meaning that, every element is assigned to one and only one zone. The set of values of x_{ik} is the whole reserve configuration as it tells which planning unit should be in each zone.

The first term in the equation (1) is the summed cost of the assignment of each planning unit to a zone within the whole configuration. The second term has been modified from Ball and Possingham (2000) to include a more sophisticated idea of connectivity, similarly to the boundary cost in Marxan. Here b is a weight to the relative importance that balances the cost of including a planning unit in the configuration for connectivity. The term $\sum_{ik}^{p}x_{ik}=1$ is the zone compatibility multiplier, where e_i is the zone which planning unit i has been assigned and e_j is the zone which planning unit j has been assigned. Typically this value will be zero if $e_i = e_j$ and take a different value otherwise. In an application where this connectivity is measured in terms of boundaries then it is the boundary cost incurred when two adjacent planning units are assigned to two different zones. While c_{ik} is the cost of zoning planning unit i as zone k and cv_{ij} is the connectivity value matrix, which controls the compatibility between zones and provides the juxtaposition pattern observed in the outputs, and the cost is given by the addition of configuration and connectivity. The last term account for the penalty for shortfall in representation, it is zero when all features meet their targets in that specific spatial configuration, and get larger with shortfall. The term was expanded from Marxan so that each feature can have a separate target for each zone. FPF and FR respectively describe the penalty factor and feature representation, FPF is also a weight that describes the importance of meeting the target for feature k. H is a step function that is zero when the term between the first parenthesis $\leq$ to zero and one otherwise. T_{ak} is the target amount for feature k, hence $-\sum_i x_{il} a_{lkl}$ is the amount of feature k in a configuration. The second parenthesis term measures the representation shortfall for feature k and is zero when representation meets the target exactly and one when the feature is not represented in the configuration.

Target Setting

In the initial scenarios we aim to represent 293 features ranging from watersheds, soils and vegetation classes to modelled species occurrences and wildlife densities in the zoning system. Features such as vegetation, watershed, freshwater eco-domains and soil types, targets were based on the minimum 20 per cent representation of the Brazilian forestry code and the rarity of these features (Lourival et al. 2009). Wildlife and cattle densities that are culturally and economically valuable were set at 30 per cent for each zone. It is important to emphasise, however, that the target values from these workshops were proposed under the assumption that the presence of a feature outside of strict reserves did not provide a contribution towards conserving that feature. Such an assumption is not sensible in the context of a BR systematic zoning since the presence and maintenance of genes, species and processes associated with biodiversity are not exclusive to protected areas (Negi and Nautiyal 2003, Wallington et al. 2005, West et al. 2006).

We aimed feature representation in three ways; first using the traditional feature targets for 293 features (e.g. species, vegetation, indigenous territories). For species representation we used the same targets as proposed by the PROBIO priority setting workshop and aim to represent the number of occurrences, or percentage of the total area, to be achieved across all zones. Then we used a "zone target" for 37 features, in which a minimum representation of a feature required in particular zone of the PBR. Finally assuming that different zones make different contributions to conserve a feature, we set the values for zone contribution to 292 features, expressing the differential capability of each zone to maintain each feature of interest in the long run (Table 6.1).

For data layers where features had spatially variable density, such as the aerial survey data for cattle, caimans, capybaras, marsh and pampas deer (Mourão et al. 1994, Mourão et al. 2000), we set specific zone targets. For example, we set targets for low and medium cattle densities specifically to buffer zones in the floodplain, assuming that they contribute to an important cultural attribute of the Pantanal. On the other hand, planning units that had high cattle densities were assigned to transition and available zones, with the purpose of achieving ranching sustainability goals, throughout the watershed.

Table 6.1 Targets set for Biodiversity (biod.), Socio Cultural (cult.) and Sustainability (sust.) Features for the Pantanal Biosphere Reserve

Features	Unit	Overall target	Biosphere Reserve objective	Zone contrib. (Available)	Zone contrib. (Transition)	Zone contrib. (Buffer)	Zone contrib. (Core)	Zone target (Available)	Zone target (Transition)	Zone target (Buffer)	Zone target (Core)
Density Cayman	L – M – H	0.3	biod-sust	---	0.6 or 1.0	0.75	1.0	---	---	---	---
Density Capybara	L – M – H	0.3	biod-sust	---	0.6 or 1.0	0.75	1.0	---	---	---	---
Density Marsh deer	L – M – H	0.3	biod-sust	---	0.6 or 1.0	0.75	1.0	---	---	---	---
Density Pampas deer	L – M – H	0.3	biod-sust	---	0.6 or 1.0	0.75	1.0	---	---	---	---
Distance to river	6 classes	0.2	biod-sust	---	1.0	1.0	1.0	---	---	---	---
Distance to road	18 classes	0.2	sust-cult	---	1.0	1.0	1.0	---	---	---	---
Freshwater domain	40 classes	0.2	biod-sust	---	0.5 or 0.7	1.0	0.25 or 1.0	---	---	---	---
Soil types	16 classes	0.2-0.3	biod-sust	---	0.5	1.0	1.0	0.1	0.32	0.2	0.18
Vegetation	38 classes	0.2-0.3	biod-sust	---	0.7	0.7	0.0 or 1.0	0.1	0.32	0.2	0.18
Vegetation subclasses	13 classes	0.2-0.7	biod-cult	---	0.7 or 1.0	0.7 or 1.0	0.0 or 1.0	0.1	0.32	0.2	0.18
Watersheds	20 units	0.2	biod-cult	---	0.5	0.9	1.0	0.1	0.32	0.2	0.18

Features	Unit	Overall target	Biosphere Reserve objective	Zone contrib. (Available)	Zone contrib. (Transition)	Zone contrib. (Buffer)	Zone contrib. (Core)	Zone target (Available)	Zone target (Transition)	Zone target (Buffer)	Zone target (Core)
Indigenous Land	26 reserves	0 or 1	cult -biod	-----	0.5	1.0	1.0	0.0	0.0	1.0	0.0
Protected Areas	19 reserves	0 or 1	Biod	-----	0.0	0.5	1.0	0.0	0.0	0.0	1.0
Deforest-ation	Pres/Abs	0	sust.	-----	-----	-----	-----	0.5	0.5	0.0	0.0
Cattle density	4 classes	0.3	cult-sust	1.0	1.0	1.0-0.75 0.75-0.0	0.0	0.7	0.7	0.0	0.0
Species models	117 species	0.3 to 1	biod.	-----	0.5, 0.6 or 0.75	0.6, 0.75 or 1.0	1.0	-----	-----	-----	-----

Note: Targets are presented in the following order: Overall representation targets, contribution of features by zone and zone targets. Targets are achieved when feature representation needs are reached per zone and when the area of a feature in each zone multiplied by the contribution of that feature in that zone reaches the overall representation target.

According to the BR model a core zones consists of areas designated as strict reserves (IUCN categories I to IV), which are nested within a buffer zones that allow traditional land use practices, mostly non consumptive. The transition and available zones allow for more intensive land use practices, transition zones however also need to comply with the sustainability goals, while available planning units are not necessarily included in such management schemes.

Based on such restrictions we set feature-by-zone contribution to targets, as mentioned above. We used this capability to systematize the level of protection that each zone provides to a feature, or defines the level of use permitted to a zone (i.e. permissibility). This feature can be used as well as a measure of the likelihood of persistence of a feature in a zone. The values attributed to each feature in each zone (Table 6.1) reflect the relation between their rarity and vulnerability (Pressey and Taffs 2001). We illustrated that using the jaguars as an example: in Table 6.1 one jaguar-territory hectare in core zone provides a contribution equivalent to one hectare to the jaguar target. In a buffer zone, one hectare of jaguar modelled distribution is worth 0.75 of a hectare, for representation target purposes, since the likelihood of their persistence in buffer zones is smaller. In transition zones this value is reduced to 0.5, meaning that 2 hectares of jaguar suitable habitat, in transition zones, are equivalent of one hectare in the core zone (Crawshaw and Quigley 1991). Basically the proportions (in Table 6.1) were multiplied by the area for the feature allocated in that zone, providing their corrected contribution to the overall feature target (Watts et al. 2009). Marxan with Zones still can assign planning units to whichever zone improves the overall efficiency, but assumes that the occurrence of features of interest in that zone represents a fraction of its occurrence in more protective zones. This is used to calculate the a_{ik} in equation (1) which is how much we conserve of feature k in zone l in planning unit i.

Species models and representation targets

We modelled the distribution of 117 species of vertebrates (Figure 6.3) using the maximum entropy algorithm in MAXENT software (Phillips et al. 2006). All species considered were red-listed either by the World Conservation Union (IUCN) or by the Brazilian Environmental Institute (IBAMA) and their point occurrences were obtained from the Conservation International database (Ministério do Meio Ambiente 2007). Targets for species were set accordance with the recommendation of experts appointed by the Brazilian government in a major prioritization workshop (Ministério do Meio Ambiente 2007). We chose MAXENT because of its advantages over other methods for presence-only data (Elith et al. 2006).

We adjusted these targets to enhance complementarity between zones and account for the differential contribution expected from each of the BR zones, using the zone contribution weights. The predicted distribution maps were based on correlation of presence-only data and the predictive variables such as elevation, vegetation, soils, distance to rivers, distance to roads, and fragility, acknowledging

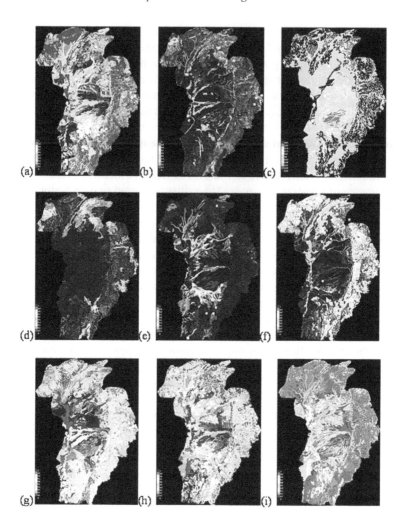

Figure 6.3   A sample of the MAXENT spatial representation model output for: (a) Ocelots (*Leopardus pardalis*),(b) Crowned eagle (*Harpyhaliaetus coronatus*),(c) Pampas deer (*Ozotoceros bezoarticus*), (d) Amphisbaena (*Bronia bedai*), (e) Giant otter (*Pteronura brasiliensis*), (f) Yellow headed parrot (*Alipiopsitta xanthops*),(g) Maned wolf (*Chrysocyon brachyurus*), (h) Hyacinth macaw (*Anodorhynchus hyacinthinus*),(i) Mamore arboreal rat (*Oecomys mamorae*). The greyscale informs the likelihood of presence of those species, the lighter the grey the most likely conditions for occurrence; based on soil, vegetation, fragility and distance to rivers, to roads and deforestation. White dots show the presence locations used for training.

however, the uncertainties associated with species modeling (Loiselle et al. 2003) and the problems associated with omission and commission errors as sources of inefficient resource allocation (Field et al. 2004).

In wetlands, species densities are often associated to climate which drives the productivity of the ecosystem (Daoust and Childers 1999, DeAngelis et al. 1998, Odum et al. 1995) and are positively correlated to ranching and ecotourism potential. Therefore, in addition to species density representation targets, we set targets for economic sustainability: in the case of caimans, capybaras, marsh-deer and pampas-deer densities (Table 6.1), they were chosen because of their ecotourism potential, data consistency and availability from aerial survey time series. Their densities were transformed in categorical variables and the targets were set for hectares of low, medium and high densities in each of the zones (Mourão et al. 2000).

For the purpose of this analysis we consider that a target is satisfactorily achieved when the selected planning units cover more than 90 per cent of the representation requirement. Hence, based on the distinct but complementary characteristics of targets we aimed to achieve explicit representation of quantitative and qualitative features and be able to illustrate the importance of target setting to the Biosphere Reserve's biodiversity, socio-cultural and sustainability objectives. Despite our efforts to set targets for as many ecological, social and economic features possible, we believe for the actual review process of the PBR, a more extensive set of features should be investigated. The current study enables BR practitioners to become aware and accountable for their successes and failures, while providing BR with a more appropriate spatial evaluation tool. It also opens up the possibility for development of quantitative indicators for an eventual monitoring programme, capable of addressing the expansion of the BR programme elsewhere.

Costs

Costs and constraints are an essential part of the optimisation process under the *minimum set* problem formulation. They are a basic component of the objective function (equation 1) under the formulation c_{ik} (= cost of planning unit i in zone k), or the "*factor to be minimised*", while the algorithm attempts to represent all features at their target levels. In Marxan with Zones, costs are calculated based on two matrices a "planning unit by feature cost", which consists of the area occupied by a cost-features in a planning units and a "zone by feature cost" (Table 6.2) which impose a weight on costs, when a planning unit is assigned to a particular zone. Altogether we used 37 cost layers going from opportunity cost to the actual land value (Lourival et al. 2008) which can be a major component of BR implementation, for core zones.

Table 6.2   Range of Values for Each Cost Layer According to the Zone of a Planning Unit

Cost layer Features	B R Objective	Unit	Zone 1 Available	Zone 2 Transition	Zone 3 Buffer	Zone 4 Core
Cattle density	cult-biod	4 classes	0	0	0	100
Deforestation	biod	pres/abs	0	0	80	100
Distance to river	sust	6 intervals	0	80	50	0
Distance to roads	sust	18 intervals	0	0	50	10
Erodability	biod	4 classes	0	20	100	100
Fire risk	biod	13 classes	0	10	50	70
Fragility	sust	6 classes	0	0	50	100
Soil types	sust	16 types	0	0-100	0-100	0-100
Vegetation subclasses	cult-sust	13 types	0	0-100	0-100	0-100
Acquisition costs	cult-sust	continuous	0	20	50	1000

Note: The values for each zone are weights that are multiplied by the area occupied by the feature cost layer

These costs are dynamic by nature, responding to changes in commodity prices and infrastructure. We integrated acquisition costs based on the equations proposed in Lourival et al (2008) for the Taquari catchment and we extrapolated acquisition costs across the entire UPRB. This extrapolation assumed that the behaviour of land value is equivalent across similar geographies in the planning region. The spatial distribution of explanatory variables; including presence of native vegetation, proximity to infrastructures, water availability, floodability and area used for ranching and agriculture, were used to extrapolate acquisition cost across the entire basin. Soil types and vegetation subclasses were used as individual cost layers due to their importance in defining current and potential land uses.

It is also important to emphasize that some of the layers used as costs, were also targeted as desirable features to be represented according to zone objective. Certain soil types, for example, are desirable features for ranching sustainability and agriculture if assigned to buffer or transition zones. Nevertheless, they also can represent costs when seen as fragile and subject to erosion, under the same land use/zone (see Tables 6.1 and 6.2). Deforestation and fire occurrences represent highly undesirable features for core zones, while "accepted" in moderation to buffer and transition zones, due to the use of native grasses.

Connectivity and Spatial Compatibility between Zones (zone juxtaposition)

The spatial configuration of reserve systems are fundamental for guaranteeing the persistence of species and ecological processes (Cabeza et al. 2004, Gerber et al. 2003). When this idea is expanded to BR zoning, the issue of compatibility between zones becomes even more important, since we are negotiating compromises between dissimilar land uses. More often than not, some land uses are incompatible and therefore need spatial separation between them, while the need for connectivity becomes essential to ecological and socio-economic processes. Control over the juxtaposition of zones is a remarkable capability of the Marxan with Zones software since it allows adjustments in the spatial compatibility of zones permitting nestedness, as originally proposed by the BR model.

We investigated the issue of compatibility between zones in the PBR case study, to provide alternative zone configurations. We used the compatibility matrix (Table 6.3) to evaluate the effect of nestedness on the spatial configuration of the PBR, in order to evaluate the flexibility of solutions as suggested in the draft of the Madrid Action Plan. We fixed parameters such as: targets, costs, boundary length in all scenarios (MAB Programme 2008) and varied the compatibility values in six scenarios, described below, evaluating their performances, on the basis of compactedness and nestedness and efficiency.

Table 6.3   Zone Compatibility Matrix Used to Explore the Compatibility between Zones and their Juxtaposition

Zone	Zone	Scenario (1)	Scenario (2)	Scenario (3)	Scenario (4)	Scenario (5)	Scenario (6)
		unlocked	locked	unlocked	locked	unlocked	locked
1	1	0.00	0.00	1.00	1.00	0.00	0.00
1	2	1.00	1.00	0.00	0.00	0.01	0.01
1	3	1.00	1.00	0.00	0.00	0.80	0.80
1	4	1.00	1.00	0.00	0.00	0.10	0.10
2	2	0.00	0.00	1.00	1.00	0.00	0.00
2	3	1.00	1.00	0.00	0.00	0.10	0.10
2	4	1.00	1.00	0.00	0.00	0.80	0.80
3	3	0.00	0.00	1.00	1.00	0.00	0.00
3	4	1.00	1.00	0.00	0.00	0.01	0.01
4	4	0.00	0.00	1.00	1.00	0.00	0.00

Note: The values under each scenario provide the spatial configuration seen in figures (4) and (5). The terms "locked" and "unlocked" signify the compulsory presence or absence, respectively, of existing reserves and indigenous land on each scenario.

Another parameter that we altered was the status of existing reserves and indigenous territories. In scenarios (2), (4) and (6), all planning units were available for allocation to any zones, while in scenarios (1), (3) and (5) the reserves were preferentially assigned to core areas and indigenous land to buffer zones. The terms "locked" and "unlocked" signify the presence and absence respectively, of existing reserves and indigenous land on each of the scenarios represented by Figure 6.4 ('a' to 'g'):

- Scenario (1) – total compatibility between planning units in the same zone, but no compatibility between zones, existing reserves and indigenous land were excluded from the analysis.
- Scenario (2) - total compatibility between planning units in the same zone, but no compatibility between zones, existing reserves and indigenous land were locked in the analysis. Protected areas were locked into core areas while indigenous lands were locked in buffer zones.
- Scenario (3) – no compatibility between planning units in the same zone and total compatibility between zones, existing reserves and indigenous land were excluded from the analysis.
- Scenario (4) – no compatibility between planning units in the same zone, but total compatibility between zones, existing reserves and indigenous land were locked in the analysis. Protected areas were locked as core areas while indigenous lands were locked as buffer zones.
- Scenario (5) – compatibility between planning units in the same zone and variable compatibility between zones. Values were set to facilitate nestedness and clumping. In this scenario existing reserves and indigenous land were excluded from the analysis.
- Scenario (6) - total compatibility between planning units in the same zone, but variable compatibility between zones. Values were set facilitate nestedness and clumping. In this scenario existing reserves and indigenous land were locked in the analysis. Protected areas were locked to core areas while indigenous lands were locked in buffer zones.

The Summed Solution output of Marxan with Zones is the frequency with which each planning units is assigned to one of the four zones, this output is commonly recognized as irreplaceability score in Marxan (Carwardine et al., 2007). We classified the selection frequency per zone using their distance from the mean measured in standard deviation units, using the same criteria described by Lourival et al. (2009).

The spatial distribution of the selection frequencies (Figure 6.3) compared using the Kappa statistic (Monserud and Leemans, 1992), measured the dissimilarities between each summed solution (i.e., planning unit selection frequency), for each of the zones. The Cohen's kappa pair-wise comparison was conducted between scenarios (1) and (2), (3) and (4), and (5) and (6) with the purpose of understanding the effect of locking in reserves, on planning unit selection frequency. We also compared (using the same statistics) the dissimilarity between solutions based on changes in the compatibility matrix, for the above scenarios (Table 6.4.).

Table 6.4   **Pair Wise Comparison of Selection Frequency between Scenarios and Zones Using Cohen's Kappa Statistics**

Scenario	Scenario	Agreement – Core	Agreement - Buffer	Agreement – Transition
1 (u)	2 (l)	low	low	very low
3 (u)	4 (l)	very low	very low	very low
5 (u)	6 (l)	low	low	very low
1 (u)	3 (u)	very low	very low	very low
1 (u)	5 (u)	low	very low	very low
3 (u)	5 (u)	very low	very low	very low
2 (l)	4 (l)	very low	very low	very low
2 (l)	6 (l)	low	very low	very low
4 (l)	6 (l)	very low	very low	very low

Note: We designate the status of reserves in a scenario as (u) for unlocked and (l) when reserves were locked in a solution. Where kappa (k) (k) = 0 means no agreement and k=1 means perfect agreement between the solutions for each of the two scenarios.

Species Models and Representation Targets

We modelled the distribution of 117 species of vertebrates using the maximum entropy algorithm in MAXENT software (Phillips et al. 2006). All species considered were red-listed either by IUCN or by IBAMA and their point occurrences were obtained from the Conservation International database (Ministério do Meio Ambiente 2007). Representation targets were based on the threat status of these lists and the targets set by the second priority setting workshop conducted by the Brazilian Government (Ministério do Meio Ambiente 2007).

We adjusted these targets to enhance complementarity between zones and account for the differential contribution expected from each of the BR zones, using the zone contribution weights. The predicted distribution maps were based on presence-only data and the predictive variables were elevation, vegetation, soils, distance to rivers, distance to roads and fragility. We chose a threshold of 70 per cent likelihood of occurrence (Figure 6.3) as our cut-off point to generate the presence maps where Marxan with Zones would search for their representation.

Comparison between ad hoc and Systematically Zoned Biosphere Reserves

We finally want to compare the gap between the resulting assignments of planning units to specific zones with the current spatial configuration of the Pantanal Biosphere Reserve, ratified by UNESCO in 2000, based on the recommendations of the PROBIO Cerrado-Pantanal priority setting workshop (Ministério do Meio Ambiente 1999). This is essential step in systematic planning since it provides the starting point for comparison in cases where reserves and indigenous lands were already designated (Pressey 1994). We used the same targets for the *ad hoc* PBR (baseline scenario) as the ones explained for scenarios (5) and (6), comparing with the best solution for 1,000 runs of Marxan, each with 1,000,000 interactions. We then illustrated the possibilities offered by Marxan with Zones to support the development of new proposals for Biosphere Reserves, and the upgrading of existing BRs, offering alternative spatial configurations based on the tradeoffs between spatially explicit objectives.

Results

The current BR design lacks a defensible framework and urgently needs a systematic review of its zonation scheme, as acknowledged by UNESCO in the Madrid Action Plan. We chose scenarios (5) and (6) to illustrate the difference in outcomes between a systematically designed PBR compared to the ad hoc designed PBR, as proposed in 2000 for the Pantanal (Figure 6.4 'e' and 'f' versus 'g'). Therefore, despite the importance, or potential of the Pantanal Biosphere Reserve to engage landowners and other local stakeholder in a conservation based development exercise, reality has showed that non inclusive planning tend to fall short of people's expectations, causing indifference and lack of effectiveness.

Target Achievement

Our results for feature targets show that 82 per cent of all 293 features in scenario (5) reached their respective targets. In scenario (6) where reserves were locked in, 84 per cent of overall targets were achieved showing no distinctive difference between scenarios (5) and (6) in overall target achievement, for the single best solution of each scenario.

We set targets in two different ways, first using the overall feature target and second using feature-by-zone targets. In both cases the representation target was considered adequately achieved when 90 per cent of the target was reached. Feature representation was fully achieved in both scenarios (5 and 6) for wildlife densities (marsh deer, pampas deer, caimans and capybaras) and watersheds. The targets were also achieved for cattle densities and distance to roads and rivers, which are associated with land value and the sustainability of properties within PBR planning units (Lourival et al. 2008).

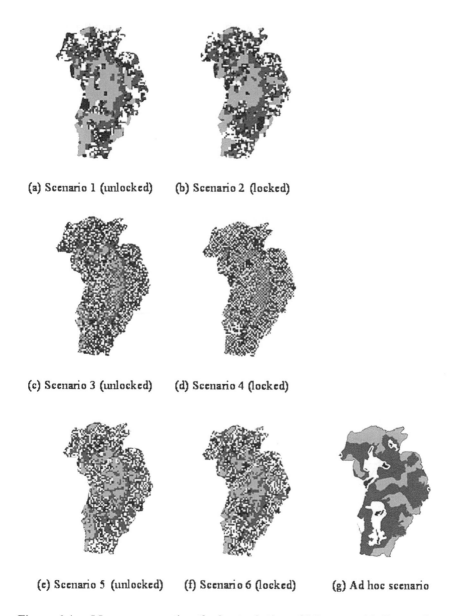

(a) Scenario 1 (unlocked)   (b) Scenario 2 (locked)

(c) Scenario 3 (unlocked)   (d) Scenario 4 (locked)

(e) Scenario 5 (unlocked)   (f) Scenario 6 (locked)   (g) Ad hoc scenario

Figure 6.4   Maps representing the best solution of Marxan with Zones, for all scenarios (i.e. 1 to 6) plus the ad hoc scenario as specified in Table 6.5. The core areas (black), buffer zones (light grey) and transition zones (dark grey). Here the transition zone planning units are split into dark grey (irreplaceable) and intermediary grey as planning units that are part of the planning area but do not contribute or were not necessary for target achievement.

In Table 6.5 we show a comparison between scenarios (5) and (6) in regards to under-representation. Scenario (5) had one vegetation class, two vegetation subclasses and one soil type under-represented (with respectively 83%, 77%, 61% and 87% achievement). While the scenario (6) under-represented one freshwater eco-domain, one vegetation subclass, and one soil type (with respectively 73%, 81% and 56% achievement). The soil type and one of the under-represented vegetation subclasses were common to both scenarios.

Table 6.5   Comparison between scenarios (5) and (6) for missed targets by zone

	Targets by Zone	All zones combined	Available	Transition	Buffer	Core
Scenario (5) [112] no reserves	Missed (n=35)	0.31	0.09	0.17	0.40	0.34
	Avrg. achievement	0.61	0.46	0.64	0.32	0.34
	Standard deviation	0.36	0.23	0.21	0.29	0.28
	Variance	0.13	0.05	0.04	0.08	0.08
Scenario (6) [114] reserves locked in	Missed (n=33)	0.29	0.12	0.06	0.33	0.48
	Avrg. achievement	0.73	0.50	0.52	0.67	0.49
	Standard deviation	0.28	0.24	0.29	0.17	0.27
	Variance	0.08	0.06	0.08	0.03	0.07

Note: The number in [ ] represents the overall number of features targeted for all zones. We show the number of targets that did not reach 90% representation, the average (avrg) proportion of their achievements and their standard deviation.

Species Models and Representation Targets

In the solutions proposed by Marxan with Zones, full representation was achieved in scenario (5) for 62 per cent of species, while in scenario (6); around 63 per cent of species were fully represented. Significant differences in representation, between the scenarios, occurred only in four species; nevertheless under-represented species to both scenarios had average achievement of respectively 78 and 77 per cent.

Zone Targets

The possibility of setting targets by zone is what sets apart Marxan with Zones from the existing systematic planning software. We used this capability to set zone targets for 32 of the 293 features under analysis, based on UNESCO-MAB guidelines. We chose to target these 32 features directly to particular zones to maximize sustainability in culturally sensitive areas (e.g. indigenous territories assigned to buffer zones in scenario 6), or to maximize use of already degraded spaces, assigning them to transition zones.

Overall, scenario (6) missed fewer targets (29%) and had higher average achievement levels (73%) for missed targets than scenario (5) with 31 per cent and 61 per cecnt respectively, for the single best solution. Results were very close in terms of their representation capabilities for zone targets, however what makes them different is the percentage of missed targets for core zones which was 34 per cent for the no-reserve scenario and 48 per cent when reserves were locked in (Table 6.5), pointing to the inefficiencies of current reserves. Despite those figures, our results for buffer and transition zones showed inverted trend.

Connectivity and Spatial Configuration Zone Compatibility (juxtaposition)

We tested Marxan with Zones' capacity to generate spatial patterns compatible with the BR model; in particular we aimed to achieve certain levels of nestedness between zones, without compromising efficiency, whilst trying to represent overall targets for the PBR. We also tested spatial compatibility between zones since we were particularly interested in clarifying the effect of *locking in* existing reserves (protected areas and indigenous territories) within the planning area.

Our results illustrate the variety of configuration options that can be achieved using Marxan with Zones. We reached high levels of spatial aggregation (clumping) in the overall best solution, for scenarios (1) and (2), when we imposed complete compatibility between planning units of the same zone and no compatibility (anti-clumping) between planning units of different zones (see Figure 6.4 'a' and 'b', and Table 6.3). The clumping process showed for each zone was reflected in the planning unit selection frequency hierarchy (Figure 6.5 'a', 'b' and 'c'), where most frequently selected planning units occupy the center of clumps, while less frequently selected planning units tend to surround them. A discrete increase in planning unit assignment (3%) was observed when reserves were *not locked in* (scenario 1) when compared to locked in reserves (scenario 2).

A checkered pattern emerged for scenarios (3) and (4) (anti-clumping), when we adopted complete incompatibility between planning units of the same zone, and total compatibility between planning units of distinct zones (Figure 6.3 'c' and 'd'). This pattern was broken only when reserves were *locked in*, resulting in planning unit clumping, where reserves existed.

When we imposed clumping to planning units (scenarios 1 and 2), the shortfalls were on average 30 per cent bigger than those in scenarios (3 and 4

– anti-clumping), however these scenarios (3 and 4) mobilized on average, 10 per cent more planning units. Nevertheless when compared, the difference in shortfalls between scenarios (1 and 3 unlocked) was reduced by 34 per cent while between (2 and 4 locked) the reduction of shortfall was 25 per cent in both cases (clumped and anti-clumped). In summary, allowing the software to choose freely, without the constraints of existing reserves, improved overall target achievement.

In scenarios (5) and (6) we defined the compatibility matrix with properties common to both sets of scenarios (1 and 2), and (3 and 4). We aimed to generate flexible nestedness by allowing planning units of the same zone to be compatible, and increased incompatibility hierarchically between zones. The pair wise compatibility matrix (Table 6.3) allowed for conditions where that core zones were more compatible with buffer zones than with transition zones. The results (Table 6.6) showed an average increase in planning unit assignment of 13.5 per cent between scenarios (1 and 2), and scenarios (5 and 6), while there is no significant increase between (3 and 4), and (5 and 6).

However, the planning units' assignment to each zone between the above sets of scenarios varied dramatically (Table 6.6). It is worth noticing that the amount of planning units assigned to core zones varied from 16 and 18 per cent of all planning units in scenarios 5 and 6, to 30 and 25 per cent in scenarios 3 and 4.

The sensitivity of this procedure can be explored through several combinations of compatibility values, which can be used to increase efficiency by assigning planning units to different zones. The scenarios proposed, engaged between 70 and 83 per cent of all planning units of the UPRB watershed, the remainder (30–17%) being composed of planning units that are not needed to contribute to achieve the representation requirements because of cost and or lack of features.

Discussion

Zoning has been the spatial mechanism to reduce conflicts in the face population growth and a way found by society to organize and manage their land use needs . Through zoning, agreed compromises amid the needs of humanity are set (Bojorquez-Tapia et al. 2004, Werner 1926). So far zoning has been used as a spatial tool in urban planning (Conway and Lathrop 2005), tourism, forestry and industry with relative success (Bos 1993). However, the conservation literature lacks examples, methods and applications of spatially explicit, multi-objective and multi-stakeholder zonation schemes (Moilanen et al. 2009).

Despite the established literature and the proven advantages of systematic planning to conservation (Bojorquez-Tapia et al. 2003, Margules and Pressey 2000, Pressey 1994), Biosphere Reserves continue to be designed using *ad hoc* configuration approaches (Dyer and Holland 1991, Pressey and Tully 1994). For that reason the capacity to demonstrate that BR can spatially represent environmental, cultural and economic sustainability requirements remains elusive. Defining the role and contribution of each of the BR zones towards transparent objectives has

Table 6.6   Summary of Planning Unit Assignment According to Zone Compatibility

Scenario	Reserve status	No. of PU in all scenarios	Proportion of total PU	PU in transition zone	%	No. PU in buffer zone	%	No. PU in core zone	%	Target shortfall in Hectares
Ad hoc	Locked	2403	0.64	1807	0.48	689	0.18	57	0.02	324,685.3
1	Unlocked	2615	0.70	1137	0.31	672	0.18	805	0.22	116,323.2
2	Locked	2704	0.73	1119	0.30	727	0.20	858	0.23	111,515.1
3	Unlocked	3104	0.83	1182	0.32	808	0.22	1114	0.30	76,113.2
4	Locked	2988	0.80	1177	0.32	866	0.23	944	0.25	83,370.0
5	Unlocked	2981	0.80	1440	0.39	929	0.25	612	0.16	104,008.6
6	Locked	3051	0.82	1413	0.38	963	0.26	675	0.18	99,659.3
Average		2907	0.78	1244	0.33	827	0.22	834	0.22	98,498.2
Stdv		199	0.05	142	0.04	113	0.03	182	0.05	15,803.9

Note: Results are shown by scenarios with the number and percentage of planning units (PU) mobilized in each scenario out of a total of 3727 planning units and their assignment by zones.

been a point of discussion for some time (Tangley 1988). Very little has been done to quantitatively account for the social, biodiversity and sustainability functions of BR (Price 1996).

We believe, however, that this is the most suitable time to advocate the case of systematic zoning for Biosphere Reserves, considering the timeframe set in the Madrid Action Plan (MAB Programme 2008) whereby all BRs need their zonation to be reviewed by 2013.

Our intention is to contribute to the evaluation process of Biosphere Reserves in general and the rezoning of the Pantanal BR. Our results showed that the current design of the PBR not only under-represents the biodiversity objectives, but also ignores the other two essential components of all BRs, (i.e. the protection of local cultures and the economic sustainability of the region). Although some studies discussed the qualitative characteristics of BR (Batisse 1990, Kellert 1986, Solecki 1994, Tangley 1988) so far few studies evaluated their quantitative characteristics (Howard et al. 1997, Lourival et al. 2009). Dyer and Holland (1991) emphasised the lack of clear definitions and tangible goals for BRs, to take into account the importance of ecological processes, complementarity, connectivity and comprehensiveness of BR networks.

We demonstrated that principles and procedures of systematic planning were met and expanded through systematic zoning, while compromises between conflicting objectives were optimized in a spatially explicitly manner (Pressey and Bottrill 2009). Biosphere Reserves proved be an interesting platform to develop systematic zoning applications, basically because they were conceived under a multi-objective, multiple-zone and multiple-cost framework.

Target Achievement in ad hoc Versus Systematic Reserve Design

Our results have confirmed that the effectiveness of a BR can dramatically increase when they are systematically designed. The PBR does not meet a large number of its targets and would need to reassign at least 300,000 hectares in planning units to achieve all targets (Table 6.5). When compared with systematically designed PBR (using the same targets and costs), shortfalls were reduced to less than 100,000 hectares, without losing efficiency.

Although increases in penalties for under-representing features could bring Marxan solutions close to full representation, we believe that spreading the current targets to reflect the response of each feature to different land use regimes, is a more efficient way to engage the private sector in conservation, and all zones would proportionally be contributing to targets, which is one of the recommendation of the Madrid Action Plan (MAB Programme 2008).

Systematically planned BRs provide guidance to the spatial allocation of new zones in order to achieve the desired feature representation, since all zones have to contribute equally towards BR objectives. This can improve the flexibility of BRs and provide opportunities to the BR review process, considered in the Madrid Action Plan (MAB Programme 2008). Enhancing and expanding the understanding of feature contributions, provided by each zone, and the casual use of floating targets (varying between zones) can allow BRs to simultaneously balance comprehensiveness, complementarity between zones as well as improve their overall efficiency, while becoming a major lesson in landscape scale planning.

Efficiency in Biosphere Reserves

Even though the tradeoffs between comprehensiveness and efficiency are not evaluated in depth here, a substantial increase in number of planning units selected was observed between *ad hoc* and systematically designed scenarios (Figure 6.4). We credited such increases to the magnitude of the targets we used. These targets derived from the legislation and governmental guidelines for species representation in protected areas *strictu sensu* (i.e. equivalent to IUCN categories I to IV) (Ministério do Meio Ambiente 2007), were never used in the *ad hoc* scenario. It is our conviction that targets need to be revisited to adequately reflect species densities and population requirements, in their relation to different land use regimes, regulated by the zone restriction procedures.

The concept of Efficiency, which is based on planning units available for selection and planning units actually selected, is commonly used in systematic conservation planning (Pressey and Nicholls 1989), but it needs to be adjusted to tackle the problems associated with zoning. A slight modification in the way efficiency is normally calculated allows planners to integrate the effects of multiple ownership in off-reserve conservation (Lourival et al. 2009), more specifically, planning units assigned to core areas (i.e. to become IUCN categories I to IV) in the Pantanal Biosphere Reserve need to be increased by more than one order of magnitude (Figure 6.6 'c' and 'd'). Moreover, if planners want to achieve the proposed targets, buffer zone areas would have to almost double, while transition zones could be reduced. Nonetheless, the results for target achievement are just an illustration of the advantages and possibilities of Marxan with Zones. If full feature representation was imperative several parameters in the software could be adjusted to do so, either via the target per zone or overall target.

Some other measures of performance regularly applied to conservation planning, can also be used for systematic zoning. As an example we used the simple system of classifying planning units according to their selection frequency (Stewart et al. 2003). This method, was expanded by Lourival et al. (2009), and can be used to classify planning units for all zones (Figure 6.5). These selection frequencies, shown in one standard deviation intervals, can be used to guide the priority of zone implementation based on their irreplaceability levels. Our approach suggests a large expansion in the core zones (see Figure 6.6 'c' and 'd') maybe politically unpalatable. Market-based methods such as legal reserve compensation defined by the Brazilian Forest code (Lourival et al. 2008) and other incentives may be the best way to achieve the expansion in core zones or at least become complementary to acquisition policies.

Zone Compatibility and Juxtaposition

The configurations generated by Marxan with Zones offer greater flexibility for the decision-making process, allowing a variety of scenarios, which can fulfill the BR objectives. Two emergent properties became important in systematic zoning, modulating the compatibility between zones. The first can be called Permissibility and involves the level of acceptable use of a planning unit that does not compromise its zone objectives. The second property is zone Juxtaposition which defines the spatial compatibility between zones. Together they enable us to isolate threatening processes from sensitive areas, create connectivity corridors and allow spreading and/or clumping of zones.

Overall, the best solution for scenarios (5) and (6) illustrates well (Figure 6.6) the principle of flexible nestedness proposed by the Madrid Action Plan. Under these new guidelines the "egg model" originally prescribed for BR (Figure 6.1) become amenable to embrace alternative spatial configurations and respond to design limitations.

Exploring the sensitivity of the compatibility matrix for intermediate scenarios (5 and 6) illustrates the capacity of the compatibility matrix to influence implementation costs and the efficiency of the BRs, by assigning fewer planning units to core zones, but compensating in other zones when pertinent.

Note for the two upper images: Limit of the Upper Paraguay River Basin-UPRB (green), map (a) represents scenario 5 where reserves are not locked in the solution and map (b) shows scenario 6 with the reserves locked in. Those are the best solutions of Marxan with Zones for the Pantanal Biosphere Reserve; in (black) are the core areas, in (light grey) the buffer zones and in (dark grey) the transition zone.

For the two lower images: the black pixels below are frequently selected planning units for core zones in scenarios 5 (a) and 6 (b). These are the necessary increments for core zones for the PBR to aim at the preset core zone targets. The overlapped figure represented in light grey (buffer) and black polygons (core) and dark grey line (transition) are placed to mask what increases are necessary in zones previously considered as transition areas.

Conclusion

We demonstrated the applicability and flexibility of Marxan with Zones to BR systematic planning. We also showed that socio-cultural objectives can be explicitly optimized and represented across all zones in the same way biodiversity and economic objectives can. In addition we illustrated, in a variety of spatial configurations, the tradeoffs between feature representation and efficiency for Biosphere Reserves.

Interestingly, an emergent property of the comparison between the *ad hoc* PBR and systematically designed alternatives was that, despite the spatial configuration

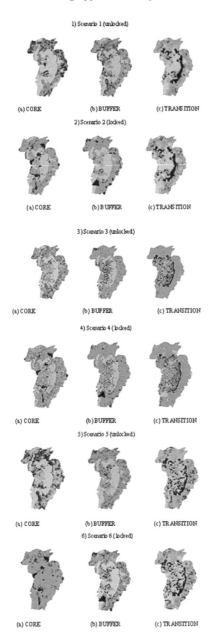

Figure 6.5　Variation in selection frequency of planning units respectively to core, buffer and transition zones of the Pantanal Biosphere Reserve classified by grades of one standard deviation unit, from low (light grey) to high (black) selection frequency for each scenario compared in Table 6.6.

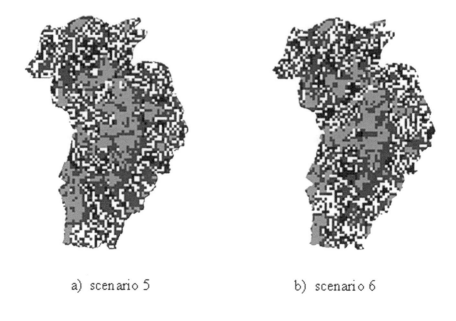

a) scenario 5   b) scenario 6

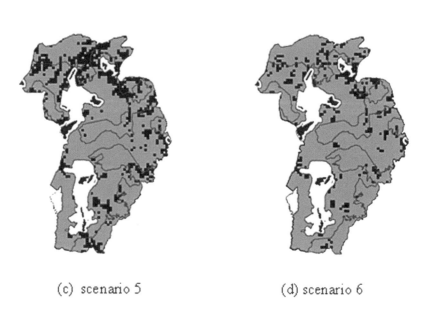

(c) scenario 5   (d) scenario 6

**Figure 6.6   Simulation of Scenarios 5 and 6 (according to the Madrid
                Action Plan)**

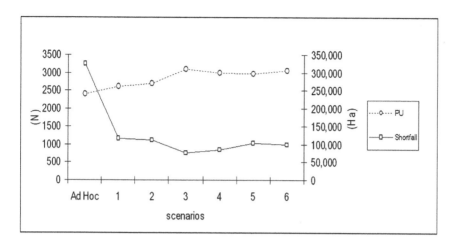

Figure 6.7 **Comparison between Ad hoc and Systematically Designed Pantanal Biosphere Reserve. The graph represents the relationship between number of planning unit engaged in each scenario and the shortfall measured in hectares for all targets.**

of zones the diversity of scenarios, the variability in the percentage of planning units assigned to each one of the three BR zones was fairly constant. On average, systematically designed core zones used 22 per cent of all planning units (SD = 5%), buffer zones also occupied 22 per cent (3%) and transition zones 33 per cent (SD=4%) of all planning units to represent feature and zone targets. While the *ad hoc* PBR had 2 per cent of planning units in core areas, 18 per cent in buffer zones and 48 per cent in transition zones (Figure 6.7).

In addition, systematically designed BRs assigned 23 per cent of planning units to *available*, either because they did not contribute to targets or because they were too expensive to be included. In the *ad hoc* PBR the class of planning units *available* accounted for 30 per cent of the watershed for no explicit reason. We chose to add available planning units to the Transition zone (summarized in Table 6.6), since they represent no acquisition cost, and provide a sense of watershed based management, which ultimately will be the source of income to the BR, through its catchment authority.

Considering that the MAB program has no guidelines for quantifying the amount of the planning region to be allocated to each zone, our systematic approach for the Pantanal Biosphere Reserve provided an interesting insight into how much of the region needs to be in different zones (i.e.: 22±5 per cent in core zones (i.e. IUCN I to IV), 22±3 per cent in buffer and 33±4 per cent in transition zones), in order to efficiently enhance target (i.e. objectives) achievements.

UNESCO conceived BRs to support sustainable societies, however, until now it was unable to quantify the contribution of BRs to their broader objectives. It is thus clear, that protected areas alone cannot provide protection to all those objectives. Systematically zoned BRs are an amenable alternative to tackle such objectives transparently.

Acknowledgements

We would like to thank The Capes Foundation (Brazilian Ministry of Education) and The Commonwealth Environmental Research Facilities-AEDA for providing the funds for this study. We also thank Dr. Gordon Grigg, Dr. Robert (Bob) Pressey and our colleagues of the spatial ecology lab from the University of Queensland for their support and constructive criticism, The Embrapa Pantanal researchers; Drs. Marta Silva and Roberto Padovani. Special thanks to Leonie Seabrook, Peter Baxter and Ché Elkin for revising the manuscript.

Acronyms

CARE - Comprehensiveness, Adequacy, Representation and Efficiency
IBAMA - Brazilian Environmental Institute
IUCN - onal Union for Conservation of Nature
MAB - UNESCO Man and the Biosphere Programme
PBR - Pantanal Biosphere Reserve
UPRB - Upper Paraguay River Basin

References

Assine, M.L. and Soares, P.C. 2004. Quaternary of the Pantanal, west-central Brazil. *Quaternary International*, 114, 23-34.
Ball, I.R. and Possingham, H.P. 2000. *Marxan v1.8.6 - Marine Reserve Design Using Spatially Explicit Annealing*. Brisbane: The University of Queensland. Available at: www.uq.edu.au/marxan
Batisse, M. 1990. Development and implementation of the Biosphere Reserve Concept and its applicability to coastal regions. *Environmental Conservation*, 17(2), 111-116.
Batisse, M. 1997. Biosphere reserves: a challenge for biodiversity conservation and regional development. *Environment*, 39(5), 7-33
Bojorquez-Tapia, L.A., Brower, L.P, Castilleja, G., Sanchez-Colon, S., Hernandez, M., Calvert, W., Diaz, S., Gomez-Priego, P., Alcantar, G., Melgarejo, E.D., Solares, M.J., Gutierrez, L. and Juarez, M.D. 2003. Mapping expert knowledge: redesigning the Monarch Butterfly Biosphere Reserve. *Conservation Biology*, 17(2), 367-379.
Bojorquez-Tapia, L.A., de la Cueva, H., Diaz, S., Melgarejo, D., Alcantar, G., Solares, M.J., Grobet, J. and Cruz-Bello, G. 2004. Environmental conflicts and

nature reserves: redesigning Sierra San Pedro Martir National Park, Mexico. *Biological Conservation*, 117(2), 111-126.

Bos, J. 1993. Zoning in forest management: a quadratic assignment problem solved by simulated annealing. *Journal of Environmental Management*, 37(2), 127-145.

Cabeza, M. and Moilanen, A. 2006. Replacement cost: a practical measure of site value for cost-effective reserve planning. *Biological Conservation*, 132(3), 336-342.

Cabeza, M., Araujo, M.B., Wilson, R.J., Thomas, C.D., Cowley, M.J.R. and Moilanen, A. 2004. Combining probabilities of occurrence with spatial reserve design. *Journal of Applied Ecology*, 41(2), 252-262.

Carwardine, J., Rochester, W.A., Richardson, K.S., Williams, K.J., Pressey, R.L. and Possingham, H.P. 2007. Conservation planning with irreplaceability: thus the method matter? *Biodiversity and Conservation*, 16(1), 245-258.

Conway, T.M. and Lathrop, R.G. 2005. Alternative land use regulations and environmental impacts: assessing future land use in an urbanizing watershed. *Landscape and Urban Planning*, 71(1), 1-15.

Crawshaw, P.G. and Quigley, H.B. 1991. Jaguar sacing, ativity and hbitat ue in a sasonally fooded evironment in Brazil. *Journal of Zoology*, 223, 357-370.

Daoust, R.J. and Childers, D.L. 1999. Controls on emergent macrophyte composition, abundance, and productivity in freshwater everglades wetland communities. *Wetlands*, 19(1), 2-275.

Dasmann, R.F. 1988. Biosphere reserves, buffers, and boundaries. *Bioscience*, 38(7), 487-489.

DeAngelis, D.L., Gross, L.J., Huston, M.A., Wolff, W.F., Fleming, D.M., Comiskey, E.J. and Sylvester, S.M. 1998. Landscape modeling for everglades ecosystem restoration. *Ecosystems*, 1(1), 64-75.

Dhargalkar, V.K. and Untawale, A.G. 1991. Marine biosphere reserves: need of the 21st century. *Journal of Environmental Biology*, 12, 169-177.

Dyer, M.I. and Holland, M.M. 1991. The Biosphere-Reserve Concept: needs for a network design. *Bioscience*, 41(5), 319-325.

Elith, J., Graham, C.H., Anderson, R.P., Dudik, M., Ferrier, S., Guisan, A., Hijmans, R.J., Huettmann, F., Leathwick, J.R., Lehmann, A., Li, J., Lohmann, L.G., Loiselle, B.A., Manion, G., Moritz, C., Nakamura, M., Nakazawa, Y., Overton, J.M., Peterson, A.T., Phillips, S.J., Richardson, K., Scachetti-Pereira, R., Schapire, R.E., Soberon, J., Williams, S., Wisz, M.S. and Zimmermann, N.E. 2006. Novel methods improve prediction of species' distributions from occurrence data. *Ecography*, 29(2), 129-151.

Field, S.A., Tyre, A.J., Jonzen, N., Rhodes, J.R. and Possingham, H.P. 2004. Minimizing the cost of environmental management decisions by optimizing statistical thresholds. *Ecology Letters*, 7(8), 669-675.

Gaston, K.J. and Rodrigues, A.S.L. 2003. Reserve selection in regions with poor biological data. *Conservation Biology*, 17(1), 188-195.

Gerber, L.R., Botsford, L.W., Hastings, A., Possingham, H.P., Gaines, S.D., Palumbi, S.R. and Andelman, S. 2003. Population models for marine reserve design: a retrospective and prospective synthesis. *Ecological Applications*, 13(1), S47-S64.

Higgins, J.V., Bryer, M.T., Khoury, M.L. and Fitzhugh, T.W. 2005. A freshwater classification approach for biodiversity conservation planning. *Conservation Biology*, 19(2), 432-445.

Higgins, J.V., Ricketts, T.H., Parrish, J.D., Dinerstein, E., Powell, G., Palminteri, S., Hoekstra, J.M., Morrison, J., Tomasek, A. and Adams, J. 2004. Beyond Noah: saving species is not enough. *Conservation Biology*, 18(6), 1672-1673.

Howard, P., Davenport, T. and Kigenyi, F. 1997. Planning conservation areas in Uganda's natural forests. *Oryx*, 31(4), 253-264.

Junk, W.J. and Nunes da Cunha, C. 2005. Pantanal: a large South American wetland at a crossroads. *Ecological Engineering*, 24(4), 391-401.

Kellert, S.R. 1986. Public understanding and appreciation of the Biosphere Reserve Concept. *Environmental Conservation*, 13(2), 101-105.

Loiselle, B.A., Howell, C.A., Graham, C.H., Goerck, J.M., Brooks, T., Smith, K.G. and Williams, P.H. 2003. Avoiding pitfalls of using species distribution models in conservation planning. *Conservation Biology*, 17(6), 1591-1600.

Lourival, R., Caleman, S.M.D., Villar, G.I.M., Ribeiro, A.R. and Elkin, C. 2008. Getting fourteen for the price of one! Understanding the factors that influence land value and how they affect biodiversity conservation in central Brazil. *Ecological Economics*, 67(1), 20-31.

Lourival, R., McCallum, H., Grigg, G.C., Arcangelo, C., Machado, R.B. and Possingham, H. 2009. A systematic evaluation of the conservation plans for the Pantanal wetland in Brazil *Wetlands*, 29(4), 1189-1201.

MAB Programme. 2008. *The Madrid Action Plan 2008 - 2013*. 3rd Wolrd Congress of Biosphere Reserves and 20th session of the International Coordinating Council of the MAB Programme. Madrid: UNESCO.

Margules, C.R. and Pressey, R.L. 2000. Systematic conservation planning. *Nature*, 405(6783), 243-253.

Ministério do Meio Ambiente. 1999. *Ações Prioritárias para a Conservação da Biodiversidade do Cerrado e Pantanal*. Brasília: MMA/PROBIO.

Ministério do Meio Ambiente. 2007. *Áreas Prioritárias para a Conservação, Uso Sustentável e Repartição de Benefícios da Biodiversidade Brasileira*. Biodiversidade 31. MMA: Brasilia. Available at: www.mma.gov.br/portalbio

Moilanen, A., Franco, A.M.A., Eary, R.I., Fox, R., Wintle, B. and Thomas, C.D. 2005. Prioritizing multiple-use landscapes for conservation: methods for large multi-species planning problems. *Proceedings of the Royal Society B*, 272(1575), 1885-1891.

Monserud, R.A. and Leemans, R. 1992. Comparing global vegetation maps with Kappa Statistics. *Ecological Modelling*, 62, 275-293.

Mourão, G., Coutinho, M., Mauro, R., Campos, Z., Tomas, W. and Magnusson, W. 2000. Aerial surveys of caiman, marsh deer and pampas deer in the Pantanal Wetland of Brazil. *Biological Conservation*, 92(2), 175-183.

Mourão, G.M., Bayliss, P., Coutinho, M.E., Abercrombie, C.L. and Arruda, A. 1994. Test of an aerial survey for caiman and other wildlife in the Pantanal, Brazil. *Wildlife Society Bulletin*, 22(1), 50-56.

Murphy, F.C. 1958. *Regulating Flood-Plain Development*. Chicago.

Negi, C.S. and Nautiyal, S. 2003. Indigenous peoples, biological diversity and protected area management: policy framework towards resolving conflicts. *International Journal of Sustainable Development and World Ecology*, 10(2), 169-179.

Odum, W.E., Odum, E.P. and Odum, H.T. 1995. Natures pulsing paradigm. *Estuaries*, 18(4), 547-555.

Padovani, C.R., Cruz, M.L.L. and Padovani, S.A.G. 2004. *Desmatamento do Pantanal Brasileiro para o Ano de 2000. Proceedings of the IV Simpósio de Recursos Naturais e Sócio-econômicos do Pantanal, Sustentabilidade Regional*. Corumbá: CPAP/EMBRAPA.

Phillips, S.J., Anderson, R.P. and Schapire, R.E. 2006. Maximum entropy modeling of species geographic distributions. *Ecological Modelling*, 190(3-4), 231-259.

Poore, D. 1995. Unesco-International-Conference on Biosphere Reserves, held in Seville, Spain, 20-25 March 1995. *Environmental Conservation*, 22(2), 186-187.

Possingham, H., Ball, I. and Andelman, S. 2000. Mathematical methods for identifying representative reserve networks, in *Quantitative Methods for Conservation Biology*, edited by S. Fersonand and M.A. Burgman. New York: Springer-Verlag, 291-305.

Possingham, H.P. 2001. Models, Problems and Algorithms: Perceptions About Their Application to Conservation Biology. Pages 1-6 in MODSIM 2001.

Possingham, H.P., Wilson, K.A., Andelman, S.J. and Vynne, C.H. 2006a. Protected areas: goals, limitations, and design, in *Principles of Conservation Biology*, edited by M.J. Groom, G.K. Meefeand and C.R. Carroll. Sunderland: Sinauer Associates, 691-700.

Pressey, R.L. 1994. Ad hoc reservations: forward or backward steps in developing representative reserve systems. *Conservation Biology*, 8(3), 662-668.

Pressey, R.L. 1999. Systematic conservation planning for the real world. *Parks*, 9(1), 1-5.

Pressey, R.L. and Bottrill, M.C. 2009. Approaches to landscape- and seascape-scale conservation planning: convergence, contrasts and challenges. *Oryx*, 43(4), 464-475.

Pressey, R.L. and Nicholls, A.O. 1989. Efficiency in conservation evaluation: scoring versus iterative approaches. *Biological Conservation*, 50(1-4), 199-218.

Pressey, R.L. and Tully, S.L. 1994. The cost of ad hoc reservation; a case-study in Western New-South-Wales. *Australian Journal of Ecology*, 19(4), 375-384.

Pressey, R.L. and Taffs, K.H. 2001. Scheduling conservation action in production landscapes: priority areas in Western New South Wales defined by irreplaceability and vulnerability to vegetation loss. *Biological Conservation*, 100(3), 355-376.

Price, M.F. 1996. People in Biosphere Reserves: an evolving concept. *Society and Natural Resources*, 9(6), 645-654.

ReVelle, C.S., Williams, J.C. and Boland, J.J. 2002. Counterpart models in facility location science and reserve selection science. *Environmental Modeling and Assessment*, 7(2), 71-80.

Rosova, V. 2001. Biosphere Reserves: model territories for sustainable development. *Ekologia-Bratislava*, 20, 62-67.

Sabatini, M.D.C., Verdiell, A., Iglesias, R.M.R. and Vidal, M. 2007. A quantitative method for zoning of protected areas and its spatial ecological implications. *Journal of Environmental Management*, 83, 68-76.

Schoemaker, P.J.H. 1991. The quest for optimality: a positive heuristic of science. *Behavioral and Brain Sciences*, 14(2), 205-214.

Solecki, W.D. 1994. Putting the Biosphere Reserve Concept into practice: some evidence of impacts in rural communities in the United-States. *Environmental Conservation*, 21(3), 242-247.

Stewart, R.R., Noyce, T. and Possingham, H.P. 2003. Opportunity cost of ad hoc marine reserve design decisions: an example from South Australia. *Marine Ecology-Progress Series*, 253, 25-38.

Tangley, L. 1988. A new era for Biosphere Reserves: Mexicos Sian Kaan shows that its hard to be everything a biosphere reserve should be. *Bioscience*, 38(3), 148-155.

Ukeles, J.B. 1964. *The Consequences of Municipal Zoning*. Washington DC: Urban Land Institute.

UNESCO. 1995. The Sville Strategy for Biosphere Reserves. *Nature and Resources*, 31(2), 2-17.

UNESCO. 2002. *Biosphere Reserves: Special Places for People and Nature*. Paris: UNESCO.

UNESCO. 2007. *Biosphere Reserves: World Network*. Paris: UNESCO.

Veneziani, P., Dos Santos, A.R., Crepani, E., Dos Anjos, C.E. and Okida, R. 1998. Map of erodibility classes of part of Taquari River Basin, based on Tm-Landsat Images. *Pesquisa Agropecuária Brasileira*, 33, 1747-1754.

Verdiell, A. and Sabatini, M.C. 2005. A mathematical model for zoning protected natural areas. *International Transactions in Operational Research*, 12(2), 203-213.

Wallington, T.J., Hobbs, R.J. and Moore, S.A. 2005. Implications of current ecological thinking for biodiversity conservation: a review of the salient issues. *Ecology and Society*, 10(1).

Wantzen, K.M. 2003. Cerrado streams: characteristics of a threatened freshwater ecosystem type on the tertiary shields of Central South America. *Amazoniana-*

Limnologia et Oecologia Regionalis Systemae Fluminis Amazonas, 17(3-4), 481-502.

Watts, M.E., Ball, I.R., Stewart, R.S., Klein, C.J., Wilson, K., Steinback, C., Lourival, R., Kircher, L. and Possingham, H.P. 2009. Marxan with zones: software for optimal conservation based land- and sea-use zoning. *Environmental Modelling and Software*, 24(12), 1513-1521.

Werner, H.M. 1926. *The Constitutionality of Zoning Regulations*. Urbana: University of Illinois Press.

West, P., Igoe, J. and Brockington, D. 2006. Parks and peoples: the social impact of protected areas. *Annual Review of Anthropology*, 35, 251-277.

Wilson, K.A., Meijaard, E., Drummond, S., Grantham, H.S., Boitani, L., Catullo, G., Christie, L., Dennis, R., Dutton, I., Falcucci, A., Maiorano, L., Possingham, H.P., Rondinini, C., Turner, W.R., Venter, O. and Watts, M. 2010. Conserving biodiversity in production landscapes. *Ecological Applications*, 20(6), 1721-1732.

Chapter 7

Organizational Complexity and Stakeholder Engagement in the Management of the Pantanal Wetland

Thomas G Safford

Introduction

The Pantanal wetland and the surrounding Upper Paraguay ('Alto Paraguai' in Portuguese) River Basin are some of the most unique hydrological and ecological systems in the world. Similarly, the watershed has diverse human populations and their varied relationships with the Pantanal environment shape the course and content of management efforts. Settlements include large urban centers as well as sparsely populated rural communities. Natural resources are central to the region's economy and way-of-life and range from large-scale monoculture agriculture in the uplands to low-density ranching and subsistence fishing in the lowlands. Just as scientific information about the Pantanal's physical, hydrologic and biological characteristics is essential for management efforts, sociological analysis of the social and organizational complexity in the region is equally important.

As already described in previous chapters, the Brazilian portions of the Pantanal are divided between the states of Mato Grosso and Mato Grosso do Sul and these states along with the Brazilian federal government are the central players in environmental management activities. State and federal authorities have recognized that the region's diverse stakeholders often have competing interests and this makes engaging these myriad groups in development and environmental planning critical. Managers have actively promoted multi-party planning approaches as a mechanism for identifying and achieving common social and environmental objectives. While achieving some success, these collaborative efforts have encountered difficulties both ensuring broad participation and balancing the competing interests of different stakeholders. Investigating how inter-group dynamics and relationships shape organizational behaviour is a central area of inquiry for sociologists. Thus, analyzing the organizational experiences of stakeholder groups linked to multi-party planning activities is an important way social scientists can contribute information to support the management of the Pantanal.

Around the globe, one of the most important trends in the natural resources field has been the emergence of collaborative multi-party approaches as a dominant paradigm for managing land, water, and fishery resources (Clark et al. 2005,

Gray 1989, Sabatier et al. 2005, Wondolleck and Yaffee 2000). Social scientists have discovered that organizational networks play an important role in shaping the relationships between public sector actors and stakeholder groups engaged in these collaborative management activities. The nature of these interactions is often a key factor determining whether these planning efforts achieve or do not achieve their objectives (Bazerman and Hoffman 1998, Morton 2008, Waganet and Pfeffer, 2007).

As an innovator in environmental governance, Brazil has been at the forefront of collaborative natural resource management efforts (Abers 2007, Abers and Keck 2006, Hochstetler and Keck 2007, Safford 2004, 2010). These activities bring together public and private sector groups to identify common needs and concerns and to craft programmes that will ensure both sustainable development and environmental protection. Collaborative multi-party planning efforts in the Pantanal have included processes linked to multi-million dollar internationally funded development programmes, such as the *Hidrovia Paraná-Paraguay* and the *Programa Pantanal*, as well as numerous smaller projects sponsored by the Brazilian and state governments. Myriad business, industry, and non-governmental groups from across Mato Grosso and Mato Grosso do Sul have either directly or indirectly engaged in these planning activities. This study draws upon conceptual insights from organizational sociology to investigate the interests and behaviour of these diverse stakeholder groups and how these attributes shape the nature of their engagement in multi-party planning and collaborative management of the Pantanal.

A Sociological Framework for Researching Organizational Dynamics

In order to investigate organizational dynamics, sociologists have studied distinct sets or "fields" of organizations. Organizational fields represent communities of organizations that have common interests and whose members interact frequently as a part of their involvement in similar activities (DiMaggio and Powell 1983, Meyer and Rowan 1977, Scott 2008). Fields can also encompass organizations that share mutual concerns and issues, such as the groups collectively interested in managing natural resources in the Pantanal (Hoffman 2001, Hoffman and Ventresca 2002, Safford 2010). The emphasis of field-level analysis on the formation of shared beliefs and the interactions between different organizational actors make this an ideal analytical unit for investigating collaborative environmental planning efforts.

Extensive research has shown that the relationships among actors within organizational fields affect the formation of commonly held beliefs about appropriate organizational practices (Astely and Van de Ven 1983; Hall and Tolbert 2005). These beliefs become "rationalized" as members of the field mutually reinforce particular ideas and perspectives as a part of their interactions. Sociologists have investigated the origins and implications of these field-level rationalized beliefs by tracking common patterns in the development of organizational structures and practices among actors engaged in shared endeavours (Scott 2008). Collectively rationalized beliefs are not static; rather they are continuously shaped by inter-organizational relations (Hoffman 2001, Hall and Tolbert 2005). Through in-depth analysis of the Pantanal organizational field, this study will show how the distinct characteristics and behaviours of interested stakeholder groups, and their inter-organizational relationships, lead to highly rationalized beliefs about environmental management and the utility of multi-party collaborative planning approaches.

Research Design and Methodological Approach

This project utilizes an embedded case study research design (Yin 1994). When implementing such an approach the initial focus is on embedded sub-units, in this instance organizations interested in the management of the Pantanal region. Analysis is then scaled-up to illustrate broader patterns, such as trends in stakeholder engagement in environmental management across the Upper Paraguay Basin. Embedded case studies are especially useful for organizational research as they can be employed to study a larger process that individual organizations are collectively engaged (Yin 1994). To support this form of analysis, data is collected about the interests and behavioural patterns of the organizational types being studied as well as the social and institutional environment in which they operate.

The first step in this research project consisted of scoping interviews with key informants involved with environmental planning in the Pantanal. These interviews helped establish the constellation of actors within the Pantanal organizational field. Based on this information, a typology of stakeholder groups linked to management efforts was constructed. This typology structured the identification of key organizational actors connected to development and environmental planning activities in the Pantanal. In addition, background material from project documents as well as key informant interviews was used to isolate critical organizational issues affecting stakeholder engagement in multi-party planning efforts. This information about organizational features was then used to develop an interview guide that queried professionals from stakeholder organizations about their interests and engagement in collaborative natural resource management efforts.

The description and analysis in this study draws primarily upon two hundred and two interviews conducted with organizational actors involved in environmental planning in the Pantanal region between1998 and 2002 (Safford 2004). Respondents came from a wide array of organizations that represented the breadth of interests linked to the use and conservation of the Pantanal. They included individuals from federal, state and municipal government agencies, environmental and social justice non-governmental (NGO) groups, indigenous tribes and community associations, business and industry federations, cooperatives and unions, research and extension agencies, and international lenders.

An effort was made to carefully census these various sectors to identify individual actors with either strong interests in natural resource management or direct involvement in multi-party planning activities. Organizational diagrams from governmental agencies, membership lists of trade associations, NGO networks, and other professional associations along with key informant interviews were used to establish a baseline characterization of organizations linked to management of the Pantanal. Although the focus here is on the involvement of business and non-governmental stakeholders, interview data from governmental actors provides important contextual information, especially related to governmental – non-governmental inter-relationships.

A snowball sampling process was used to develop the list of potential interviewees, where respondents provided the names and contact information for other actors involved in management and planning efforts in the Pantanal (Babbie 2001). At multi-party events sponsored by both governmental and non-governmental entities, it was also standard practice to compile attendance lists. These became critical sources for identifying organizational participants and developing a database of potential respondents. Although the Upper Paraguay River Basin covers a large area, the number of organizations directly involved in natural resource issues is relatively small by comparison. This enabled interviews to be conducted with individuals from virtually all of the prominent organizations concerned with management activities in the area.

Because understanding how inter-group interactions influenced organizational actors' views about environmental issues and management approaches, observing organizational behaviour first-hand was an important aspect of the data collection for this study. During research trips to Brazil in 1998, 2000–2001, and 2002 observation data was collected at forty-five different multi-party meetings linked to the planning of the *Programa Pantanal* and other natural resource management projects. These included events open to any interested stakeholder, as well as others limited to specific sub-sets of organizational actors such NGO networks or farmers cooperatives. A concerted effort was made to attend as many different types of multi-party activities as possible in order to build a solid empirical foundation for assertions about inter-organizational processes and actors' beliefs about collaborative management. Finally, throughout this field research any and all available written materials produced by organizations involved in natural resource management endeavours were collected. In total, 291 reports, pamphlets, scientific studies, and memos were catalogued for use in informing this study.

Data gathered using these three research methods were coded and analyzed using a pattern-matching technique that identified common trends in the beliefs and behaviours of actors in the Pantanal organizational field (Bitektine 2008, Trochim 2000, Yin 1994). In this instance, patterns in both the views and behaviour of actors from different stakeholder groups were matched against theoretical constructs regarding organizational relationships and the formation of collectively rationalized beliefs (Scott 2008). This approach enabled both intra- and inter-group comparisons of different actors' conceptualizations of collaborative management and appropriate ways for interacting with the range of interested actors in the Pantanal region. The pattern-matching methodology is an effective tool for examining levels of consensus between distinct groups involved with shared endeavours, such as organizational actors involved with the management of the Pantanal (Bitektine 2008, Trochim 2000).

The Pantanal Organizational Field

Development and environmental planning in the Pantanal intersects with the interests of a wide range of stakeholder groups. For the purposes of this research, two overarching categories of organizational actors are used to structure sociological analysis of their engagement in collaborative multi-party planning efforts: 1) the business and production sector and 2) the civil society and non-governmental sector.[1] These categories are empirically derived based on similarities in the roles particular types of organizations play within environmental management activities in the Pantanal. The subsequent sections provide description of key organizational actors within these two sectors as well as analysis of their interest in environmental concerns and engagement in multi-party planning. This study does not provide a comprehensive assessment of every organization within these two sectors; rather it focuses on representative sets of organizational actors that can be used to make inferences about the way different members of the Pantanal organizational field perceive collaborative multi-party environmental planning efforts.

One of the most intriguing elements of Pantanal region as a location for investigating collaborative approaches is the range of organizational actors involved in natural resource management and the distinct beliefs and practices they bring to multi-party processes. The region does not have an extremely large number of stakeholder groups, nor the most successful participatory activities in Brazil. Nonetheless, the constellation of interested actors in the Upper Paraguay Basin represents key organizational types that are pivotal players in natural resource management efforts across Brazil and the globe. Thus, insights from the organizational experiences of stakeholder groups in the Pantanal not only have implications for the future management of this unique wetland, but also for collaborative environmental management efforts in other regions and countries.

---

1   Other key organizational entities involved with the management of the Pantanal include municipal, state, and federal natural resource and environmental agencies and universities and extension groups. These actors play pivotal roles in management efforts, however, because business and civil society organizations are the focus of this study, analysis of governmental and research agency engagement in multi-party planning is not addressed in this chapter.

The Business and Production Organizational Sector

Table 7.1   Key Members and Organizational Features of the Business and Production Sector of the Pantanal Organizational Field

	Business and Production Organizations (Mato Grosso)	Business and Production Organizations (Mato Grosso do Sul)
Key Members of Organizational Field	*Federação das Indústrias* – State Industry Federation *Sindicato Rural* – Rural Landowners Union FAMATO – Association of Agricultural Producers and Processors Nature tourism operators	*Federação das Indústrias* – State Industry Federation *Sindicato Rural* – Rural Landowners Union FAMASUL – Association of Agricultural Producers and Processors Nature tourism operators
Organizational Features	<ul><li>Business and production sector actors rarely participate in multi-party planning activities and exhibit little support for collaborative approaches.</li><li>Industry and agricultural federations effectively use political lobbying to forward both their individual organizational interests as well as those of their members.</li><li>The *Sindicato Rural* is the most powerful industry group; its affiliated actors use both lobbying and direct engagement in multi-party planning to achieve the union's objectives.</li><li>Agricultural producer associations oppose multi-party approaches and collaborative planning, seeing them as a challenge to the status-quo which favors their interests.</li><li>Traditional tourism-related groups use lobbying in similar ways as other business sector actors, but nature tourism operators show a greater openness to collaboration.</li></ul>	

Groups and associations representing business entities have strong interests in the management of natural resources in and around the Pantanal. These organizations range from trade federations promoting particular industries to unions of agricultural producers. One of the defining characteristics of actors from this sector is their political power and ability to use lobbying to forward their interests within environmental programmes. Although they share similar concerns, business and production groups both individually and collectively have chosen distinct ways to engage in multi-party planning activities. The relationships between organizations from this sector and governmental and civil society actors, as well as the particular tactics they have used to influence multi-party planning processes, have important implications for broader efforts to promote collaborative management in the Pantanal region.

Actors from Business and Industry Associations

Natural resources are the economic engine for the Pantanal region. Trade groups and industry associations work to ensure unfettered access to water, land, and fishery resources that these operations depend upon. The business community is well organized and the industrial federations (federações das industrias) in both states actively forward the interests of the business community within governmental arenas. At the time of this study, the president of this federation in Mato Grosso owned a water extraction and well-drilling firm, while in Mato Grosso do Sul the leader operated an agricultural processing business. The fact that the two leaders were from resource-related industries reflects both federations' membership at large and the importance of these types of activities in the two states.

Given these links, organizational actors affiliated with the state industry federations would be affected by a variety of environmental management programmes in the Pantanal. Nonetheless, actors from these federations had minimal contact with the state environmental agencies and had little or no interaction with other types of organizations involved with natural resource management activities. Analysis of attendance lists from multi-party planning events linked to the *Programa Pantanal* project illustrate that in only two instances had individuals from the industrial federations in either state participated. Similar patterns appear in analyzing the engagement of these groups in other regional multi-party planning efforts. Having identified the marked absence of these groups, it is critical to establish why they chose to not collaborate in these environmental management activities.

Interview data from respondents representing the state industry federations, as well as their member organizations, highlights that these actors made strategic decisions to focus on lobbying elected officials rather than engaging in multi-party planning to forward their interests. In Mato Grosso do Sul, an actor from a prominent agro-business indicated that he had worked closely with the assemblyman who chaired the natural resource committee in the state legislature and he was confident agriculturalists would receive the necessary support at the time of implementation of regional development projects. In addition, interview data gathered from state assemblymen in Mato Grosso do Sul involved with environmental planning show a clear link between the business community and these politicians. As one assemblyman indicated, as an elected official it was his job to ensure that the needs of his constituents were met and that businesspeople had better things to do than attend multi-party meetings convened by government environmental agencies.

In Mato Grosso, the head of the industry federation indicated that he had attended a number of multi-party events in the past and found they were completely unproductive. He perceived them as forums where activist NGOs complained about environmental problems associated with development and state agency officials bemoaned the lack of funding for addressing them. With strategic campaign contributions, he indicated that he gained access to the governor and key legislators. Through these interactions he garnered the support he and other members of the federation desired.

This indirect approach to influencing the course of natural resource management activities achieved his and other industry stakeholders' objectives. Nonetheless, by not participating in multi-party activities he limited his interactions with other organizational actors. This constrained his ability to contextualize his interests within those of other groups concerned about the Pantanal and impeded the development of a collaborative management approach. Interview data from other actors from the business sector show a consistent frustration with rhetoric surrounding environmental issues in the Pantanal. Business sector respondents perceived civil society and NGO groups as inherently anti-business and anti-development and this discourse impeded collaborative interactions that might have led to shared beliefs and understanding across these distinct types of organizations.

These individual accounts provide important insights into why industry and business sector organizations have historically been peripheral players within the multi-party planning efforts in the Pantanal. Nonetheless, for managers seeking to address resource-related concerns, the beliefs of actors from this sector will be important to consider. Even though they may not directly participate in multi-party planning, they shape collaborative processes through their political clout and ability to externally alter the course of management activities to forward their interests. The success of multi-party planning depends not only on the nature of interactions among participants, but also hinges on who is not participating and how they may be influencing management activities outside of formal collaborative processes.

Actors from Agricultural and Ranching Associations

Developing alliances with politicians to forward their interests was also the primary way actors from the agriculture and ranching communities engaged in the management of the Pantanal. However, these actors chose to press their interest not only by using political tactics, but also through directly participating in multi-party planning activities. This two-pronged approach was a pragmatic effort to use all possible tools of influence. It also illustrates recognition that collaborative processes play an increasingly important role is shaping environmental management in the region. A number of respondents from agricultural associations suggested that ten or fifteen years earlier there would have been no need to attend multi-party environmental planning events. These actors were pragmatic and understood that governmental agencies and international lending institutions looked to these multi-party activities for guidance and thus it was critical for ranching and agriculture groups to participate.

A number of respondents from agricultural and ranching organizations also pointed out that the interests and objectives of these groups were not homogenous. This meant that organizations with parochial interests could use multi-party forums to forward their concerns if they were not met through the agricultural sector's broader political lobbying activities. After analyzing a range of documents produced by agricultural actors, content analysis data show consistent support for promoting agro-pastoral development in the Pantanal region. However, an extremely wide range of methods were cited as ways to achieve this shared objective.

This finding highlights a highly rational shared belief about the use of natural resources to spur development, but divergence in these actors' beliefs about specific management strategies and approaches. For example, ranchers from lowland Pantanal communities promoted expanding low impact "green" production to differentiate Pantanal beef, while producers in upland areas of the basin sought increased intensive production and the development of feed lots. Because these objectives are so different, it is logical that each group would independently seek to promote their interests within management programmes.

The most powerful organization within the production sector is the *Sindicato Rural* (Rural Landowners Union). This organization brought together both ranchers and farmers from across both states. Each municipality in the basin has a local affiliate which operates under state and national conglomerations of the *Sindicato Rural*. Because the union worked as both a lobbying group and ran independent agricultural extension activities, the organization was highly decentralized. These different elements of its operations were one of the reasons respondents from production sector organizations cited as the source of heterogeneity in the union's positions and activities related to natural resource management. Actors from *Sindicato Rural* affiliates in the uplands of the basin shared a belief in the importance of promoting technically-advanced agriculture, while those in lowland Pantanal portions emphasized sustainable use.

Like the industry federations, the rural landowner union played key roles in state politics. This influence enabled them to mobilize both legislators and governors to support their interests within environment-related programmes. Interview data gathered from all types of actors in both states highlights a consistent reference to the power of the agricultural and ranching lobbies in both Mato Grosso and Mato Grosso do Sul and their ability use political influence to shape natural resource management in the Pantanal.

Assemblymen linked to these groups blocked legislation perceived as contrary to their interests and politicians supporting these agro-pastoral groups pressured state environmental agencies on their behalf. As this was always a cloud over multi-party activities, state management actors regularly sought input from the *Sindicato Rural* and other groups in this sector to ensure their programmes would not be opposed. This created a power imbalance, but it also increased interactions among governmental and production-oriented actors. Although the rural landowners union often applied coercive pressure, they did interact more frequently with government natural resource management professionals than those from industry associations. These interactions provided opportunities to develop more collaborative inter-organizational relationships and aided in the formation collective beliefs about issues concerning the development and protection of the Pantanal.

Interview data from individuals associated with the *Sindicato Rural* highlight that actors who backed collaborative management approaches rationalized this support based on their ability to secure funding for their members through these activities. However, these respondents also indicated that participating in multi-party processes had helped them recognize the value of input from other interested actors in achieving broader management objectives for the Upper Paraguay region. This finding suggests that multi-party planning activities not only brought individual benefits to agriculturists but fostered more positive inter-group relationships that could be the foundation for a broadly-supported collaborative management approach for the Pantanal.

The *Sindicato Rural* was not the only organization linked to agriculture wielding influence in the Pantanal region. State trade associations of agricultural producers, known as FAMATO in Mato Grosso and FAMASUL in Mato Grosso do Sul, had staff working directly with government officials on development projects, as well as lobbyists seeking to influence elected officials. FAMATO in particular worked aggressively to promote its interests within the governmental sphere and opposed any management processes that might undermine the value of their influence with politicians in Mato Grosso.

When actors from this agricultural trade association participated in multi-party activities they actively vetoed programmes that required consensus, and where representatives of both state associations did not like the decisions reached, they went to court to challenge the legitimacy of multi-party bodies themselves. Respondents from FAMATO were highly rational about using these obstructionist tactics. They felt agriculture's preeminent position in the region's economy meant it needed to be supported using all possible methods. Interview data show that individuals representing members of both agricultural federations shared a strong belief that maximizing natural resource use was essential to their states. They viewed multi-party approaches and collaborative environmental planning as counter to this objective. Conversely, actors from public sector agencies and NGOs cited the presence of actors from both FAMATO and FAMASUL as extremely disruptive for collaborative planning efforts. The nature of their engagement undermined support for a broad multi-party approach. This illustrates the critical importance of establishing a shared belief in the benefits of inter-organizational collaboration among all interested stakeholders in order for these processes to function effectively.

Unlike the *Sindicato Rural*, which worked directly with ranchers and farmers throughout the Upper Paraguay Basin, FAMATO and FAMASUL did not have the same decentralized structure. These associations were linked to agro-businesses and processors as well as producers. Respondents from affiliate businesses interviewed for this study felt that given the political power of these associations it was not necessary for their individual organizations to participate in multi-party environmental planning activities. This should not suggest that actors from member groups all had the same views about natural resource-related issues in the Pantanal. Nonetheless, they appear to have felt that the state associations sufficiently represented the broad interests of agriculture. Because FAMATO and FAMASUL worked independently, they did not have experience collaborating or drawing upon input from outside groups. This may explain in part why actors from these groups were unwilling to compromise or to consider the views of groups not aligned with these associations. This disconnect between agricultural groups and other interested actors is one of the greatest challenges confronting managers attempting to promote more collaborative management in the Pantanal region.

The final business and production sector actor linked to multi-party environmental planning efforts is the tourism industry. Although state-wide federations of tourist operations exist in both states, these associations rarely played a direct role in natural resource management activities. As was the case with the *Sindicato Rural*, regionally important tourism-related organizations, linked to nature and eco-tourism in particular, wielded influence within the planning of development projects. Respondents from these organizations stated that tourism federations represented hoteliers and that in order for the special needs of nature tourism operators to be included in government programmes they needed to be more directly involved in multi-party planning activities.

This nascent type of tourism did not have access to private sector funding in the way traditional tourism operations did nor did they have extensive political connections or clout like agro-business organizations. This led to a highly rationalized belief in the need to seek out other forms of assistance and new alliances. As protecting flora and fauna was essential to their operations, an affinity emerged between actors from nature tourism organizations and those from other non-governmental groups focused on conservation in the Pantanal. In this way the nature tourism sector represents a divergence in the patterns of organizational behaviour exhibited by business and production sector groups. This illustrates that rather than a fundamental divide between business actors and other non-governmental groups, it is inter-organizational relationships and the formation of common beliefs that ultimately determine whether wide-ranging stakeholder groups can successfully forward a collaborative approach to the management of the Pantanal.

The Civil Society and Non-Governmental Organizational Sector

Table 7.2   **Key Members and Organizational Features of the Civil Society and Non-Governmental Sector of the Pantanal Organizational Field**

	Civil Society and Non-governmental organizations (Mato Grosso)	Civil Society and Non-governmental organizations (Mato Grosso do Sul)
Key Members of Organizational Field	FORMAD – State NGO Forum ICV – Institute for Life Ecotrópica FETAGRI – Agricultural Workers Union Fisher cooperatives and community associations International environmental organizations (WWF, CI, TNC)	FORMADES – State NGO Forum ECOA – Ecology and Action IBISS – Brazilian Institute for Social Health Indigenous tribes – Guató, Terena, Kadiweú FETAGRI – Agricultural Workers Union Fisher cooperatives and community associations International environmental organizations (WWF, CI, TNC)
Key Organizational Features	• Institutional problems and internal divisions are a barrier to collaboration among NGOs in Mato Grosso. • In Mato Grosso do Sul, NGOs value collaborative interactions and these lead to common beliefs about both environmental concerns and management approaches. • IBISS establishes health-environment links aiding in the development of an integrated perspective on natural resource management in Mato Grosso do Sul. • Indigenous tribes' interactions with NGOs foster collaboration, but deep-seated mistrust of outsiders inhibit their engagement in multi-party planning activities. • Labour unions are not major players in the Pantanal organizational field, in large part due to practical decisions by these actors to focus on other regions of the two states. • Commercial and sport fishers share concerns about fishery problems, but intra-state divisions and debates about management approaches limit collaboration. • International NGOs collectively focus on environmental conservation, but choose to forward these interests outside of multi-party planning activities.	

Non-governmental groups representing different segments of civil society are the most complex type of organizations within the Pantanal organizational field. They range from small community associations to large internationally-recognized non-profits. Actors from these groups share a common desire to influence public policy and promote alternative approaches to natural resource management, but they are far from homogenous and often have different positions on local versus regional environmental concerns. A number of organizations within this sector are also part of broader coalitions that bring together these disparate groups to forward a common front within multi-party planning processes. Non-governmental networks in both Mato Grosso and Mato Grosso do Sul play key roles in organizing interested civil society groups and creating space for these types of actors to participate in environmental management activities. Understanding both the inter-relationships among actors from this sector and the nature of their engagement in multi-party planning provides important insights that may aid managers in broadening civil society engagement and advancing a more collaborative management approach.

Actors from Non-governmental Organizations in Mato Grosso

In Mato Grosso, non-governmental organizations have struggled to develop common strategies for confronting resource-related issues in the Pantanal region. This has meant that these actors have neither the power nor the influence that business sector groups have on environmental management. Respondents from both governmental and non-governmental organizations indicated that the difficult, and often adversarial, relationship with the state government had inhibited the growth and effectives of NGOs across Mato Grosso.

Without stable sources of funding or institutional support, these groups remain fragmented and disorganized. Analysis of interview data from actors from this sector show that many believed that promoting a new development paradigm for the Pantanal region was futile given the state government's centralized control over policies related to natural resource management. These patterns are in stark contrast to agricultural associations and landowners unions who effectively navigated the state bureaucracy and lobbied political actors to forward their interests. These divergent experiences illustrate both challenges and opportunities for those promoting collaborative management in the Mato Grosso portions of the Pantanal.

At the state level, FORMAD, the state NGO Forum for Development and the Environment, was the principle organization coordinating the work of civil society organizations interested in environmental and social concerns. Its members included influential environmental groups such as ICV, *Ecotrópica*, and *Ecopanta*. The Forum had strong ties to labour organizations such as the Federation of Agricultural Workers, FETAGRI, and fishing cooperatives from across the Upper Paraguay Basin. Affiliates also included community health organizations and actors from ecumenical groups such as the CPT, the Pastoral Land Commission, which provided social services to under-served rural communities. In theory, this diverse group of organizations gave FORMAD strength in numbers and brought a diversity of perspectives to discussions of resource-related concerns in the Pantanal. Nonetheless, the Forum had more members in name than in practice.

Inter-organizational relations were at the core of FORMAD's challenges. The difficult relationship between the Mato Grosso government and NGOs at the time of this study made collaborative interactions extremely challenging. A number of respondents suggested that the Forum could never have a significant impact given the power of politicians in defining the objectives of environmental programmes. Most importantly, respondents from organizations affiliated with FORMAD voiced little support for multi-party planning processes, as their previous experience with these activities had left them sceptical about their utility. As one respondent indicated, there was no use spending time participating in planning activities sponsored by state environmental agencies when the governor would ultimately decide the priorities for natural resource programmes. Although organizational actors associated with FORMAD collectively believed in the need for a coordinated effort to change the development paradigm in Mato Grosso, they shared a highly rationalized scepticism about the value of collaborative planning.

The lack of internal consensus among groups associated with the Forum was an additional factor undermining support for collaborative approaches. Actors from environmental NGOs questioned the value of working with public health organizations and vice-versa. Reports and letters from FORMAD to public officials illustrate a collective lack of focus on solutions, with most documents merely listing each group's concerns. FORMAD meeting minutes highlight environmental problems in the Pantanal, but do not include substantive consideration of management alternatives or ways to use collaborative inputs to offer alternative approaches to the use or conservation and natural resources. With these institutional difficulties, and the powerful centralizing force of the state government, FORMAD was in a weak position to forward the cause of civil society groups or promote increased NGO involvement in natural resource management. Respondents from the Forum's member organizations shared a deep concern about social and environmental conditions across the state. Nonetheless, the strength of these core beliefs did not naturally translate into an effective coordinated strategy for addressing these concerns.

As FORMAD was limited in its ability to represent civil society within environmental management programmes, powerful NGOs in Mato Grosso like ICV and *Ecotrópica* became the dominant non-governmental actors involved in multi-party planning efforts. Rather than banding together through FORMAD, smaller community groups and associations aligned themselves with these NGOs, unions, or production associations to achieve their aims. This created a scattered approach without regular lines of communication or interaction that would have likely fostered shared beliefs about management approaches and established a foundation for broader collaboration. In some cases, groups also worked through the more organized civil society associations and networks in Mato Grosso do Sul. NGO respondents suggested that the effectiveness of actors from Mato

Grosso do Sul in representing civil society in broader regional planning efforts for the Pantanal was one of the principle reasons why a strong network of non-governmental organizations never emerged in Mato Grosso.

In the state capital of Cuiabá, problem-specific community organizations also emerged around pressing local issues such as water pollution and public health, but they rarely attempted to combine their efforts or link up with groups concerned about similar issues in other cities. In the areas beyond the capital, small regional NGOs played roles in other narrow local issues. In the municipality of Barão de Melgaço, a number of non-governmental groups were involved in conservation efforts and with protection of the local fishery. Similar organizations emerged in the town of Cáceres which also faced fishery declines. While the importance of nature tourism in Poconé mobilized preservation-oriented actors in this municipality. NGOs from all of these areas individually contributed to multi-party planning activities related to environmental management in the Pantanal.

However, documents produced by these organizations as well as interview data highlight a pattern where these regional groups narrowly focused on particular core natural resource-related concerns rather than considering the links between their issues and broader environmental management efforts. Interestingly, these actors were also often unaware of parallel activities linked to the same resource issues in other parts of the state. For example, the organizations in Barão de Melgaço and Cáceres were both concerned about fisheries declines, yet neither their reports nor interview responses pointed to the possibility of combining efforts to achieve shared objectives regarding fisheries management in the Pantanal.

When asked directly about collaboration, respondents from these NGOs voiced a consistent belief that other groups were either unaware of, or uninterested in, their problems. Thus, they did not see the benefit of developing partnerships. This vividly exemplifies how the nature of inter-organizational relations and the lack of a viable coordinating network limited the ability of NGOs in Mato Grosso to establish collaborative bonds that may have enabled them to effectively press their collective interests in the management of the Pantanal.

A number of NGOs in Mato Grosso did attempt to raise public awareness about pressing environmental problems and implemented small projects that addressed natural resource-related issues in the Pantanal. However, when asked to characterize these groups, governmental actors did not focus on these aspects of NGOs work. They viewed these actors as activists who merely questioned public policies and criticized management programmes rather than contributing concrete solutions to environmental concerns. These responses exemplify a pattern of beliefs that severely constrained interactions between civil society groups and governmental actors in Mato Grosso. Findings from this study illustrate that in order for collaborative management efforts to be successful in the Pantanal, managers must diagnose the social forces shaping both the relationships among NGO groups as well as these organizations' links with public sector actors. Clearly the nature of existing inter-organizational relationships makes this a challenging setting for promoting collaborative management. Nonetheless, the experiences of

NGOs in the neighbouring state of Mato Grosso do Sul illustrate that this type of intra and inter-organizational collaboration can occur.

Actors from Non-Governmental Organizations in Mato Grosso do Sul

Non-governmental groups in Mato Grosso do Sul exhibit marked differences in their organizational characteristics compared with their counterparts in Mato Grosso. In particular, FORMADES, the state NGO Forum for Environment and Development, is a much more cohesive and an active player in development planning and natural resource management in the Pantanal. FORMADES attracted a broad range of members, had a formal institutional structure with officers and duties, and held regular meetings. Virtually every civil society organization linked to natural resource management concerns in the state was either formally or informally associated with this NGO coalition. Its members included environmental groups, public health organizations, indigenous tribes, student associations, and a variety of community groups. Over fifty organizations were officially affiliated with FORMADES. To ensure that one group did not over-determine the Forum's activities, the coordinator rotated annually and decisions required consensus before FORMADES formally acted on an issue.

Because FORMADES was well organized, it adapted quickly to its members' needs. The Forum was extremely effective in responding to emerging issues and mobilizing actors from affiliated groups to participate in management and planning efforts that necessitated civil society engagement. ECOA, a regional environmental and social justice NGO based in the state capital Campo Grande, played a central role in coordinating the activities of FORMADES. Although the president of ECOA wielded considerable influence because of his organization's leadership among NGOs, he and other actors from ECOA made concerted efforts to ensure that no one individual or organization dominated the Forum's multi-party processes. This pattern of behaviour exemplifies a strong core belief in the value of collaborative inputs among actors from ECOA and FORMADES more broadly.

Although the majority of organizational actors affiliated with FORMADES were from groups located in Campo Grande, the Forum actively sought out organizations from outlying areas and brought them into broader civil society efforts to promote alternative approaches to natural resource management. Actors from community groups in the remote Pantanal towns of Porto Murtinho, Corumbá, Bonito, and Nioaque regularly attended meetings, and also took on leadership positions in the Forum. This was in marked contrast with the experience of the NGO forum in Mato Grosso which was never able to achieve this type of regional participation.

As more and more groups joined FORMADES, collaboration increased, but these new members looked to the Forum's long-standing leaders for guidance on how to address regional natural resource issues that were beyond the purview of their particular organization. Interview data from respondents from FORMADES' smaller affiliates show that while most of these actors had strong core beliefs about their own organization's central concern, such as over fishing, sedimentation, or bio-diversity protection, they had little knowledge about broader regional issues. Given this knowledge gap, these actors logically followed the lead of FORMADES and influential members like ECOA when considering management issues that were beyond their core areas of concern. However, an even more intriguing finding was that not only were the views of actors from these new groups affected by these exchanges, but also those of long-time participants. Patterns in the interview and content analysis data show that the beliefs of actors from ECOA and other organizational leaders were not static and were strongly influenced by their interactions with regional members of FORMADES who brought novel viewpoints on the management of the Pantanal.

Another group that had considerable power and influence within FORMADES was IBISS. This NGO focused on public health and social justice issues, especially those linked to the condition of women, children, and indigenous peoples. They brought a new perspective on environmental concerns, in particular highlighting issues relevant to women natural resource users and the impacts from large development projects on small communities such as increases in prostitution and HIV/AIDS. By talking about these broader social problems, and linking them to environmental concerns, IBISS helped FORMADES articulate a more holistic approach to management in the Pantanal region. Their efforts had an especially powerful impact on international donors and federal government agencies in conjunction with multi-party planning efforts. These affinities and the articulation of the broader implications of development projects helped strengthen the overall position of civil society groups within planning and management processes related to the Pantanal

Interview responses from actors linked to IBISS show that most of these individuals knew little about environmental issues before joining FORMADES. Through their association with the Forum they learned about environmental problems and recognized the links to their priority health-related concerns. Similar organizational learning took place among environmental actors who gained increased understanding about public health issues by working in partnership with IBISS. These collaborative exchanges marked differently from the more narrow parochial approach of groups associated with FORMAD in Mato Grosso.

The positive nature of these interactions should not suggest that the work of FORMADES was easy or entirely successful. The diversity of member groups, and the contentiousness of many of the issues under discussion, meant that consensus was not always achieved and there were often disputes between its affiliates. Nonetheless, after formalizing interactions among actors within the Forum, strong inter-organizational bonds were formed and these solidified a collective belief in the value of collaborative organizational approaches. FORMADES' experience illustrates the benefits that can result from efforts to cultivate inter-organizational relationships and collaborative exchanges. The changes in the behaviour of groups associated with the NGO Forum in Mato Grosso do Sul demonstrates that beliefs about different approaches to natural resource management can shift if multi-party processes are structured to encourage organizational learning and a common sense of purpose.

Actors from Indigenous Tribes

The Pantanal region is home to a number of different indigenous peoples and Mato Grosso do Sul has one of the largest native populations of any state in Brazil. The vast majority of these tribal communities are clustered in the southwest corner of the Upper Paraguay Basin near where the borders of Paraguay, Bolivia, and Brazil meet. Because of the remoteness of this location, these groups rely heavily on natural resources for their survival and have been important conservators of the Pantanal. However, these communities are also some of the most underdeveloped in the region and indigenous peoples have had difficulties interacting with Brazilian migrants who have increasingly encroached on their homelands.

Historically, indigenous groups have had limited involvement in environmental planning and they have relied on state government officials or those from the federal Indian agency, FUNAI, to articulate their interests. Nonetheless, increased interactions with a number of non-governmental groups changed this approach. The links between leaders from the Guató and Terena peoples and ECOA was a key factor that encouraged indigenous actors to more directly forward their own demands within multi-party planning activities. Although indigenous participation increased following these interactions, their ability to engage in collaborative management was constrained by native peoples' troubled relationships with governmental authorities.

Each of the state governments had staff assigned to work with indigenous groups. However, these departments were poorly funded and largely ineffective in addressing the problems facing native communities. Interview data from indigenous leaders and other actors outside of the government characterized them as mismanaged and corrupt. In Mato Grosso, the indigenous population located in the Upper Paraguay Basin is extremely small, especially in comparison with the large numbers in the state's northern Amazon region. This meant that state actors involved with indigenous issues mostly ignored the small tribes in its portion of the Pantanal, focusing instead on Amazonian groups.

Because Mato Grosso do Sul had a much larger indigenous presence in the Pantanal, the state attempted to develop a variety of natural resource-related programmes in its indigenous areas. Nonetheless, actors from the state government suggested that the hierarchal leadership structure among Pantanal tribes, and a deep distrust of outsiders, made it nearly impossible to work in these communities. In addition, state managers consistently cited indigenous groups' insistence on working directly with high-level federal officials and international donors as a barrier to effective collaboration within multi-party environmental programmes.

Respondents from all types of organizations interviewed for this study suggested that issues related to the involvement of indigenous groups created complicated organizational questions for natural resource managers. They consistently cited tribal actors as the most difficult to engage in multi-party planning. The only non-indigenous actors that worked effectively with native groups were social and environmental groups linked to FORMADES. The public health group IBISS had worked well with the Kadiweú and ECOA collaborated closely with the Guató and Terena tribes. These connections slowly brought indigenous actors into the broader organizational field, but they remained some of the most independent organizational actors in the Pantanal region.

In analyzing interview data from indigenous actors themselves, one pattern that emerges is a consistent belief among these respondents that indigenous communities were fundamentally different from other groups in the basin. They also strongly believed that native people themselves, rather than government officials, should design programmes to assist their tribes. These actors also voiced a pattern of resentment about the historic maltreatment of indigenous communities by governmental authorities, and the depth of these feeling will likely make them difficult to overcome. Both the focus on local autonomy and distrust for outsiders and governmental authorities is a serious barrier for broader indigenous engagement in collaborative management of the Pantanal. Nonetheless, the incremental success members of FORMADES had partnering with these groups suggests that progress can be made if relationships of trust and respect can be cultivated and tribal issues are prioritized within multi-party planning efforts.

Actors from Labour Unions and Cooperatives

Although Brazil has a well-organized labour sector, unions have played only minor roles in the planning and implementation of natural resource programmes in the Pantanal. Because agricultural production dominates the region, FETAGRI (the Brazilian National Federation of Agricultural Workers) was active in both states. In Mato Grosso, actors from FETAGRI occasionally worked with FORMAD as well as with some of its member NGOs, but the union was rarely involved with multi-party environmental planning activities. When respondents from the Federation were asked about this limited engagement, they indicated that the union had scarce human and financial resources and they felt that their efforts would be better spent on issues in the northern Amazonian areas of the Mato Grosso. Actors from NGOs and governmental agencies also suggested that the union was not well organized in the Pantanal municipalities and chose to work with strong affiliates in the northern part of the state.

Following a similar pattern, FETAGRI also chose to work in parts of Mato Grosso do Sul outside of the Upper Paraguay Basin and did not actively participate in management programmes linked to the Pantanal. However, labour organizations in Mato Grosso do Sul did have strong ties to FORMADES. Interview data from these respondents highlight a pattern of belief that the NGO forum could sufficiently lobby for agricultural workers in the Pantanal region and that the union could then focus on issues in other areas of the state. Respondents from groups linked to FETAGRI also suggested that the union was not extremely interested in environmental issues. Thus, concerns linked to protection of the Pantanal were not a priority. In both states organized labour was conspicuously absent from multi-party planning activities. Because their lack of engagement was more a pragmatic decision than one based on opposition to collaborative approaches, opportunities exist to increase the participation of unions in the future.

As fishing is an important economic activity in the Pantanal, cooperatives among small commercial fishers are key players in management and planning efforts. These cooperatives regularly convened meetings among fishers, and their representatives were present at many multi-party planning events. Sport fishing is an equally important form of fishing, but actors involved in these activities did not form networked organizations and normally worked in conjunction with broader tourism groups. Their involvement in environmental management was similar to other business and production actors in the region. They were politically active and lobbied the two governors and state assemblymen to forward their interests in the Pantanal.

Both sport and commercial fishers severely criticized the fish and wildlife agencies in Mato Grosso and Mato Grosso do Sul for failing to coordinate fisheries management. The two states had different fishing regulations and commercial fishers were not allowed to cross state lines. Confusion in fisheries policies and programmes is one of the most contentious resource related problems in the Pantanal region. Respondents from both fisher groups and state actors attributed these failures to the inability of governmental agencies and interested NGO groups to recognize the necessity of a coordinated effort to improve management activities. They shared a belief in the core issue of fishery declines, but they did not agree on a utility of a collaborative management approach for addressing them.

Although competing interests, and cross-state rivalries, limited fishers' ability to develop a common front on fishing issues, individual fisher cooperatives were important contributors to multi-party planning related to the Pantanal. In Mato Grosso do Sul, actors from these cooperatives worked effectively with other organizations on regional environmental management. Respondents from fisher groups indicated that they found their interests in fisheries conditions were closely aligned with other non-governmental actors and this facilitated collaboration. In Mato Grosso, the fishing cooperatives were also keenly interested in the management of the Pantanal, but they did not have sufficient numbers or political clout to alter state natural resource management programmes dominated by agricultural interests.

Where advances were made by fisher groups, it was through collaboration with actors from other organizational sectors rather than through a coordinated effort among fishing-related organizations. The experiences of both agricultural unions and fisher organizations in the Pantanal illustrate the challenges faced when encouraging collaboration among organizations that have strong interests outside the region or are in competition with other groups seeking similar access to scarce natural resources.

Actors from International Non-Governmental Organizations

Domestic non-governmental organizations were not the only NGO groups interested in the management of the Pantanal. A wide-array of international NGOs either directly engaged in programmes addressing environmental issues in the region or funded projects supported by local partners. Environmental organizations such as Conservation International (CI) and World Wildlife Fund (WWF) established local affiliates in Mato Grosso do Sul, enabling them to closely follow environmental concerns in the Pantanal. Other organizations such as The Nature Conservancy (TNC) and International River Network (IRN) chose more indirect involvement through their national offices located in Brasília and São Paulo.

Interestingly, these organizations were not consistent participants in multi-party planning activities. In systemizing the attendance lists from planning meetings linked to the Programa Pantanal and other multi-party processes, only one individual from WWF was regularly present. Interview data gathered from a variety of actors from these organizations highlight a pattern where respondents believed that their efforts were better spent cultivating relationships with international donors and national policy makers than working through multi-party activities in the region. Although they wanted a visible local presence, these groups shared a belief that multi-party planning activities were of limited value in forwarding their organizational objectives.

International NGOs technical credentials and financial resources gave them considerable clout and access to state and federal policy makers. These actors' ability to purchase land or fund projects also provided them with a variety of ways to unilaterally promote their interests in protecting the Pantanal environment. These NGOs were not necessarily opponents of collaboration, and actors from these organizations lauded the efforts of groups such as FORMADES and fisher cooperatives within government sponsored multi-party planning activities. Nonetheless, interview data from international NGO respondents highlight a highly rationalized belief that their organization's role was to provide a strong technical rationale for conservation in the Pantanal. These respondents consistently voiced a belief that this objective could be best achieved through strategic partnerships directly with governmental actors and international donors, rather than through multi-party activities. Where they did promote a collaborative approach was in the implementation of their own projects, not through government-sponsored processes that focused on regional environmental planning in the Pantanal.

Finally, there was also considerable competition among international NGOs. Although they shared a collective interest in preserving the Pantanal, their views on how to achieve this goal were not homogenous. Conservation International and The Nature Conservancy focused on buying land and promoting the creation of new protected areas. WWF worked more with existing landowners and sought to include human populations within broader conservation programmes. These competing approaches created challenges for collaboration among international environmental NGOs and this contributed to their collective disengagement in multi-party planning activities.

Like many business and production groups in the region, international NGOs recognized that lobbying and direct interactions with government officials was an effective way to influence environmental planning in the Pantanal. However, actors for international NGOs were not opponents of multi-party processes or collaborative input. Their lack of engagement was strategic rather than based on a shared opposition to more participatory approaches. This suggests that if future multi-party processes in the region can establish their relevance for environmental conservation these groups could become valuable contributors to the collaborative management of the Pantanal.

Discussion and Conclusions

The Pantanal region has complex environmental challenges that are linked to myriad anthropogenic influences across the Upper Paraguay Basin. Effectively tackling these problems will require harmonizing natural resource use and conservation. To forward management approaches that can achieve this objective, government authorities will need to better understand the diverse interests of stakeholder groups in the region and the factors that both facilitate and impede their engagement in collaborative management efforts. Two of the overarching insights from this study are the importance of defining the members of the organizational field and understanding not only the perspectives of stakeholder groups that regularly participate in multi-party processes, but also those that are absent. This research points to both opportunities and challenges for managers seeking to promote collaborative planning and illustrates the important role sociological analysis can play in supporting these efforts.

Although environmental management is often viewed as a strictly technical activity, one of the key insights from the Pantanal organizational experience is the central role politics play in these endeavours. The tactics used by business and production groups show that among this organizational sector there is a highly rationalized belief that political lobbying is the most effective mechanism for ensuring these actors' interests are addressed within natural resource management programmes. Business groups appear to have accomplished their objectives using these methods, but their absence from multi-party processes has impeded the development of common understandings about production-related issues across all members of the organizational field. Similarly, international environmental NGOs have chosen to work directly with government officials and international donors rather than contribute to multi-party planning to promote their conservation goals. Although many of these groups have exhibited tacit support for collaborative approaches, their lack of consistent participation has limited these actors' ability to understand other stakeholders' interests or identify common themes that could be used to promote a more integrated approach to the protection of the Pantanal.

The experience of the *Sindicato Rural* illustrates that political lobbying and direct engagement are not completely incompatible, but there first needs to be a shared recognition about the value of collaborative input. If this does not occur, groups like the agricultural trade association FAMATO, which consciously sought to impede multi-party planning from proceeding, will continue to limit the implementation of multi-party planning approaches in the Pantanal. Environmental management is inherently a political as well as technical endeavour and efforts to eliminate politics from these efforts will likely be unsuccessful. Findings from this study suggest that managers might be better served by acknowledging the political nature of management activities and illustrating to politically engaged actors that contributing to multi-party planning does not necessarily undermine their interests. Such an approach would facilitate more collaborative inter-organizational exchanges that could promote the formation of shared beliefs and greater consensus about the use and conservation of natural resources in the Pantanal.

Although civil society and non-governmental groups have more openly embraced and participated in multi-party planning, the mixed experiences of actors in Mato Grosso and Mato Grosso do Sul illustrate the ongoing challenges faced when promoting collaboration within a complex organizational context. The centralizing political climate in Mato Grosso is clearly an external factor that has discouraged civil society engagement in environmental planning activities in the Pantanal. Nonetheless, internal divisions and the lack of a shared recognition of the utility of collaborative inputs among NGO groups is also a barrier to wider use of multi-party approaches.

Conversely, the experience of the NGO forum in Mato Grosso do Sul illustrates that diverse civil society groups can overcome internal differences and achieve common understandings that in turn can increase their effectiveness in conveying civil society's interests within multi-party planning activities. In addition, the diffusion of knowledge related to linked health and environmental issues illustrates a process of organizational learning that, if replicated, could help bridge the perceived divide between social, economic, and environmental issues in the Pantanal. The organizational strength of FORMADES was a key factor leading to these successes. Managers as well as funding agencies need to invest in developing the organizational capacity of both individual civil society groups as well as NGO networks. With a strong organizational foundation these groups will be able to better articulate their interests and substantively contribute to collaborative management efforts.

Finally, balancing the individual versus collective interests of different stakeholder groups continues to be an obstacle to the implementation of a collaborative approach to the management of the Pantanal. The experience of agricultural and industry groups, as well as unions and fisher cooperatives, shows the difficulty individual actors have in transcending their parochial interests to promote broader long-term natural resource management objectives. In addition, the unique needs and troubled past of indigenous groups adds a set of organizational issues and concerns that will require innovative engagement techniques if these key actors are to be incorporated into multi-party planning activities. Findings from this study point to strained inter-organizational relationships as the social forces that are inhibiting the formation of shared beliefs and understandings. Bringing these groups together and applying conflict resolution and mediation techniques may be one way managers can diffuse the tension between these actors and begin building a foundation for greater consensus and recognition of these groups collective interest in the effective management of the Pantanal.

Social scientists have not been regular participants in applied inter-disciplinary research projects exploring the practice of environmental management. The Pantanal management experience illustrates opportunities for sociologists to contribute increased understanding of the organizational complexity surrounding environmental problems and in particular how these factors both facilitate and impede efforts to promote more collaborative approaches to managing natural resources. Although findings from this case study are specific to the Pantanal, they illustrate the utility of organizational analysis within broader scientific examination of environmental problems.

Acronyms

CI – Conservation International
CPT – Comissão Pastoral da Terra (Pastoral Land Commission)
ECOA – Ecologia e Ação (Ecology and Action)
FAMASUL – Federação da Agricultura e Pecuária de Mato Grosso do Sul (Federation of Agricultural and Livestock Producers – Mato Grosso do Sul)
FAMATO – Federação da Agricultura e Pecuária de Mato Grosso (Federation of Agricultural and Livestock Producers – Mato Grosso)
FETAGRI – Federação dos Trabalhadores na Agricultura (Federation of Agricultural Workers)
FORMADES – Fórum de Organizações Não-governamentais para o Meio Ambiente e Desenvolvimento Sustentável Sul-Matogrossense (Forum of Non-governmental Organizations for the Environment and Sustainable Development - Mato Grosso do Sul)
FORMAD – Fórum de Organizações Não-governamentais para o Meio Ambiente de Mato Grosso (Forum of Non-governmental Organizations for the Environment – Mato Grosso)
FUNAI – Fundação Nacional do Índio (Brazilian National Foundation for Indians)
IBISS - Instituto Brasileiro de Inovações pró-Sociedade Saudável (Brazilian Institute for Innovations for a Healthy Society).
ICV – Instituto Centro da Vida (Institute for Life)
IRN – International River Network
MS – Mato Grosso do Sul (The State of Mato Grosso do Sul)
MT – Mato Grosso (The State of Mato Grosso)
NGO – Non-governmental Organization

TNC – The Nature Conservancy
WWF – World Wildlife Fund

References

Abers, R.N. 2007. Organizing for governance: building collaboration in Brazilian river basins. *World Development*, 35(8), 1450-1463.
Abers, R.N. and Keck, M.E. 2006. Muddy waters: the political construction of deliberative river basin governance in Brazil. *International Journal of Urban and Regional Research*, 30(3), 601-622.
Astely, W.G. and Van de Ven, A.H. 1983. Central perspectives and debates in organization theory. *Administrative Science Quarterly*, 28(2), 245-273.
Babbie, E. 2001. *The Practice of Social Research*. Belmont, CA: Wadsworth Thomson.
Bazerman, M.H. and Hoffman, A.J. 1999. Sources of environmentally destructive behavior: individual, organizational, and institutional perspectives. *Research in Organizational Behavior*, 21, 39-79.
Bitektine, A. 2008. Prospective case design: qualitative method for deductive theory testing. *Organizational Research Methods*, 11(1), 160-180.
Clark, B.T., Burkardt, N. and King, M.D. 2005. Watershed management and organizational dynamics: Nationwide findings and regional variation. *Environmental Management*, 36(2), 297-310.
DiMaggio, P.J. and Powell, W.W. 1983. The iron cage revisited: institutional isomorphism and collective rationality in organizational fields. *American Sociological Review*, 48, 147-160.
Gray, B. 1989. *Collaborating: Finding Common Ground for Multi-Party Problems*. San Francisco: Jossey-Bass.
Hall, R.H. and Tolbert, P.S. (eds.). 2005. *Organizations: Structures, Processes and Outcomes*. Upper Saddle River: Pearson-Prentice Hall.
Hochstetler, K. and Keck, M.E. 2007. *Greening Brazil: Environmental Activism in State and Society*. Durham: Duke Press.
Hoffman, A.J. 2001. Linking organizational and field-level analyses: the diffusion of corporate environmental practice. *Organization and Environment*, 14(2), 133-156.
Hoffman, A.J. and Ventresca, M.J. 2002. Introduction. In *Organizations, Policy, and the Natural Environment: Institutional and Strategic Perspectives*, edited by A.J. Hoffman and M.J. Ventresca. Stanford: Stanford Press, 1-40.
Meyer, J.W. and Rowan, B. 1977. Institutionalized organizations: formal structure as myth and ceremony. *American Journal of Sociology*, 83, 340-363.
Morton, L.W. 2008. The role of civic structure in achieving performance-based watershed management. *Society and Natural Resources*, 21, 751-766

Safford, T.G. 2004. Collaboration and Changing Paradigms: Organizational Relations and Natural Resource Management in Brazil's Alto Paraguai Watershed. PhD Dissertation, Cornell University.

Safford, T.G. 2010. The political-technical divide and collaborative management of Brazil's Taquari Basin. *Journal of Environment and Development*, 19(1), 68-90.

Sabatier, P.A., Focht, W., Lubell, M., Trachtenberg, Z., Vedlitz, A. and Matlock, M. (eds.). 2005. *Swimming Upstream: Collaborative Approaches to Watershed Management.* Cambridge: MIT Press.

Scott, W.R. 2008. *Institutions and Organizations*. 3rd edition. London: Sage Publications.

Trochim, W. 2000. *The Research Methods Knowledge Base*. 2nd edition. Cincinnati: Atomic Dog.

Wagenet, L.P. and Pfeffer, M.J. 2007. Organizing citizen engagement for democratic environmental planning. *Society and Natural Resources*, 20(9), 801 – 813.

Wondolleck, J.M. and Yaffee, S.L. 2000. *Making Collaboration Work: Lessons from Innovation in Natural Resources Management*. Washington D.C: Island Press.

Yin, R.K. 1994. *Case Study Research: Design and Methods*. Newbury Park: Sage Publications.

Chapter 8

Reassessing Development: Pantanal's History, Dilemmas and Prospects

Antonio A R Ioris

Introduction

The Pantanal is a geological depression in the Upper Paraguay River Basin (UPRB) which extends for around 148,000 km^2 where, due to the long residence of water in the flat landscape, a large tropical wetland was naturally formed. Its territory is shared between Brazil, Bolivia and Paraguay and – as explained elsewhere in this book – the local ecosystems comprise a complex mosaic with influences from the Amazon, Cerrado and Chaco biomes. Because of its location on the line of expansion of the Portuguese and Spanish empires and, later, of the three independent nations, major historical events have taken place in the Pantanal, such as wars, rebellions and geopolitical disputes. The Pantanal is also internationally famous for its abundant biodiversity, intricate ecological functions and picturesque cultural expressions (i.e. the '*pantaneiro*' way of life derived from the close contact with the unique hydro-ecology in relatively isolated conditions; see Chapter 2 above). The region is nowadays recognised as one of the most important tropical wetlands in the world and was designated as a Biosphere Reserve by UNESCO in 2000. At the same time, it is also one of the most threatened socio-ecological systems in the planet, particularly due to water pollution, loss of biodiversity, high sedimentation, modification of natural cycles, large-scale projects and the lack of conservation units (Swarts 2000; for the international repercussion of ecological threats, see the 2009 documentary by the BBC). The wetland ecosystems are affected by urban and agro-industrial expansion in the surrounding plateaus (i.e. areas higher than 200 m in altitude), as well as by the intensification of production and infrastructure interventions in the floodplain itself.

Most of the pressures and developmental conflicts have occurred in the Brazilian section of the Pantanal, which represents around 60 per cent of the UPRB territory (with a total of 600,000 km^2) and is inhabited by more than two million people (ANA 2005). The majority of this population lives in the towns and cities established in the higher altitude terrains that surround the Pantanal wetland (the plateaus), especially in the metropolitan area of Cuiabá. Urbanisation and agribusiness are the main drivers of change in the plateaus, while the floodplain is managed for cattle production (the most widespread economic activity, which involves around four million animals), fishing (including amateur, professional and subsistence fishers) and tourism (either rural tourism or eco-tourism). Native vegetation in the floodplain has been increasingly removed to make space for artificial pastures, which happens together with illegal hunting, poaching and the use of fire as a land management technique (Alho et al. 1988, Junk and Nunes da Cunha 2005). Around 130 new hydropower schemes are under construction or being planned in the Brazilian side of the UPRB, although the local and the cumulative impacts of such structures are still not properly understood (even so, there are preliminary evidences of serious disruption caused by the operation of hydropower dams, cf. Lucianer 2010). Sewage treatment is still only available in the minority of urban settlements and treats just a small proportion of the total effluents. All those pressures prompted Junk and Nunes da Cunha (2005: 392) to declare the Pantanal at a crossroads, in the sense that "increasing economic and political pressure requires fundamental decisions to be made in the near future".

The importance of this major transboundary wetland and the complex interconnections between lowlands and uplands have placed serious demands upon the agencies responsible for the regulation and conservation of natural resources. To be sure, the growing rate of impacts has attracted substantial attention from multilateral agencies and scientific organisations in recent decades. The prevailing discourse makes reference to problems such as the uncontrolled expansion of economic development, lack of scientific information and the inadequacy of environmental regulation. However, the implementation of responses has been a matter of serious disagreement, as demonstrated by the tensions between locations, sectors and scales of policy-making. That has resulted in a serious conflict of interests between stakeholder groups, which has been aggravated by the fact that government responses have been limited and only marginally successful. The most comprehensive assessment of the Brazilian Pantanal to date is the Conservation Plan of the Upper Paraguay River Basin (PCBAP), published in 1997, which identified the main environmental problems of soil erosion (throughout the river basin and with particular severity in some specific locations), agriculture mechanisation (leading to soil compactation and sedimentation), pollution and devastation caused by gold diggers (*garimpo*), deforestation (in farmlands and in riparian areas) and the use of agriculture pesticides. Despite the issues described in the various volumes of the PCBAP, it included only a narrow consideration of the historical and sociological causes of conservation problems. A more recent study commissioned by the Brazilian government (in Palermo et al. 2003) aimed to establish the specific causes of eco-hydrological impacts in the Pantanal. The publication concentrated on the negative consequences of the human presence (water pollution, soil degradation and biodiversity loss), on alterations of hydrological fluxes (frequency of critical events, conflicts around water use and socio-economic losses) and on the implementation of the new water legislation (institutional fragility). Again, the last analysis dedicated scant attention to the socio-political basis of ecological problems.

Taking on board previous assessments and similar international experiences (e.g. Dixon 2003), this chapter will consider the legacy from the past, the controversies of regional development and the perspectives for the future. The following pages will contend that the majority of the existing analyses offer only a partial and superficial examination of the conservation challenges and environmental threats posed to the Brazilian section of the Pantanal. On the one hand, many scholars (e.g. Neves 2009) maintain an unhelpful schematic contrast between the traditional, low impact agriculture and the destructive elements of modern forms of production. In this case, the authors underestimate the link between earlier socio-economic arrangements and the contradictions of present development. The mystification of the past makes more difficult to identify the failures of regional development strategies (see below). On the other hand, political leaders and most of the mass media support, and commemorate, the conversion of large areas of central Brazil into crop production. For instance, The Economist (2010) describes the expansion of agribusiness in the regional uplands as a truly 'economic miracle', regardless of the scale and the negative externalities of agriculture expansion. A more careful examination of the dynamic, and profoundly politicised, interpenetration between the 'new' and the 'old' in the Pantanal is still uncommon in the literature. The majority of the publications continue to neglect the long-term intricacies and the scalar connections between local practices and wider politico-economic pressures.

We will argue here that, instead of a crude contrast between a bucolic, romanticised Pantanal dominated by extensive cattle production and the disruption associated with modern day agribusiness, there exists is a continuum of socio-ecological problems that crosses the centuries (see also Rossetto 2009). The differences between past and present are mainly in terms of the magnitude and the speed of social and ecological impacts, but in the end it is the same overall model of development that has been responsible for the double exploitation of society and the rest of nature. Before returning to that central contention of the chapter, it is necessary to examine the main milestones in the history of the Pantanal.

Dealing with the Past

The historical and geographical evolution of the Pantanal reflects the gradual transition from a colonial regime of production into new economic activities increasingly connected to global markets. Since the early days of colonisation, the economy of the Pantanal have been intimately associated with the territorial expansion of capitalism, but in a way that maintained the Pantanal in a subordinated condition in relation to the stronger political and economic centers (Ioris 2004). The Iberian influence started in the early 16th Century, when the region was then inhabited by semi-nomadic indigenous tribes that largely relied on the use of the floodplain's biodiversity (Wilcox 1992). The first Europeans, including Jesuit missionaries, introduced cattle in the UPRB in order to benefit from the extensive areas of nature pastures. Cattle herds adapted well to the local circumstances with seasonal floods that systematically fertilise the fields. Gold extraction started in the second decade of the 18th century in the area surrounding the city of Cuiabá, recognised as the capital of the province of Mato Grosso by the Portuguese crown in 1726.

The formal occupation of the wetland only occurred in the last decades of the 18th Century, when military fortifications were established by the imperial powers to oversee the uncertain borders. The Brazilian Pantanal became a supplier of meat, leather, calves and steers to the regional economy and, with better means of transport, to the State of São Paulo. During the Paraguay War – the most violent South American conflict that took place between 1864 and 1870 and was primarily associated with the control of the Pantanal (the hostilities started with the Paraguayan invasion of the Brazilian cities of Corumbá and Cáceres) – farmers were forced to supply the troops with meat and leather, which led to a disorganisation of cattle production for some years. The continental war decimated the few established farms and the small number of cattle that survived was dispersed and unmanaged. After the Paraguay War, there was a gradual return of settlers and a slow recovery of the farms and of fluvial navigation in the Paraguay River system. Cattle ranching were restored as, again, an extensive system of production dependent on marketplaces located outside the floodplain. At the turn of the 20th century, only 12 ranches were recorded in the Pantanal (against 3,500 in the 1970s), according to Wilcox (1992). Some sugar mills were opened along the Cuiabá River to supply regional demand, but the mills disappeared in the 20th Century after the construction of an interstate road that facilitated the commercialisation of cheaper sugar from São Paulo.

The unique Pantanal geography did not serve as mere substratum for social interactions, but was appropriated as an integral element of the process of production. As a result, farmers had to develop a high proportion of self-sufficiency in terms of resources, technology and equipment. Traditional cattle farming was characterised by extensive pasture areas and was largely adapted to the local environment. Due to the low quality of the native pastures the average density of cattle is still only 0.25 head per hectare in the Pantanal, which was nonetheless sufficient to maintain the viability of the sector. The low profitability of traditional cattle ranching in the Pantanal needs to be contrasted with the vigour of the regional culture, which is evident in the vibrant music, poetry and articraft. Yet, it is important to avoid romanticising the relationship between land owners and peons (farm workers), which were evidently marked by class differences and a rigid social hierarchy.

Cattle production in the Pantanal started to change in the second half of the 20th Century with the introduction of Indian breeds, especially Nelore, to replace the original herds with Iberian origin. Production techniques also gradually intensified with the use of industrialised inputs (i.e. machinery, artificial pastures, wire fences, veterinary medicines, mineral supplements, etc.) and new business strategies, such as the certification of the Pantanal meat as a prime quality product (the so-called 'Pantanal calf' or '*vitelo pantaneiro*'). Different than traditional cattle ranching, modern farms are more closely associated with higher rates of ecological impacts because of deforestation, the mechanical cultivation of soils and application of chemical inputs. At the same time that the floodplains were experiencing technological and social changes, additional threats to the Pantanal ecology started to come from alterations in the surrounding uplands. The main pressures in the uplands (plateaus) have been the mushrooming urbanisation process (typically with no land use planning, sewage treatment and adequate waste disposal) and the constant expansion of intensive agriculture (crop and cattle production, as well as ethanol produced from sugar cane). The conversion of the uplands into plantation farms started in the 1960s, with the arrival of a great number of migrants from the southern and southeastern states of Brazil. The so-called 'agriculture frontier' benefited from heavy farming subsidies and governmental investments in commodity storage and distribution. Government agencies were responsible for providing credit, conducting agricultural research and disseminating technologies through rural extension. Transport and communication infrastructure had to improve to allow the connection of the region with the rest of Brazil and the international markets. This expansionist process continued and, since the neoliberal reforms of the Brazilian State in the 1990s, has followed an even more distinct market-oriented approach (i.e. with less subsidies, private sources of credit, more integrated production, professional farm management, etc). The initial years of the 21st century have been a period of economic and political reorganisation regarding the use and conservation of the Pantanal, as well as the search for new business alternatives and higher integration between the Bolivian and Brazilian economies.

It is relevant that a series of development programmes were formulated by the Brazilian State specifically for the Pantanal since the 1970s. The most representative initiatives included CIDEPAN in 1971 (Consortium of Municipalities for the Development of the Pantanal), PRODEPAN between 1974 and 1978 (Programme for the Development of the Pantanal that introduced engineering works, agriculture production incentives and flow regularisation schemes) and EDIBAP between 1978 and 1981 (Geographic Study and Development Plan). The Pantanal also benefited from investments and incentives made available through regional and thematic development plans, such as POLOCENTRO, PRODEGRAN, POLONOROESTE and PRODEAGRO. The aim was the intensification of the use of natural resources and regional integration, which mainly served to stimulate agroindustrial development in the cerrado (savannah) areas around the Pantanal. With the emergence of environmental legislation in the 1980s, the conservation of the Pantanal also became a matter of particular concern for the national government. That was translated into several governmental initiatives, such as the aforementioned PCBAP (1997), the Integrated Management Programme for the Pantanal (introduced in 1999 and funded by the Global Environmental Facility) and the short-lived Programme for the Sustainable Development of the Pantanal (2001–2004), carried out by the Brazilian Ministry of the Environment.

It should be observed that the intensification of agribusiness affected the entire region, transforming the rural and urban landscapes alike with the production of new spaces and social relations. That resulted in emerging disputes between urban and rural, old and new, landlords and employees, institutions and citizens, environment and society, and so forth. Such phenomena are not exclusive to the Pantanal, but have been recurrent in the history of agricultural expansion in Brazil, as well as in Bolivia and Paraguay. There exists an increasing number of new players involved in the debate about the direction of regional development, such as scientists, environmentalists, landless groups, indigenous peoples, energy suppliers (natural gas and hydropower), navigation companies and tourism operators.

Apart from agribusiness, hydropower generation, road construction and navigation expansion have also come under fire in recent years due to the associated level of negative environmental impacts. Large-scale navigation, in particular, is strongly supported by grain producers that intend to intensify use of the Paraguay and the Plata Rivers to reduce exportation costs. Nonetheless, environmentalists have denounced the likely threats posed by intensive navigation such as changes in water velocity, discharge, surface elevation, and sediment load. On the geopolitical scale, the Bolivian government has expressed its intention to construct a transcontinental motorway through the Pantanal to offer Brazilian exporters an easier access to Pacific ports in Chile and Peru. Likewise, an international pipeline was recently built, which crosses the Pantanal to convey gas from Bolivian reserves to satisfy industrial and urban demands in the southeast of Brazil. Ultimately, these various projects have meant a transition from one elitist economic system, dominated by traditional Pantanal farmers, to another elitist system, now controlled by industrial interests and plantation farmers. Crucially, for the majority of the population in the cities and in the countryside, the recent modernisation process has offered scarce improvements in living standards and in terms of wider political opportunities.

It means that, with the complexification of the management of the Pantanal, the transformations in production activities have raised fierce controversy and have demanded responses that are often beyond the existing scientific knowledge and the current environmental policy-making framework. Because of the more explicit recognition of the economic importance of the ecosystem services provided by the Pantanal, calls for economic efficiency and market exposure have occupied centre stage in the agenda of regulatory reforms in the three countries. It has represented a move in favour of hybrid mechanisms of environmental governance and away from the divisions between state-market-society that allegedly caused most of the mistakes during the period of state-led development (i.e. 1970s and 1980s). The application of market-based solutions to environmental problems is expected to foster economic rationality and promote management efficiency (Ioris 2010). However, in the countries where the so-called governance paradigm has been applied, the outcomes of the reforms have been restricted to some bureaucratic improvements and, at best, the removal of isolated, circumstantial problems (vis-à-vis the modest contribution of environmental legislation regarding the impacts of agriculture, cities and industries). The regional experience of the Pantanal is a case in point of the inherent limitations of the institutional reforms and the contradictory influences of neoclassical economics on the ongoing reorganisation of environmental regulation. Although the new water legislation in Brazil delegated to catchment committees the approval of plans and the reconciliation of spatial differences, the core instrument of regulatory framework has been the expression of the monetary value of water through environmental charges and the internalisation of external costs (see more below). It suggests that the official initiatives continue to subject socio-natural systems to an economic rationality and unfair distribution of opportunities. In the same way, newly formed decision-making mechanisms have been dominated by the same political groups that always controlled economic and social opportunities in the region. It means that, instead of promoting a genuine change in public policies, the new approaches have largely preserved the hegemonic interests of landowners, industrialists, construction companies and real estate investors, at the expense of the majority of the population and the recovery of ecological systems.

The contemporary reality of the Pantanal is certainly too complex for any simplistic, schematic explanation. On the contrary, the transformations taking place in the Pantanal and its surrounding areas are constantly challenging established scientific patterns and conventional policy-making. The social and political consequences of those transformations have provoked unpredicted reactions and fostered new alliances and novel disputes. That is the reason why the management of the Pantanal wetland requires, first of all, improvements in the representative democracy regarding the negotiation of contrasting interests and in the mediation of socio-ecological conflicts. Innovative solutions need to consider the wider spectrum of demands both within the Pantanal region and in connection with broader environmental pressures operating at the national and international scales.

Probably the most emblematic example of the controversy between development, environmental conservation and social well-being is the degradation of the Taquari River Basin (already mentioned in other chapters). The Taquari catchment covers around 78,000 km^2 that are split between two very distinct areas, the uplands and the lowlands. The dramatic changes that took place in the fragile upland soils since the 1970s led to massive erosion and sediment deposition in the floodplain. There has been an abrupt increase in the rate of soil erosion and sediment deposition due to deforestation and inadequate soil management (Jongman 2005). Vieira and Galdino (2003) affirm that the rate of sedimentation reached the staggering rate of 36,000 tonnes per day (in 1997) with an increase in the flooded area by 370 per cent in relation to the period 1925–1975. Instead of the regular pulse of flooding (that essentially defines the Pantanal's eco-hydrology), the anthropogenic impacts on the Taquari have resulted in an area of 11,000 km^2 under permanent inundation. Erosion and siltation have decreased water transport capacity, which results in the destabilisation of the river channel and the breaching of riverbanks (river avulsions) and accelerate the creation of oxbow lakes (the so-called *arrombados*). The end result is the formation of artificial oligotrophic lakes that follow the collapse of the riverbanks.

Regrettably, the processes of deforestation and soil erosion in the upstream of the river basin ended up impacting the only area of small farms in the Pantanal floodplain, namely the Colônia São Domingos (established in the 1920s by the famous explorer Marshal Rondon), Colônia Bracinho and communities Rio Negro, Cedro and Miquelina. Many families have completely lost their land to the permanent floods and had to migrate to the cities of Corumbá and Ladário, where entirely new neighbourhoods were formed to accommodate the environmental refugees. Those that still have some agriculture had to reduce the number of cattle and subsistence production because of the inundated land. At the same time, the remaining families have struggled to cope with the significant decline in fish capture in recent years (Curado 2004). In addition, farmers and professional fishers have tried to manage the geomorphological changes with small interventions in the *arrombados*, but normally the two groups adopt conflicting strategies: while farmers try to restore the riverbanks, fishermen insist that the lakes produced by the *arrombados* are new fishing areas. It is true that the severity of the Taquari condition has attracted some attention from the media and forced the prparation of technical assessments and recommendations. Moreover, the cost of structural solutions (e.g. dredging, construction of sediment storage dams, physical reconstruction of river channels) is highly prohibitive, reaching hundreds of millions of dollars and without any guarantee that the problems can be effectively solved (Jongman 2005).

The 'tragedy' of the Taquari River is highly significant, not only due to the ecological and geomorphological disruption, but also because of its economic and social consequences and the failure of conventional response strategies. There is a notable level of scepticism among farmers and other groups of stakeholders about the repeated promises made along the years that have achieved only negligible results (Ioris 2004). To help to understand those processes of change and disputes is necessary to examine questions related to environmental values and resource scarcity, as discussed in the next section.

Pantanal's Values and the Politics of Scarcity

There are important ideological and political causes behind the growing environmental degradation and the related social impacts in the Pantanal mentioned above. First of all, it is possible to situate the long chain of changes and the acceleration of impacts as consequence of the imposition of Westernised forms of interaction between society and the rest of nature. The intensification of the use of territorial resources in the Pantanal region has raised increasing controversy and divisiveness, which suggests the existence of major contradictions in the management and conservation of resources both within the floodplain and in the surrounding plateaus. Those conflicts about the exploitation of resources and landscape changes are normally underpinned by fierce disagreements over the interpretation of the value of ecosystem features. Competing forms of valuation have informed the design of government interventions, the execution of technical assessments and the formulation of regulatory procedures. At the same time, the idiosyncratic forms of valuing nature reflect the specificity of the interactions, and deep interdependencies, between social groups and their ecological conditions. Dissimilar manifestations of values correspond not only to different subjectivities projected onto the world, but serve to expose the imbalance of power and subtle positions of authority. The challenges to improve environmental management are not restricted to reverting impacts, but above all constitute a clash of competing valuation approaches seeking legitimisation. The contested basis of the environmental values of the Pantanal system – that is, the contrasting meanings, preferences and priorities among resource users, other interested parties and government officials – is further reinforced by the fact that not all interpretations are equally accepted. Through institutionalised forms of valuation, hegemonic approaches ignore the complex relations between social inequalities and environmental degradation (Scruggs 1998).

The examination of environmental values can, therefore, offer a helpful entry point into the intricacies of public policies and environmental management approaches. Values are contingent assessments of worthiness that emerge out of multiple socio-ecological relations and follow particular demands, legacies and opportunities. As observed by Kovel (2002: 195), "ecological politics can be translated into a framework of values" and, by and large, the values that ultimately prevail are those sponsored by the more powerful in society. Despite its crucial importance, the disputes around the valuation of nature are not always properly recognised, but, in many cases, the vast universe of values is reduced to a simple dichotomy between economic (modern) and traditional (anti-modern) interpretations. In particular, mainstream assessments concentrate on the techno-monetary aspects of ecological systems and neglect the plurality of values forged through socio-spatial experiences. The priority of government interventions is often to guarantee the circulation of commodities, the accumulation of capital and the reproduction of the labour force, which expose the economic nexus that characterises mainstream valuations of nature. Values, thus, become abstract, unmediated and impersonal in a way that breaks the multiple connections between the social and the material dimensions of environmental management. The most common results of the poor incorporation and integration of stakeholder values come in the form of management failures and the perpetuation of political conflicts (Ananda and Herath 2003).

Instead of narrow approaches, the assessment of the values of nature requires a critical investigation that embraces the intersection between concrete and abstract, local and general, personal and social experiences (Hoffman 2005). Values are described as positioned constructions expressed through mechanisms of cooperation and conflict that contribute to the social production of space. In other words, values cannot be dissociated from the unity of action and reflection, that is, the praxis of valuation. Ecological values are ensembles of experiences, subjectivities and spatialities shared by a social group in specific circumstances. In that regard, economists working with nature valuation can certainly benefit from the contribution of social anthropologists and other social scientists, who emphasise the cultural intricacy of the relationship between nature and society (Treitler and Midgett 2007). According to anthropologists, management problems and conflicts are as much the result of biophysical and socio-economic conditions as the product of cultural values. Through movement, circulation and consumption, it should be possible to clarify the relation between things, cultural identities and human agency. Anthropologists have an interest in material culture and examine how things, made or modified by humans, reflect beliefs, ideas, attitudes, assumptions and, ultimately, values. Kopytoff (1986) defines such interconnection as the 'cultural biography' of things or the transformation of the meaning and value of goods across time and space. Similarly, Appadurai (1986) describes the complex and unpredictable confrontations between different regimes of valuation as 'tournaments of value'. These tournaments are complex events removed from the routines of economic life or situations when the disposition of the cultural tokens of value is at stake. Values are therefore politicised notions that involve contested relations of power and knowledge within and across societies.

Anthropologists, thus, reject the dualism between valuing subjects (individuals who interpret values) and valuable objects (things that potentially hold value). Valuation is then considered as a dynamic process related to the conceptions of the world around the speaker, cast in a moral frame of reference. Expanding from the notion of 'tournaments' (Appadurai 1986), we could claim that the multiple values of the Pantanal represent a direct indication of how the different social groups are able to interact and establish socio-ecological relations. The valuation of nature follows the belief patterns of groups or individuals and, by extension, the larger society of which these individuals are a part. In order to understand the formulation of values, one has to almost inevitably deal with issues of visibility and invisibility and has to re-examine notions about power, exchange and the human person (Graeber 2001).

Social anthropologists certainly make an important contribution to questioning the idea of nature as cultural universal and to understanding values at the intersection between humans and things. However, the common claim among those scholars that nature is a social construction – that is, the natural world as the construction of our concepts of nature – presents the serious risk of moving away from the materiality of nature and towards a relativistic stance. In this case, the meanings and values of nature become unhelpfully tangled in an ambiguous, uncertain ontology of constructed nature. The consequence is that the values of nature held by a particular social group are seen as always unique, specific and without any possibility of association or interaction with the values of other groups. Against this extreme fragmentation of valuation approaches, Strang (2005) argues that the engagements between society and nature are experienced and interpreted within specific cultural contexts, but at the same time the particular ecological qualities and the cognitive processes are universal and persist over time and space. In the same direction, Ingold (2000) reproves the anthropological claim of perceptual relativism (i.e. people from different cultural backgrounds would perceive reality in different ways due to alternative frameworks of belief), first and foremost because such view actually reinforces the Western dichotomies between nature and culture. Rather than a cultural construction of the environment that implies human cognition outside the world of nature, Ingold calls for a sentient ecology in which knowledge emerges out of feelings, sensitivities and skills developed through long experiences in particular environments.

Going beyond classical economics and the relativistic biases of some anthropological studies, the values of nature need to be described in more interconnected and dynamic ways. Environmental values are evolving conceptions that operate at the dialectics between personal preferences, group trajectories and broader socio-economic connections. The values associated with the Pantanal are qualified attributes of the metabolism between individual and collective preferences, market and non-market demands and local and higher scales of interaction. Furthermore, the valuation of nature is a political manifestation of the achievements and insufficiencies of individuals, communities and societies. Values are ultimately the enduring outcomes of past experiences that precipitated, and are stored, in the discourse, morality and imaginary of human societies. It means that the valuation is neither neutral nor purely subjective, but encapsulates accumulated knowledge, material sensibilities, socio-economic disputes, as well as fulfilled or unfulfilled aspirations. The failure to comprehensively address the dynamic genesis and interpersonal mediation of ecosystem values is a central cause of the intensification of problems and conflicts in the Pantanal. Since the last quarter of the 20th century, the official policies and management programmes mentioned above have increasingly described the Pantanal's ecology through an economic language and, at the same time, promoted new environmental regulation through the narrow lens of economic values. The main problem is that monetary valuation tends to cover and disregard the complex associations between social inequalities, environmental degradation and the imposition of new institutions.

The distortions caused by the imposition of a hegemonic valuation approach over other, more spontaneous considerations of value have important parallels with the interplay between the scarcity and abundance of resources. In effect, more than technical and managerial dilemmas, the alleged 'crossroads' of the Pantanal are directly related to the controversial oscillation between abundant and scarce of resources. The geography of scarcity provides another privileged entry point into the, kaleidoscopic interlinkages of the Pantanal throughout its history. In the first moment of the colonisation of South America, the main resource available to the coming settlers was the territory of the floodplain itself and its vast extension of native pastures. More recently, the abundance of water, biodiversity and workforce played a more central role with the regional development. Those manifestations of abundances evolved together with various forms of (human made) scarcity, not only caused by the degradation of the resource base but also the lack of technology, infrastructure, financial capital and the precarious enforcement of the existing environmental legislation. In the last half a century, the pattern of allocation and use of natural resources, primarily following the priorities of the agribusiness sector, led to an increasing scarcity of even water, land and ecosystem services. The inversion of money and technology in the regional economy, according to the centralised policies funded by the national government, has certainly failed to offer a solution to the constant emergence of new forms of scarcity. On the contrary, the shortage of resources and opportunities presents itself as a totality of relations: at the same time that some groups in the periphery suffer from the manifestation of scarcity, the insertion of the Pantanal into continental and global market transactions encourages the acceleration of environmental disruption. The creation and evolution of scarcity ultimately demonstrates the synergistic connection with political manipulation and the selective accumulation of capital.

Scarcity must be considered as a fundamental influence in the organisation of socio-spatial relations. It is precisely because of a shared sense of scarcity that the market-based economy was able to flourish and dominate European society, with its distant repercussions felt in the middle of South America and in the Pantanal. At the centre of Western modernity lies this persistent realisation of scarcity (Xenos 1989), which nurtures itself from the priority of profit and accumulation that follows the imposition of capitalist relations of production. Capitalism, at the same time that multiplies the circulation of commodities and accelerates technological improvements, relies on a set of values created out of scarcity and fulfilled through market transactions. There is, thus, an intrinsic connection between the structure of market incentives and environmental degradation, in the sense that scarcity derives from the inability of capitalism to relate with nature without disrupting the ecological balance (Perelman 1996). The core features of capitalism – the exploitation of labour power, the alienation of those that create value and the intrinsic need to expand – are all based on scarcity and inherently depend on its constant reinforcement. Particularly because of the expansionist basis of capitalism, new areas of consumption and waste are permanently being added to already existing patterns of socio-natural exploitation. In the end, capitalist society is permeated by a 'dialectic of scarcity' that arises from the constant economic growth and social privileges at the expense of subordinate social classes and the environment (Panayotakis 2003).

The material and symbolic production of scarcity has been predicated upon practices of social exclusion and discrimination that defined the history of development in the Pantanal. The intensification of economic production and the associated processes of ecological modernisation in the recent past have been advanced by governmental agencies sensible, first of all, to the needs of the stronger private interests (agribusiness in particular). Those disparities have only reinforced the inequalities established in previous centuries. Agriculture expansion, urban development and natural resource management have operated within the hegemonic asymmetries that continue to dominate the political scene of South American countries. Contrary to the orthodox public policies promoted at the federal and state levels in Brazil, the inefficiency of public services is less the result of state inability and more the convergence of powerful private interests with the containment of social demands. Scarcity is not maintained because of structural pressures and limited resources, but it is instrumental for the emergence of only a handful of circumstantial abundances. The fragility of such abundances inescapably led to a reaffirmation of the long-established social and political scarcities. For instance, those affected by the allocation, use and conservation of resources have been virtually excluded from strategic decisions, as in the case of the degradation of the Taquari River Basin. The wetland region that offers the promise of apparent abundance is also a '*megera*' that nurtures itself from socio-natural exploitation and hypertrophied scarcities. The disparate experiences of recent years demonstrate that, while public policies aim to reconcile socio-economic development with the conservation of biological systems, in practice those attempts often results in an escalation of disputes and ecological impacts.

The turbulent expansion of regional development is a compelling indication of the specificity of capitalist production in the Pantanal, which is more and more characterised by artificially created scarcities and an increasing abundance of financial resources available to exploit the resource base. Mainstream public policies are the offspring of the 'dialectics of scarcity' (as suggested by Panayotakis 2003), given that scarcity serves as powerful justification not only for the exploitation of resources, but also for the adoption of market-based environmental conservation (as established in the new Brazilian legislation). Particularly the regulatory toolkits developed for contemporary environmental management – such as user licences, user fees, decision-support systems and the payment for ecosystem services – are all rationalised in relation to rising levels of scarcity (Ioris 2010). By limiting the analysis of environmental management problems to the (utilitarian) balance between supply and demand, conventional policies neglect the relative and contested basis of scarcity. Official approaches operate within the narrow episteme of techno-bureaucratic responses (Ioris 2008) in which scarcity emerges as a 'meta-narrative' that justifies controversial solutions and allows for simplistic portrayals of property rights and resource conflicts (Mehta 2007). Especially in the last three decades, the neoliberal influence on environmental management has amplified the sense of scarcity, to the extent that market-based transactions now reach both the physical stocks of resources and the ecosystem 'services' provided by nature. Watts (2000: 197) observes that scarcity "is part of the genealogy of liberal governance", while Harvey (2009: 114) argues that the conceptualisation of scarcity "only takes on meaning in a particular social and cultural context. (...) In sophisticated economies scarcity is socially organised in order to permit the market to function. (...) This is achieved by appropriative arrangements which prevent the elimination of scarcity and preserve the integrity of exchange values in the market place".

Overall, to help our reflection about the future of the Pantanal, two main observations can be drawn about the meaning and the ramifications of scarcity. First, scarcity is always plural and multiform. It is never a single phenomenon but, on the contrary, sectoral scarcity develops into unavoidably associations with other manifestations of shortage and deprivation. These multiple facets of scarcity are not only interconnected, but are interdependent and mutually reinforcing. Scarcities are ultimately the amalgamation of material and social processes that self-reinforce each other through the perpetual creation of new demands and the persistence of a wasteful relation to nature. The satisfaction of some demands happens at the expense of novel scarcities elsewhere, for example, if democracy and representations are scarce, the tendency is that other forms of material and social scarcities will emerge and persist. In that sense, the compound ontology of scarcity is established according to specific historic-geographical circumstances and, in the market-based society, it is based on processes of social differentiation that serve primarily the accumulation of capital. Following the capitalist process of value extraction, resources need to be made scarce to expose and crystallise exchange-value or surplus wealth forced upon nature. Scarcity is carefully controlled in order to guarantee social reproduction, as well as to open new avenues for capital accumulation and curb the growth of the organic composition of capital. It is through the manipulation of scarcity that ecological systems are brought to the centre of the intricate commodification of everything.

Second, in a market-based society, such as the one that is increasingly encroaching upon the Pantanal, scarcity never exists without its dialectical opposite, that is, abundance. The realisation of abundance presumes the preservation and, what is more important, the multiplication of scarcity. The colonisation of scarcity through the promises of abundance operates as a powerful ideological tool for the legitimisation of socio-spatial inequalities. In the capitalist context, the creation of exclusive forms of abundance is predicated upon the perpetuation of the double imposition of scarcity over nature and society. In the case of environmental management, for the advance of economic and political forms of socio-natural control, abundance needs to be realised through the constant promotion of scarcity. The multiple scarcities associated with nature are always contained and recycled through social and geographical dislocations that bring new forms of abundance. The dialectical unity of abundance and scarcity becomes even more apparent in the processes of capitalist urbanisation and agribusiness. Considering that capital itself posits the conditions for its realisation (i.e. it presupposes what it is not yet in being but merely in becoming, cf. Marx 1973: 459), the model of regional development presupposes the persistence of the apparatus of scarcity: the financial and technological investments in nature follow the signposts of abundance, creating a chain of scarcities that is functional for uneven accumulation of material and social opportunities.

Reassessing the Processes of Change: Nature, Society and the State

The previous pages discussed critical aspects underpinning socio-economic trends and governmental initiatives related to the exploitation and conservation of the Brazilian Pantanal, in particular the contested meanings of ecological values and resource scarcity. For more than three centuries, the Pantanal has been gradually incorporated into the expansion of international markets and the consolidation of the three nation states. The regional processes of change have reflected idiosyncratic mechanisms of capital circulation through the conquest of territories, the appropriation of resources and the associated exploitation of the working force. Although it is evident that the main patterns of economic production evolved from cattle ranching, fishing and hunting into a combination of agribusiness, agro-tourism and urbanisation, it is also possible to identify a clear line of continuity between past and present. The Pantanal remains a marginal, remote area in the Brazilian and South American geography conventionally related to the abundance of resources that can be mobilised through the exploitation of low paid labourers. Different than claimed by most available interpretations, socio-economic development in the Pantanal has been primarily defined by the creation and reaffirmation of socio-spatial inequalities.

The politicised complexity of environmental conservation requires an explanatory framework that comprehensively captures the multidimensionality of the relations between society and the rest of nature in the Pantanal region. As emphasised by Swyngedouw (2004), society and nature co-evolve together as a socio-natural hybrid that describes processes that are simultaneously material, discursive and symbolic. Even so, the politico-economic intricacy of ecological management issues has been continuously neglected by policy-makers and senior authorities. It is also the case that most environmental management approaches nowadays emanate from, or are informed by, multilateral agencies beyond the realm of national politics (e.g. World Bank, UNESCO and the Inter-American Development Bank). The mainstream doctrine of environmental management has included calls for integration and public engagement, which are summarised under the expression 'environmental governance', as already mentioned above. The paradigm of governance insists on a transition to more flexible procedures that connect the action of both the state and a multitude of organisations and movements that constitute the non-state (Conca 2006). Instead of the conventional exercise of authority, better governance is expected to create lasting and positive changes according to goals such as openness accountability, effectiveness and participation (Batterbury and Fernando 2006). Yet, behind repeated references to governance, environmental policy-making around the world remains controlled by centralised government agencies and subject to the influence of powerful sectoral interests (Blomquist and Schlager 2005).

In the Brazilian Pantanal, despite the apparent changes in discourse, today's governance approaches have largely replicated the technocratic, top-down procedures that characterised infrastructure projects and development programmes in the 1970s and 1980s. The main differences are the use of new strategies based on the economic (monetary) value of nature and on instrumentalised forms of public participation. Contemporary environmental legislation has nominally recognised the importance of integrated responses, but in practice it has mainly secured long-established advantages, increased the circulation of capital around environmental restoration and paved the road for the formation of public-private partnerships (Ioris 2007). Recent institutional reforms – which spread through laws, guidance and government programmes – have essentially failed to confront the sources of social injustice that underpin environmental degradation. The Pantanal experience of environmental governance has significant parallels with the international practice, where there has been limited attention to questions of social 'power and to the consolidation of multiple injustices through everyday struggles fought over the material and symbolic elements of the lived environments (Loftus and Lumsden 2008). The results are successive failures in the recent history of environmental policy-making, which unfolds in a succession from one vantage point to another, as a "parallax movement" (a notion suggested by Žižek 2006: 26) that never reaches a definitive solution, nor reconcile, in a satisfactory manner, social and natural demands.

To a large extent, the oversimplification of environmental problems and the reduction to superficial calls for environmental governance derive, primarily, from the denial of the biased interventions and political commitments of the state. Kalyvas (2002: 105) observes that in recent years the state seems to have "retreated from the realm of social sciences", whilst there is a growing emphasis on globalisation and on postmodernist discourses (which partially deal with the connections between global and local scales) that tends to overlook the sociological significance of the national state. Even political ecology scholars have concentrated their studies on popular mobilisation, moral economy and the contradictions of capitalism (Mann 2009), but not dedicated enough attention to the evolution and the scalar linkages of environmental management problems promoted by, or in relation to the state. Instead of being one among other socio-ecological players (as conceptualised, in an otherwise interesting book, by Bryant and Bailey 1997), the state is in effect the embodiment of political hegemonies and the immanence of social relations inscribed in the management of nature. If the resolution of environmental management problems depends on how citizens perceive their claims and also on how they are able to collectively negotiate their demands through identity, economic activity and spatial location (Anand 2007), the state remains the central protagonist of the assessment of socio-natural systems and coordination of responses.

Consequently, there is a common analytical gap, even among critical approaches, in the way that the state is nowadays increasingly called to exercise a strong procedural power to govern not just property, rights and knowledge, but state bureaucrats ultimately have to administer "the sky, the climate, the sea, viruses, or wild animals" (Latour 2004: 204). Instead of simply being the apparatus of government, the structure of the state is effectively a continuation of civil society that, because of its political commitments, performs the ambivalent role as agent of both reform and stability. Whilst containing the demands of the broad society through a combination of coercion and consent, the action of the state primarily reflects "back its prestige upon the class upon which it is based" (Gramsci 1971: 269). Particularly the capitalist state constitutes the main power instrument of the dominant groups, which nonetheless operates in a continuous process of formation and superseding of an unstable socio-ecological equilibrium.

Against both the assumption that the state is a neutral entity promoting the common good (as implicit in the theory of environmental governance) and the limited discussion of the class-based tendencies of state action (as still implicit in the work of many critical scholars), Jessop (2007) has argued that state power combines centralised and diffuse authority in conformity with the fundamental features of political economy and profoundly embedded in social relations. The state is an institutional ensemble that does not exist in isolation of the balance of political forces, but these forces are in effect responsible for shaping – at least in part – the structure and intervention of the state. Yet there is never a full correspondence between the capitalist state and the interests of the dominant classes, but the appropriation, pressure and colonisation of the public administration by the hegemonic groups are not absolute. It means that to a certain extent the capitalist state remains politically separated from the circuits of capital and accumulation (Offe 1996) and, therefore, needs to be understood as a dynamic institution that offers unequal opportunities to different social groups to achieve their specific political purposes. This selectivity of the state is not given in advance, but is the result of the interplay between state priorities and socio-political contestation within and beyond state institutions (Jessop 1990). For those reasons, the political construction of state authority requires analytical tools that should be capable of exploring the inscribed asymmetries of state action, as well as the social forces that enjoy different capacities to pursue strategies that are more or less adapted to its selective functioning.

In addition, the study of the state also requires a more systematic treatment of spatial dynamics, as well as of the relations between nature and society. As observed by Swyngedouw and Heynen (2003: 912-913), "...socioecological processes give rise to scalar forms of organisation – such as states, local governments, interstate arrangements and the like – and a nested set of related and interacting socioecological spatial scales. (...) These territorial and networked spatial scales are never set, but are perpetually disputed, redefined, reconstituted and restructured in terms of their extent, content, relative importance and interrelations". The need to embrace the socio-ecological and socio-spatial properties of the state is even more justified under the current pressures of neoliberalism, because since the accumulation crisis of the capitalist system in the 1970s the national state has been under intense pressure from the twin forces of globalisation and localisation (Neumann 2009). A spatialised analytical framework should be able to link the general form of the capitalist state with the contingent historico-geographical experiences in the course of capitalist development (Brenner 2004). Jessop et al. (2008) have taken into account the four key dimensions of socio-spatial relations employed by social scientists in the last decades (i.e. territory, place, scale and networks) as an approach that extends through interconnected scales and tries to integrate the multiplicity of fields of state operation.

Nonetheless, the spatialised solution advanced by Brenner (2004) and Jessop et al. (2008) still seems rather like a prescriptive categorisation of spatial interrelations, given that it is built around inconvenient binarisms (i.e. the examples given are the linkages between place and territory, scale and territory, and network and territory). According to the last authors, those 'duets' are expected to become the entry points into the assessment of complex, multidimensional systems. However, such formulation has also the negative consequence of limiting the analysis of socio-spatial relations to a sort of checklist of paired associations, instead of embracing the full extent of interscalar and multisector state interventions. The conclusion is that, if spatialisation represents an important contribution towards recognising the socio-ecological complexities of the state, it is still necessary to incorporate the political ecological dimension as a more integral component of the evolving relations between state and society. Instead of schematic constructions, a more comprehensive political ecology should understand environmental management through the multiple interconnections between state interventions, socio-natural interactions and power disputes. Whilst rejecting any claims of geographical determinism or ecological fetishism, we submit that the biophysical characteristics of managed ecological systems are not only shaped, but directly influence social relations and political disputes within and beyond the state. These are particular socio-physical phenomena through which symbolic formations are forged, social groups enrolled, natural processes and things entangled, and political power relations expressed and reconstituted (Swyngedouw 2007).

This brief discussion about the configuration and the functioning of the state has important consequences for understanding the challenges faced by the various groups involved in regional development and in the conservation of the Pantanal. From a critical perspective, the fundamental problem with managing of the Pantanal is the authoritarian attempt by the state to incorporate the geographical periphery into mainstream, hegemonic development strategies. The Pantanal was on the periphery of the Portuguese and Spanish empires, which later became the border of the independent South American countries. Suddenly, because of the availability of land and natural resources, it was brought to the centre of the projects of national development of the three countries (mainly because of gas in Bolivia, navigation in Paraguay and agribusiness expansion in Brazil). There was no shortage of international development banks willing to finance such megalomaniac projects, as well as politicians eager to associate their names with the incorporation of the Pantanal into traditional development strategies. There were additional plans to connect navigation through various river systems in the continent, from the Amazon to the Plata, as a regional infrastructure scheme to place the core of South America at the centre of global trade (Barkin 2009), but always leaving regional society and ecology subordinate to the main political centres and stronger economic interests.

Based on the above, the evolution of public policies and state interventions in the Brazilian section of the Pantanal can be reinterpreted in three main phases with distinctive, but also common features. The first phase lasted from the early 18th century to mid-20th century (the moment of national industrialisation and transference of the capital to Brasília), when the economy was essentially based on the production of calves and young adults that were later commercialised outside the Pantanal. This established form of production persisted in the floodplain, but its regional importance was superseded by the introduction of plantation farms in the surrounding plateaus, heavily supported by governmental subsidies and public investments in infrastructure (roads, warehouses and communication) since the 1970s. The neoliberal reform of the Brazilian State in the 1990s, together with processes of international integration and economic diversification, led to the introduction of new economic activities championed by the government and with growing participation of private corporations (e.g. the construction of gas pipelines and large dams, the increase of tourism and eco-tourism, the adoption of high-input agriculture technologies in both the floodplain and in the upper areas). Because of local and international pressures, the planning process changed from a purely technocratic and centralised style to a formal, but largely bureaucratised, attempt to gauge public perception and listen to differing stakeholder voices (moreover, not necessarily sharing decision-making power). The Pantanal became associated with a peculiar symbolism that combines ecological vulnerability, a unique culture and uneven economic opportunities (Vargas 2006). It should be possible to perceive that during those three phases (see Table 8.1) there persisted a trend of socio-natural exploitation that was primarily advanced by the national state according to its political priorities, spatial strategies and political commitments.

Table 8.1   Evolution of State Policies and Environmental Management in the Pantanal

Criteria	Traditional management	Techno-economic management	Market-based management
Period	18th to mid-20th cent.	1960s-1990s	since the 1990s
Focus and ideological basis	Extensive use of resources	Engineering infrastructure, intensification	Diversification, sustainable development
Main sources of change	Sparse population; military priorities	Central government	Various government authorities and private stakeholders
Political direction	Territorial consolidation	Economic growth	Economic growth and societal gains of efficiency

Conclusions

This chapter intended to offer a critical conceptualisation of the deep causes of environmental disruption and of the growing conflicts between sectors and locations in the Pantanal. The discussion was situated in the broad panorama of historical and geographical transformations that gradually brought the Upper Paraguay River Basin to the centre of national development and cross-country integration (more recently, due to the export of natural gas from Bolivia to Brazil and the plans to increase fluvial navigation). As pointed out by Lukács (1971), it is crucial to address the entirety of relations, given that the category of totality determines not only the object of knowledge, but also the action and position of the subject. That has major repercussions for the discussion about the trends and prospects of the Pantanal, because to a large extent the problems of environmental degradation and poverty are directly associated with the discriminatory pattern of policy-making and the elitist configuration of regional development. A more sustainable management of the Pantanal can only be achieved with the reversal of such tendencies and the effective inclusion of those more directly affected by the negative consequences of regional development. The conservation of the Pantanal is not teleological, but certainly mediated by conflicts about the access to territorial resources and the opportunity to influence the state.

So far, the overall process of economic expansion and the incorporation of the Pantanal into national development have been marked by socio-spatial differentiation and political differences. Uneven opportunities and asymmetrical development have not only characterised production units (e.g. farms), but also the differences between locations and between countries. The main development pressures continue to come from the eastern border of the Pantanal: in the past, it used to follow the demands of the Portuguese and Spanish crowns, and nowadays it is the industrialised and heavily urbanised areas of Brazil that exert the most serious pressures on the Pantanal. In that context, the severe degradation of catchments, like the Taquari, represents that most dramatic chapter in the long history of intensification of agriculture production in the surrounding plateaus at the expense of the ecosystems and low-income social groups. On the other hand, it is important not to base the critique of current trends on the romanticising of the past. Different than the conventional discourse about the management of the floodplain, it is not accurate to say that traditional forms of economic production were wholly sustainable. Long-term ranching has had direct impacts on fauna and flora (due to changes in land use, the spread of diseases and competition for forage between cattle and undomesticated herbivores, cf. Taboco 1990, quoted in in Wilcox 1992) and has also facilitated the activities of external fishermen, hunters and poachers. It means that the floodplain was preserved after nearly three centuries mainly because of the low profitability of conventional cattle production. Wilcox (1992: 255) aptly observes that "ranching has not 'saved' the Pantanal. (...) "If anything, the Pantanal has saved itself" due to its remoteness and the availability of land elsewhere.

In the present, as much as in the past, scarcity and abundance of resources in the Pantanal remain a contested and relative condition. There are discrepant opinions about the ecological value of the Pantanal's features, which can be linked with the limited understanding about the underlying cause of environmental problems (Marchini 2003). Most assessments to date identified as the main structural problems of environmental management the lack of information and research, the use of soil beyond the carrying capacity, perverse agriculture incentives, lack of regional infrastructure and losses in the commercialisation of commodities. However, this list seems to include the outcomes of the problems rather than dealing with the underlying drivers of social injustice and geographical inequalities. Both traditional economic production and agribusiness activities have, in actual fact, been based on the twin exploitation of landscape resources and the workforce. In other words, the long interaction between social groups and nature has been mediated by the uneven balance of power within the Pantanal and in relation to the hegemonic centres of decision-making. Crucially, the affirmation and maintenance of power imbalances has been the main outcome of public policies formulated and enforced by state agencies. The national state remains the most ambiguous, enigmatic and influential player in the regional affairs. Government interventions created the international boundaries that fragment and threaten the Pantanal; it also supported and promoted a pattern of economic production that subordinates the peoples of the Pantanal to exogenous priorities. Yet, the reform of the state apparatus has introduced new mechanisms of environmental regulation that paradoxically offer opportunities for public participation and organised protest. The final conclusion is that the future of the Pantanal will continue to depend on the interplay between multiple interests manifested primarily in relation to, and via the democratisation of, the national state.

References

Alho, C.J.R., Lacher, T.E. and Gonçalves, H.C. 1988. Environmental degradation in the Pantanal ecosystem. *BioScience*, 38(3), 164-171.

ANA. 2005. *Strategic Action Program for the Integrated Management of the Pantanal and the Upper Paraguay River Basin: Final Report*. Brasília: ANA/ GEF/PNUMA/OEA.

Anand, P.B. 2007. Capability, sustainability, and collective action: an examination of a river water dispute. *Journal of Human Development*, 8(1), 109-132.

Ananda, J. and Herath, G. 2003. Incorporating stakeholder values into regional forest planning: a value function approach. *Ecological Economics*, 45(1), 75-90.

Appadurai, A. 1986. Introduction: commodities and the politics of value, in *The Social Life of Things: Commodities in Cultural Perspective*, edited by A. Appadurai. Cambridge: Cambridge University Press, 3-64.

Barkin, D. 2009. The construction of mega-projects and the reconstruction of the world. *Capitalism Nature Socialism*, 20(3), 6-11.

Batterbury, S.P.J. and Fernando, J.L. 2006. Rescaling governance and the impacts of political and environmental decentralization: an introduction. *World Development*, 34(11), 1851-1863.

BBC. 2009. Brazil's huge wetland under threat. *BBC News* [Online, 3 July] Available at: http://news.bbc.co.uk/1/hi/world/americas/8130261.stm [accessed: 10 August 2010].

Blomquist, W. and Schlager, E. 2005. Political pitfalls of integrated watershed management. *Society and Natural Resources*, 18(2), 101-117.

Brenner, N. 2004. *New State Spaces: Urban Governance and the Rescaling of Statehood*. Oxford: Oxford University Press.

Bryant, R.L. and Bailey, S. 1997. *Third World Political Ecology*. London and New York: Routledge.

Conca, K. 2006. *Governing Water: Contentions Transnational Politics and Global Institution Building*. Cambridge, Mass. and London: MIT Press.

Curado, F.F. 2004. *Caracterização dos Problemas Relacionados aos "Arrombados" na Bacia do Rio Taquari*. Relatório Final projeto ANA/GEF/PNUMA/OEA. Corumbá: Embrapa Pantanal.

Dixon, A. B. 2003. *Indigenous Management of Wetlands: Experiences in Ethiopia*. Aldershot: Ashgate.

Graeber, D. 2001. *Toward an Anthropological Theory of Value: The False Coins of our Own Dreams*. New York: Palgrave.

Gramsci, A. 1971. *Selections from the Prison Notebooks*. Trans./Edit. Hoare, Q. and Smith, G.N. London: Lawrence and Wishart.

Harvey, D. 2009. *Social Justice and the City*. Revised edition. Athens and London: Georgia University Press.

Hoffman, J. 2005. Economic stratification and environmental management: a case study of the New York City Catkill/Delaware watershed. *Environmental Values*, 14(4), 447-470.

Ingold, T. 2000. *The Perception of the Environment: Essays in Livelihood, Dwelling and Skill*. London and New York: Routledge.

Ioris, A.A.R. 2004. Conflicts and contradictions on the occupation of the Pantanal space, in *The Pantanal: Scientific and Institutional Challenges in Management of a Large and Complex Wetland Ecosystem*, edited by D. Tazik, A.A.R. Ioris and S.R. Collinsworth. Washington DC: USACE, 26-38.

Ioris A.A.R. 2007. The troubled waters of Brazil: nature commodification and social exclusion. *Capitalism Nature Socialism*, 18(1), 28-50.

Ioris, A.A.R. 2008. Regional development, nature production and the techno-bureaucratic shortcut: the Douro River Catchment in Portugal. *European Environment*, 18(5), 345-358.

Ioris, A.A.R. 2010. The political nexus between water and economics in Brazil: a critique of recent policy reforms. *Review of Radical Political Economics*, 42(2), 231-250.

Jessop, B. 1990. *State Theory*. University Park, PA: Pennsylvania State University Press.

Jessop, B. 2007. From micro-powers to governmentality: Foucault's work on statehood, state formation, statecraft and State Power. *Political Geography*, 26(1), 34-40.

Jessop, B., Brenner. N. and Jones, M. 2008. *Theorizing sociospatial relations. Environment and Planning D*, 26(3), 389-401.

Jongman, R. 2005. *Pantanal-Taquari: tools for decision making in Integrated Water Management*. ALTERRA Special Publication 2005/02. ALTERRA: Wageningen.

Junk, W.J. and Nunes da Cunha, C. 2005. Pantanal: a large South American wetland at a crossroads. *Ecological Engineering*, 24(4), 391-401.

Kalyvas, A. 2002. The stateless theory: Poulantzas's challenges to postmodernism, in *Paradigm Lost: State Theory Reconsidered*, edited by S. Aronowitz and P. Bratsis. Minneapolis and London: University of Minnesota Press, 105-142.

Kopytoff, I. 1986. The cultural biography of things: commoditization as process, in *The Social Life of Things*, edited by A. Appadurai. Cambridge: Cambridge University Press, 64-94.

Kovel, J. 2002. *The Enemy of Nature: The End of Capitalism or the End of the World?* London: Zed Books.

Latour, B. 2004. *Politics of Nature: How to Bring the Sciences into Democracy*. Trans. C. Porter. Cambridge, Mass. and London: Harvard University Press.

Loftus, A. and Lumsden, F. 2008. Reworking hegemony in the urban landscape. *Transactions of the Institute of British Geographers*, NS 33, 109-126.

Lucianer, B. 2010. Este rio está morrendo. *Correio do Estado, Campo Grande*, 29 June, 4-5.

Lukács, G. 1971. *History and Class Consciousness: Studies in Marxist Dialectics*. Trans. R. Livingstone. London: Merlin Press.

Mann, G. 2009. Should political ecology be Marxist? A case for Gramsci's historical materialism. *Geoforum*, 40(3), 335-344.

Marchini, S. 2003. *Pantanal: Opinião Pública sobre Meio Ambiente e Desenvolvimento*. Tefé: Instituto de Desenvolvimento Sustentável Mamirauá.

Marx, K. 1973. *Grundrisse*. Trans. M. Nicolaus. London: New Left Review and Penguin Books.

Mehta, L. 2007. Whose scarcity? Whose property? The case of water in western India. *Land Use Policy*, 24(4), 654-663.

Neves, A.C.O. 2009. Conservation of the Pantanal wetlands: the definitive moment for decision making. *Ambio*, 38(2), 127-128.

Neumann, R.P. 2009. Political ecology: theorizing scale. *Progress in Human Geography*, 33(3), 398-406.

Offe, C. 1996. *Modernity and the State: East, West*. Cambridge: Polity Press.

Palermo, M.A., Alho, C.J.R. and Tonet, H.C. 2003. *Implementação de Práticas de Gerenciamento Integrado de Bacia Hidrográfica para o Pantanal e Bacia do Alto Parguai*. Relatório Final projeto ANA/GEF/PNUMA/OEA. Brasília: ANA.

Panayotakis, C., 2003. Capitalism's 'dialectic of scarcity' and the emancipatory project. *Capitalism Nature Socialism*, 14(1), 88-107.

PCBAP. 1997. *Plano de Conservação da Bacia do Alto Paraguai*. Brasília: MMA.

222 *Tropical Wetland Management*

Perelman, M., 1996. Marx and resource scarcity, in *The Greening of Marxism*, edited by T. Benton. New York: Guilford Press, 64-80.

Rossetto, O.C. 2009. Sustentabilidade ambiental do Pantanal Mato-Grossense: interfaces entre cultura, economia e globalização. *Revista NERA*, 12(15), 88-105.

Scruggs, L.A. 1998. Political and economic inequality and the environment. *Ecological Economics* 26(3), 259-275.

Strang, V. 2005. Common senses: water, sensory experience and the generation of meaning. *Journal of Material Culture*, 10(1), 92-120.

Swarts, F.A. (ed.) 2000. *The Pantanal: Understanding and Preserving the World's Largest Wetland*. St Paul, Minnesota: Paragon House.

Swyngedouw, E. 2004. *Social Power and the Urbanization of Water: Flows of Power*. Oxford: Oxford University Press.

Swyngedouw, E. 2007. Technonatural revolutions: the scalar politics of Franco's hydro-social dream for Spain, 1939-1975. *Transactions of the Institute of British Geographers,* NS 32, 9-28.

Swyngedouw, E. and Heynen, N.C. 2003. Urban political ecology, justice and the politics of scale. *Antipode*, 35(5), 898-918.

The Economist. 2010. The miracle of the cerrado. *The Economist*. [Online, 26 August] Available at http://www.economist.com/node/16886442 [accessed: 10 September 2010].

Treitler, I. and Midgett, D. 2007. It's about water: anthropological perspectives on water and policy. *Human Organization*, 66(2), 140-149.

Vargas, I.A. 2006. *Território, Identidade, Paisagem e Governança no Pantanal Mato-Grossense: Um Caleidoscópio da Sustentabilidade Complexa?* PhD thesis, Curitiba, Brazil: UFPR.

Vieira, L.M. and Galdino, S. 2003. *A Problemática Socioeconômica e Ambiental da Bacia do RioTaquari*. Corumbá: Embrapa Pantanal.

Watts, M.J. 2000. The great tablecloth: bread and butter politics, and the political economy of food and poverty, in *The Oxford Handbook of Economic Geography*, edited by G.L. Clark, M.P. Feldman and M.S. Gertler. Oxford: Oxford University Press, 257-275.

Xenos, N. 1989. *Scarcity and Modernity*. London: Routledge.

Wilcox, R. 1992. Cattle and environment in the Pantanal of Mato Grosso, Brazil, 1870-1970. *Agricultural History*, 66 (2), 232-256.

Žižek, S. 2006. *The Parallax View*. Cambridge, Massachusetts: MIT Press.

Chapter 9

Management and Sustainable Development of the Okavango

Cornelis Vanderpost, Natalie Mladenov, Michael Murray-Hudson,
Piotr Wolski, Lapologang Magole, Lars Ramberg, Joseph E Mbaiwa,
C Naidu Kurugundla, Nkobi M Moleele, Gagoitseope Mmopelwa,
Eben Chonguiça, Donald L Kgathi, Mpaphi C Bonyongo,
Sibangani Mosojane, Susan Ringrose

Background

The Okavango Delta constitutes one of the world's most pristine subtropical wetlands with high biodiversity of all life forms and a very high abundance of African wildlife. The Okavango River terminates in an alluvial fan known as the Okavango Delta in the flat, semi-arid landscape of the Kalahari (Mendelsohn et al. 2010). The wildlife abundance of the Delta has made it an international tourism destination, renowned for its tranquility, natural beauty and near pristine environment in a world in which most wetlands have become heavily transformed. The Okavango Delta also presents a unique combination of hydro-ecological dynamics and socio-economic settings.

Recent pressures generated by population growth, urbanization, tourism and other development initiatives have demonstrated the need for development and management plans. The Okavango Delta was listed under the Ramsar convention in 1997 as a wetland of international importance and this resulted in the formulation of the Okavango Delta Management Plan (ODMP), which has just entered its implementation phase. But pressures and management issues are not confined to the Delta alone, as it is part of a larger river basin that spans the three countries of Angola, Namibia and Botswana. For this larger basin, the Permanent Okavango River Basin Commission (OKACOM) was created by the three riparian states to oversee water developments. The relative inflow contributions (Table 9.1) of each riparian country reflect issues related to water supply, scarcity, and demand. For Angola, the Cubango is only one of the many (and somewhat peripheral) rivers in this well-watered country. For much drier Namibia, the Okavango is a rare perennial river with irrigation potential, while for Botswana the Okavango is the 'Jewel of the Kalahari', an oasis of plants and animals, which are assets for people's livelihoods and for the modern tourism industry.

Table 9.1   Contribution of Basin Countries to Annual Inflow in the Okavango Basin

Country	Average annual river inflows (Mm³)	Annual inflow %	% Basin area contributing to annual inflows	% Basin area not contributing
Angola	9,320.5	94.5	38.7	0.9
Botswana	256.4	2.6	3.8	36.7
Namibia	286.1	2.9	4.1	15.8
Total	9,863.0	100.0	46.6	53.4

Note: Adapted from Ashton et al. (2003)

Namibia has long had plans to develop the Eastern National Water Carrier as part of its National Water Master Plan and take advantage of the Okavango River's perennial water in an otherwise water scarce region (Pinheiro et al. 2003). To promote conservation of the Okavango Delta, Botswana ratified the Ramsar Convention on Wetlands of International Importance, listing the Delta as a Ramsar site. This vast water body in a predominantly dry land has created a unique wetland environment, justifying its status as one of the largest Ramsar sites in the world.

The Okavango (also known as Cubango) originates in the Angolan Highlands, where average rainfall of 1,300 mm per season gives the Okavango river system its major push. The many tributaries running to the lower southeast eventually converge into two major river branches, the Cubango and the Cuito (Figure 9.1). On the Angola-Namibia border, the two streams unite to continue as the Okavango River into Botswana only to spread out again into several distributary channels upon reaching the geological fault lines that control the shape of what is known as the Okavango Delta (Figure 9.1).

In the southern part of the basin, the climate is characterized by occasional and sometimes persistent drought episodes. The Okavango River and Delta that flood during the local dry season are of extreme importance to plant and animal life and human livelihood as the annual flooding of the Delta takes the form of a single large pulse that is out-of-phase with the local rainy season. This has important consequences for wildlife and plant ecology as it makes water available throughout the year – from local rains during the rainy season to flooding during the dry April-October months. Base-flow supplemented by local rains supports 3,500–6,000 km² that are semi-permanently inundated. The flood pulse causes annual expansion and contraction of the inundated area by a further 3,000–6,000 km². The sum of local rainfall – both current and antecedent – and rain in the Angola catchment determine the level and duration of the flood of the Delta, and the amount of outflow from it to Lake Ngami, the Boteti River, the Selinda spillway and the Mababe depression. The high flood of 2010 served to remind us that all these often dry systems are in fact interconnected when in flood.

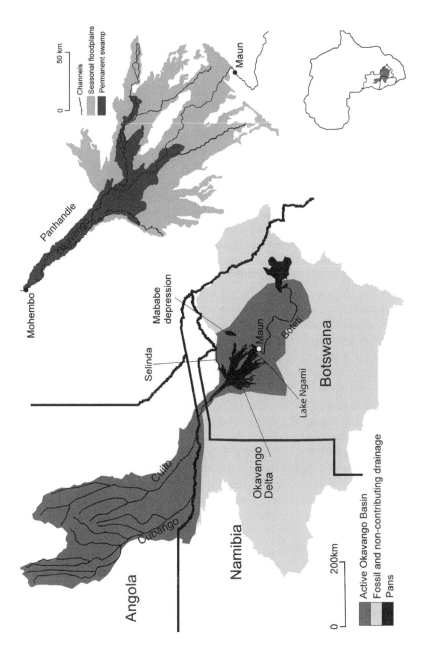

Figure 9.1 Okavango Basin and Okavango Delta.

The Delta wetlands (Figure 9.1) encompass 12,000 km² of inundated land punctuated with over 100,000 islands, creating an extensive dryland–wetland interface that supports biodiversity. Significantly, the Delta is a freshwater wetland even though 98 per cent of its waters evaporate, leaving behind some 380,000 tons of salts per year. Perhaps most importantly, the basin has so far escaped damming, large scale agricultural and water abstraction schemes, or regulation of its irregular flooding levels and distribution.

Three main factors make the Delta a unique wetland: its position within the Kalahari geological setting, its climate, and its transboundary character (Ramberg and Wolski 2008). Geologically, what we know as the Okavango Delta is part of a large 22,000 km² alluvial fan deposited by the Okavango River in an active tectonic trough located in a vast flat basin known as the Kalahari (McCarthy et al. 1992). Well-sorted medium grained sands make the basin highly permeable, which is crucial for maintaining the Delta's fresh water nature. Transport of sandy sediment determines the dynamics of channel migration and shifts in flood distribution. The associated process of the drying of old floodplains and the formation of new ones is an important ecosystem disturbance factor leading to rejuvenation of the system and to the release of nutrients stored in formerly inundated peat beds. Low concentration of solutes (including nutrients) in the flood water makes the Delta an oligotrophic wetland. Importantly, the influx of solutes does not exceed the capacity of processes responsible for salt removal, which maintains the system as a freshwater one. It also causes the soils to be rather poor, which is one reason for the limited development of intensive agriculture and, hence, the absence of fertilizer-derived nutrient/pollution loads in the river.

The dynamic interaction of inundation with geomorphology and biology causes self-organization of the system into a mosaic of islands and floodplains that support biodiversity. Tree-covered islands are also instrumental in the process of immobilizing and removing dissolved chemicals from surface waters. Through a groundwater recharge process (Ellery and Ellery 1997, Ramberg et al. 2006a, Ramberg and Wolski 2008) (see Figure 9.2) riparian trees fringing islands pump underground water at a rapid rate such that the water table beneath the islands is significantly lower than the surrounding swamps. As water is transpired by the trees , it is replaced by fresh water from the surrounding floodplains and swamps. The saline groundwater ultimately sinks to the bottom of the aquifer in density-induced flow, preventing the salts from accumulating at the surface and, thus, keeping the surface water fresh.

The striking spatial arrangement of vegetation communities (Ellery and Ellery 1997) supports a high diversity of wildlife (Ramberg et. al. 2006b). This ecosystem is mainly engineered by hydrology through shifts in flooding patterns over a number of time scales and through seasonal flooding. In, addition, animals (especially large herbivores) and plants exert their influence on the distribution of water and cause the removal and redistribution of nutrients .

Floodplains are key components of the Delta's ecological functioning. When the water level rises in May–June, the trampled and grazed floodplain grasslands

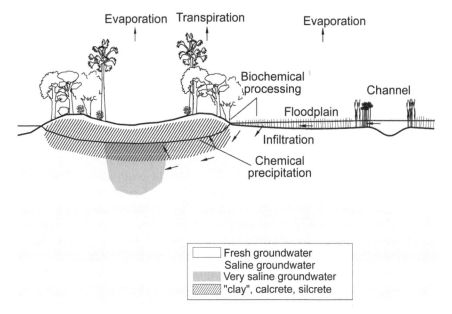

Figure 9.2   An Island Cross-section Showing Vegetation at the Island Fringe and Evapotranspiration-driven Groundwater Flow (adapted from Ramberg and Wolski 2008).

are inundated, nutrients are released and the primary production of phytoplankton, algae and macrophytes explodes, enticing aquatic herbivores from zooplankton to fish to migrate in to feed and spawn. When the water recedes in August through September, large terrestrial ungulates move in to feed on the grasses that sprout in the drying floodplain margins. Variable flooding levels determine not only the phenology of growth but also the accessibility and quality of the grasses. Floodplain grasses function as a reliable grazing resource during the dry season. In the process of grazing, the animals release large quantities of dung on the floodplains, thus contributing significantly to the nutrient cycle (Lindholm et al 2007; Ramberg et al. 2011). This seasonal floodplain ecology strongly influences the amount of fish in the perennial swamps and rivers as well as fish-dependent predators such as crocodiles, fish eagles, herons, otters, and man. Similarly, the number of herbivores is regulated by the volume of high quality grazing produced on the seasonal floodplains after water has receded (Ramberg et al. 2006b) during the dry season when other grazing resources are scarce.

Frequency and duration of inundation and flow action create environmental gradients, leading to the development of distinct vegetation communities, and result in the zonation of vegetation away from the primary channels with channel-flanking vegetation being more productive relative to floodplain vegetation (Bonyongo et al. 2000). Channel blockages, notably by *Cyperus papyrus, Phragmites australis* and *Vossia cuspidata,* influence water distribution and in some cases the formation of new channels and floodplains. Mostly, however, they are formed by a decline of flow and accumulation of sediment in the channel bed that causes a rise of the water surface and water loss out of the channel (McCarthy et al. 1992). Because the apparent channel blockages are symptoms and not the cause of channel failure, efforts at channel-clearing have failed. Pathways created by hippopotamus (*H. amphibius*) also play a key role in the development of new channels and changes in the direction of water flow.

The entire river is also a lifeline for the people making a living along and from it. Population density is generally low: the 600,000 people living in the river basin are mostly concentrated in the upper catchment in Angola, along the Namibian side of the Okavango river banks and on the south-western fringes of the Delta in Botswana. In the Angolan section of the basin, people are poor and the country derives limited benefits from the river. In Namibia the Okavango waters are of higher value because of prospects for irrigated farming and hydro power. In Botswana the waters are of high value on account of the multi-million dollar international tourism industry. In each country, nevertheless, the majority of the people are poor, even in Botswana where tourism income is unevenly distributed. Throughout the basin people hope to better their lives by utilizing or harnessing the resources of the river, making the striking of a balance between the requirements of the three different countries a major challenge. If Angola and Namibia were to harness the Okavango waters to the extent that flow and flood conditions were permanently altered (for example through dam construction or substantial water abstraction), prospects for nature-based tourism in Botswana could be hampered. Alternatively, if the three countries could agree on an equitable development formula based on a conservation/tourism scenario for this last unaltered African river-basin, they could combine forces for mutual benefit.

Management concerns for the river basin and the Delta revolve around a number of broad categories that include trans-boundary hydrological issues such as water abstraction, pollution, potential dam construction in Angola and Namibia or artificial channel clearing in Botswana with cascading effects on biological productivity and biodiversity. There are also other concerns such as the spread of invasive species like Salvinia molesta and damage to trees caused by the increasing elephant population. This latter is part of the broader conflict between conservation/tourism interests and the traditional users of the Delta resources. This raises issues relating to the distribution of tourism benefits across the three countries and among social groupings within them. Finally, there are challenges brought about by expected regional climate change, especially with respect to predicted rainfall regime changes.

Lessons Provided for Basin Development by the Okavango Delta

Tourism

The Botswana part of the basin has the most developed and flourishing tourism industry. In the last 20 to 30 years, the Okavango Delta has become a major international nature-based tourism destination, receiving between 50,000 and 100,000 tourists every year (Mbaiwa 2005). The industry is more developed in Botswana than in Angola and Namibia partly due to recent civil strife in the latter two countries but also due to the scenic nature of the Okavango Delta. Tourism products include the rich wildlife diversity and scenic beauty of the area, while cultural tourism based on local (ethnic) cultures, is also beginning to take shape (Mbaiwa and Sakuze 2009). Tourism development in the Okavango Delta provides employment opportunities to local communities and is a significant source of foreign exchange for Botswana, although the majority of the region's residents remain poor in spite of tourism revenues. The growth of tourism has stimulated the development of infrastructure and facilities such as hotels, lodges and camps, airport and airstrips in wildlife areas as well as permanent "service villages" for the considerable number of staff working in the tourism industry. These latter are located inside the protected areas and are illegal but tolerated. Through backward linkages, wholesale and retail businesses and banking services have also grown in gateway tourism towns such as Maun, which now have tarred roads and other communication facilities such as internet, partly due to tourism development.

Tourism and conservation have been afforded significance in the development space created and agreed upon in the basin. Management is essential because tourism may impact negatively upon conservation targets if not adequately controlled. Tourism generates environmental impacts such as pollution of fresh water bodies, the level depending on the number of visitors and how they are managed. But a possibly greater challenge is the balance that is required between the business demands of the industry and the need for local residents to share in the benefits. For subsistence farmers the increase in the number of elephants in northern Botswana, which is testimony to the success of nature conservation measures and is a bonus to wildlife based tourism, becomes a threat when the severity and number of negative human-elephant encounters rises.

Tourism Impacts on the Environment

Positively, tourism development has contributed to the development of policies such as the Tourism Policy of 1990, the Wildlife Conservation Policy of 1986 and the Community-Based Natural Resource Management Policy of 2007, which aim to promote tourism development and conservation. Negatively, tourism development in the Okavango Delta has been criticized for increasing environmental impacts. Such impacts include the disruption of sensitive ecological areas due to driving outside prescribed trails, noise pollution, and poor waste management in lodges and camps (Roodt 1998, NRP 2000, Mbaiwa 2003). Noise pollution, for example, is generated by small engine airplanes carrying tourists to and from the lodges and camps and by the movement of motor boats in the area (Mbaiwa 2002, Roodt 1998). Fast moving motor boats also create wakes which disturb nesting birds, mammals and reptiles which live in or along the water. Crocodiles and hippos seek undisturbed areas and the presence of too many boats disturbs them. Roodt (1998) mentions that hippos, which were present in large numbers seven years ago, have already moved out of the Xakanaxa lagoon due to human disturbance.

Poor waste water and sewage management is another issue facing tourism development in the Okavango Delta. For instance, septic tanks for wastewater collection in camps may pose a threat to local groundwater (McCarthy et al. 1994, Aqualogic 2009). Many tourist camps in the Okavango Delta rely on borehole water to supply camp needs, and, due to the small size of many islands, boreholes may be located in close proximity to sites of effluent discharge into the groundwater (McCarthy et al. 1994).

Ecosystem Value from Tourism

As a famous international tourist destination, the Okavango Delta has huge tourism value in terms of non-consumptive tourism (wildlife viewing, filming, photography, canoeing, and recreational fishing) and through consumptive tourism (safari hunting). In addition to benefits from bush products such as thatching grass, reeds and papyrus, palms, wild foods and medicines, woody resources, wild animals, birds and fish, which have been valued at more than BWP 12 million/ annum (Turpie et al. 2006), local communities also benefit from employment opportunities created by the tourism business and income from community based tourism (Mbaiwa and Darkoh 2006). Mladenov et al. (2007) surveyed tourists who had visited the Delta and calculated that tourists spent a total of BWP 437 million/ annum on Okavango Delta safaris during 2001–2002. Mmopelwa and Blignaut (2006) estimated the direct and non-consumptive tourism values at BWP 4 million and BWP 68 million/annum, respectively.

The economic value of the Delta as a wildlife refuge, one of its ecosystem supporting services, has been estimated by Turpie et al. (2006) to be approximately BWP 77 million/annum, estimated in terms of hunting and ecotourism values for different flooding regimes. Using survey data of tourist travel costs and hypothetical donations for the preservation of the Delta, Mladenov et al. (2007) estimated a value for ecosystem protection of the Delta from the tourism sector at BWP 110 million/annum and recognized that this could have been higher if the sampled population had included those who had not (yet) visited the Okavango Delta but might have assigned value to this ecosystem without ever having seen it.

Community-Based Natural Resource Management (CBNRM)

Animals and people share the same Okavango waters and river banks. Management is therefore necessary to avoid conflict. The real challenge is to create community benefits for residents by involving people in nature conservation and in the tourism industry. Community-Based Natural Resource Management (CBNRM) is one of the global leading themes in conservation since the 1990s (e.g. Ramberg 1994). CBNRM is an incentive-based conservation philosophy that links conservation of natural resources with rural development (Swatuk 2005, Twyman 2000, Thakadu 2005, Blackie 2006, Mbaiwa 2010). The basic assumption of CBNRM is that for a community to manage its natural resource base sustainably, it must receive direct benefits arising from its use. These benefits must exceed the perceived costs of managing the resources. In the Okavango Delta, CBNRM has had mixed results. Some of the CBNRM projects have collapsed due to poor management of finances and environmental resources whilst a few have succeeded and significantly benefited participating villages economically. For example, the Sankoyo village CBNRM project (involving photographic and safari hunting) has been successful in generating benefits such as income and employment opportunities (Arntzen et al. 2003, 2007, Mbaiwa and Stronza 2010). The Sankoyo Village CBNRM project also raised income which financed the construction of houses for the needy, funeral insurance and expenses for all members, scholarships and household dividends (Mbaiwa and Stronza 2010). There are, however, no studies showing that conservation of natural resources has improved. In contrast, the Gudigwa eco-lodge project was not successful (Magole and Magole 2005). Despite the failures of CBNRM in the Okavango Delta, CBNRM has potential as a tool to achieve both biodiversity conservation and poverty alleviation. CBNRM based activities, however, need to be considered adequately and in conjunction with the commercial tourism sector in the integrated management plans for the Okavango that are on the drawing board or already in operation.

Control of Invasive and Problem Species: Salvinia molesta

Salvinia molesta, a floating water fern native to south-eastern Brazil in South America (Forno and Harley 1979), has become invasive in regions of the world where it has been introduced due to lack of co-evolved enemies. It invaded Botswana in 1948 (Edwards and Thomas 1977) and was discovered in the Okavango Delta in the 1970s and 80s (Forno and Smith 1999). The negative impacts of *S. molesta* in the Okavango Delta include blockage of streams and channels, choking backwater bodies such as ponds and lagoons, elimination of indigenous vegetation, impairing the access of wildlife to drinking water and disruption of navigation and recreational activities such as fishing and wildlife tourism.

Despite the heavy use of herbicides, mainly paraquat from 1972 to 1976, *S. molesta* was not brought under control until a host-specific bio-control agent, the weevil *Cyrtobagous salviniae,* was released (Schlettwein 1985). This biological control has proven to be effective against *S. molesta* in Botswana (Naidu et al. 2000) resulting in salvinia control at several sites in the Moremi Game Reserve (Figure 9.3) (Kurugundla 2003). The involvement of local communities in salvinia control along the Santantadibe River at Ditshipi and the strengthening of private sector salvinia control and monitoring capacity in collaboration with the Biokavango Project in the Moremi Reserve are among the achievements of stakeholder participation (Kurugundla et al. 2010). Despite such successes, hippopotami and boats continue to spread the infestation to new areas of the Delta.

The Government of Botswana invests annually in the monitoring and control of alien invasive species in the country, which includes regulating the movement and importation of boats and fishing gear to prevent the transportation and spread of aquatic weeds (Smith 1993). These efforts can not be relaxed if the currently low overall level of invasive species infestation of the Okavango basin is to be maintained.

Elephant Conflict and Management

Like elsewhere in Africa, Human Elephant Conflict (HEC) is prevalent in Northern Botswana due to the increase in both human and elephant populations. Evidence from aerial surveys (Figure 9.4) suggest that the population of elephants in northern Botswana has increased to its present level of around 150 000 at a growth rate of between 4 and6 per cent per annum (Calef 1990, Cumming 1981, Craig 1989, and Gibson et al. 1998). There is evidence suggesting that the elephant range was fairly restricted in northern Botswana prior to 1984, and that the range has since expanded substantially, with most elephants now located outside protected areas (Gibson et al. 1998, Mosojane 2004). This expansion is due, to some extent, to the fact that the Government of Botswana banned the hunting of elephants for ten years starting in 1980, while from 1989 an international ban on the ivory trade was imposed by the Convention on International Trade in Endangered Species (CITES). This international ban facilitated a steady increase in the elephant population which started expanding to its former range, which, however, became inhabited simultaneously by more people. Thus, increases in both human and elephant populations contribute to HEC conflict in Northern Botswana (Gibson et al. 1998, Craig 1991, DWNP Botswana 2002, CSO Botswana 2002), particularly because a growing human population results in increased fragmentation of elephant habitat through land-clearing for agriculture and for homesteads (Tchamba 1996, Barnes 1996).

There is remarkable temporal variation in the elephant range between wet and dry seasons in northern Botswana. The wet season range is estimated to be 100,000 km^2, extending almost throughout northern Botswana, while in the dry season this shrinks to about 60,000 km^2, mainly as a result of the drying up of seasonal water holes and intermittent streams. During this period elephants tend to concentrate near permanent water sources such as the Chobe, Linyati, and Kwando rivers in Northern Botswana and in the Okavango Panhandle and Delta. It is estimated that elephant densities during the dry spells can reach seven

Figures 9.3   Biocontrol of Salvinia in one of the Hippo Pools in Moremi Game Reserve (left- August 2009, right- January 2010), Okavango Delta.

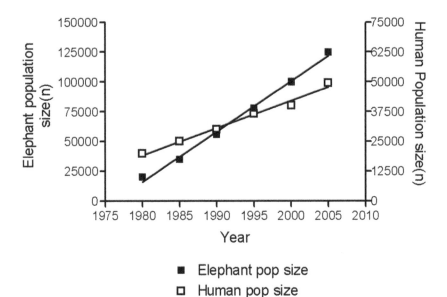

Figure 9.4   **Population Estimates for People and Elephants Living in Northwest, Botswana. Data on human numbers were kindly provided by CSO Botswana (2002). Elephant population size in northern Botswana was based on aerial counts extracted from records of Botswana's DWNP and Gibson et al. 1998).**

individuals/km^2 near water sources. For the Okavango Panhandle, aerial surveys indicated that the density of elephants in the area was nearly three times higher during the dry season than in the wet season (Mosojane 2004). The dry season influx of elephants into the Okavango Panhandle results in more elephants crossing the narrow landscape along the Okavango River, which is densely inhabited by humans and has many subsistence agricultural fields (Mosojane 2004). Therefore, the primary conflict between local communities living around the Okavango Delta and elephants is through crop raiding by elephants in subsistence arable fields. Almost all agricultural activities in the Okavango Delta are non-commercial and are subsistence in nature, with crops harvested being an absolute necessity for survival. Agricultural productivity loss resulting from elephant damage ranges up to 40 per cent, with serious implications for the livelihoods of many of the people living in affected areas.

Actions that may be taken to mitigate conflict between people and elephants differ from one country to another and may alleviate the conflict temporarily (Thouless 1994). Some of the traditional non-lethal methods practiced in Africa and also in the Panhandle of the Okavango Delta include drum beating, throwing fire sticks at elephants and making fires around cultivated fields (Thouless and Sakwa 1995, AFESG 1999). All these deterrents are only temporary measures, as elephants become habituated to them (AFESG 1999).

The Botswana Government compensates farmers for crop losses resulting from elephant damage (Table 9.2). The compensation is based on a standardized formula of Pula BWP250 (~US$36) per hectare damaged, irrespective of the kind of crops planted or damaged. Although compensation appears to meet the expected revenue from harvest (Table 9.2), it fails to recognize the loss in revenue and quality of life of farmers when they have to spend sleepless nights beating drums and making fires and decoys around cultivated fields in an effort to scare elephants away. Unlike other African regions, where HEC incidences occur primarily along the periphery of protected areas (Barnes 1996, Thouless and Sakwa 1995), humans and elephants living along the Okavango Panhandle and other parts of northern Botswana share the same space and water sources. This poses a management issue for Botswana but also for the other riparian states as they open up as elephant habitats. Finding solutions that cater for both the ecological needs of the elephants and the socio-economic needs of subsistence farmers and the tourism industry is likely to remain a challenge for many years to come.

Table 9.2   **The Financial Implications of Crop Damage by Elephants along the Okavango Panhandle (fiscal amounts calculated in Botswana Pula-BWP)**

Months	Affected Fields (n)	Total Expected Harvest (BWP)	Field Raided (%)	Income Before Compensation (BWP)	Compensation (BWP)	Actual Income with Compensation (BWP)
Reported						
March	11	714	29	485	140	626
April	17	652	36	414	155	570
May	20	918	37	572	250	822
Total	48	2284		1471	545	2018
Random[a]						
March	40	593	12	543	102	556
April	40	503	45	495	185	578
May	40	481	45	470	187	555
Total		1577		1408	474	1689

[a] *Fields* in the study site were allocated numbers from 1 to 111, then 40 fields were randomly selected. Elephant incidences were checked twice a week in these 40 fields.

Source: Mosojane 2004 (Data collected in October 2003 to May 2004).

Tsetse Fly Management

The presence of tsetse flies, which cause human and animal trypanosomiasis, in many parts of Africa has long been a major constraint to agricultural development. Botswana has a long history of attempting to control tsetse flies through different methods. Host-game elimination was undertaken between 1944 and 1949, while bush clearing to remove habitats for tsetse flies was carried out during the 1954–1967 period (Bonyongo and Mazvimavi 2007). Due to undesirable environmental effects, game elimination and bush clearing were abandoned in the late 1960s and replaced with spraying of dieldrin and DDT, endosulphan, and a cocktail of endosulphan and synthetic pyrethroids between 1960 and 1991. Due to the toxic effects of endosulphan on non-target species, particularly fish, sequential aerial spraying was stopped in the early 1990s in favour of odour baited targets which were used from 1991–2000. These baited targets were successful in suppressing tsetse fly populations but could not achieve eradication. Moreover the targets were continuously ravaged by elephants and baboons, resulting in high maintenance costs and incomplete fly control. Thus, trypanosomiasis still affected at least 300 cattle around the Okavango Delta by 1999. In the year 2000, the African Heads of States Meeting of the Organization of African Unity resolved that the eradication of tsetse fly should be a collective responsibility of all countries. Subsequently, a Pan African Tsetse and Trypanosomiasis Eradication Campaign (PATTEC) was formulated that aimed to eradicate tsetse and trypanosomiasis within the shortest possible time. Following this, the Botswana government successfully eradicated tsetse flies from the Okavango Delta and the Kwando-Linyanti areas, using sequential spraying of deltamethrin, a pyrethroide of relatively low toxicity.

In the early days, concerns for the fate of non-target insects and the dangers posed by insecticides to the environment were largely ignored. But, following robust environmental awareness campaigns linked to new international conventions and the increasing need to manage and maintain biodiversity, aerial spraying for the eradication of tsetse in the new millennium was preceded by an environmental impact assessment (EIA) and accompanied by a comprehensive environmental monitoring programme. Components that were monitored included terrestrial invertebrates, aquatic micro-invertebrates, fish, birds, tourism-impacts and socio-economic issues (Perkins and Ramberg 2004, Ramberg et al 2006b). The major conclusion from environmental monitoring was that short-term reductions were observed in aquatic and terrestrial invertebrate abundance and species diversity, but that in most cases recovery of populations occurred shortly after spraying. Although changes were observed with respect to invertebrate species composition and bird and fish behavior, these could not be attributed with certainty to spraying. Over the past 10 years, the Okavango region has been virtually free of tse-tse flies, allowing expansion of cattle husbandry as well as tourism.

Climate Change and the Future

As with many aquatic ecosystems around the world which experience pressures of water resource allocation, management of the Okavango will be strongly influenced by climate change. The effects of climate change on the Okavango can conceptually take the form of: i) changes in hydroperiod caused by change in rainfall and evaporation with secondary effects on plants, animals and human socio-economic systems; and ii) increased CO_2 levels and temperatures, affecting the biochemistry of plants and the ecology of animals with secondary effects propagating through the ecosystem and resulting in socio-economic impacts. Although both are potentially important, only the first has been studied in the Okavango basin through hydrological models.

Our current understanding of the hydro-ecology of the Okavango suggests that the annual flood pulse is the dominant driver of the wetland's ecosystems (Murray-Hudson 2009). The hydrological processes generating this pulse integrate rainfall and evaporation conditions in the catchment at a longer time scale than that of an individual rainfall event. The Okavango region is characterized by semi-cyclic long term variability in rainfall and temperature, resulting in multi-annual sequences of drier and wetter years of approximately 40-years duration (McCarthy et al. 2000). Historic rainfall time series (1920-to date) from the region show no evidence of change in total annual rainfall and inter-annual rainfall variability beyond the range of this periodic variability (Mazvimavi and Wolski 2006). These findings do not exclude the possibility of anthropogenic, GHG (green-house-gases)-driven change, but recognize that such changes have not yet been stronger than and are thus not distinguishable from natural rainfall variability. Temperature records, however, show an increase in temperatures towards the end of the 20th century that exceeds the range of variability observed in the earlier period (Wolski 2009).

Projected Climate Change

Projections of future temperature and rainfall vary between various GHG emission scenarios and between different General Circulation Models (GCMs) used to predict future climate conditions. Recent assessments of climate change in the Okavango region incorporate results of several GCMs (Murray-Hudson et al. 2006, Bauer 2006, Milzow, 2009) or 18 GCMs available from the World Climate Research Program (WCRP) Climate Model Intercomparison Project (CMIP3) multi-model dataset (www-pcmdi.llnl.gov). This multi-model work produced projections of both wetter and drier conditions than those observed before, with higher likelihood associated with drier conditions. In the most recent work (Wolski, 2009), the results of nine GCMs were downscaled to local conditions. The downscaled projections shift towards wetter conditions, but still encompass the drier end of the spectrum (Wolski 2009).

Raw GCMs consistently simulate increases in the mean, minimum and maximum temperatures in the Okavango region during the 21st century, although they differ in the magnitude of the increases. Results of statistical downscaling of projected temperature changes are consistent with the results of raw GCM analyses, suggesting an increase of 2.3–3°C relative to the 1960–1990 reference period. Results of statistical downscaling of GCMs predominantly suggest an increase in annual rainfall: by the mid-21st century, rainfall is projected to increase by 0–20 per cent in relation to rainfall in the 1960–1990 reference period.

Projected Hydrological Change

Results of hydrological modeling using outputs of statistical downscaling of projected future climate, indicate that runoff from the Okavango catchment responds more strongly to change in rainfall than to change in temperature (evaporation). In contrast, the flooding in the Okavango Delta responds more strongly to change in evaporation than to change in rainfall. This is understandable as evaporation has limited impact on the generation of runoff in the catchment, but plays a dominant role in determination of flood extents in the endorheic Okavango Delta where 98 per cent of water inputs evaporate. In the overall catchment, the projected increase in rainfall overcompensates for the projected increase in evaporation resulting from increased temperatures. Therefore, an increase in runoff (total and monthly) is projected with the increase in peak flows projected to be stronger than the increase in low flows.

In the Delta proper, under "dry" scenarios (low rainfall increase and high temperature increase within the inter-model range), the increase in evaporation and transpiration may exceed the increase in local rainfall and inflow from the catchment, with a resulting decrease in the system's overall wetness. However, under "wetter" scenarios (high rainfall increase and low temperature increase within the inter-model range), an increase in overall wetness is projected. The "dry" conditions will manifest as a decrease in frequency and duration of inundation throughout the Delta and by a reduction of low flows in the rivers draining the system. The "wet" conditions will manifest as an increase in duration and frequency of inundation throughout the Delta, and an increase of high and low flows in the rivers draining the system. Importantly, no significant changes in interannual variability of hydrological conditions are projected. This means that interannual variability, under low or high floods, will continue to be present. Under "wet" scenarios, high floods are expected to be larger than and low floods will not be as low as those observed in the past. For "dry" scenarios, high floods will not be as high, and low floods will be lower than those observed in the past.

Results of hydrological analyses based on raw GCM data indicate conditions qualitatively similar to these projected by the "dry" downscaling scenarios, i.e. a decrease in frequency and duration of inundation throughout the Delta and by a reduction of low flows in the rivers draining the system, but of generally larger magnitude.

Hydrology as an Ecological Determinant

The primary production of the Delta wetlands and riparian zones is fundamentally dependent on the effects of flood-pulsing and rainfall. Recent work by Wolski et al. (2010) suggests that while the seasonally flooded grasslands show the highest CO_2 uptake rates in summer, the *Cyperus papyrus*-dominated communities of the perennially flooded parts of the Delta have the highest carbon uptake over the course of the year. Likewise, plant community composition is controlled by hydroperiod. Murray-Hudson (2009) found that frequency of annual flooding was the strongest determinant of Delta floodplain plant species composition, while McCarthy and Ellery (1994) have shown that the riparian trees in the Delta utilize shallow groundwater derived from floodplain infiltration. When flooding drops to a frequency of less than about 1 in 10 years, a terrestrial succession begins on floodplains, with a progressive encroachment by woody species (Heinl et al. 2008, Murray-Hudson 2009, Ramberg et al. 2006b); this succession is effectively arrested by renewed flooding of sufficient duration.

Given these relationships, climate change-induced shifts in hydroperiod clearly will affect both the production and community ecology of Delta vegetation. Changes in proportions of perennially, seasonally and intermittently flooded areas will result in changes in primary production and, thus, changes in habitat availability for the different components of aquatic and terrestrial biota. Mosepele et al (2009), for example, found a strong relationship between fish catch per unit effort and discharge into the Delta. Larger incoming floods increase fish catches and reduced flows result in declining catches. Such changes are likely to be mediated through a combination of increased primary production by aquatic plants in the increased area of flooding, and the availability of detritus derived from the flooding of terrestrial grasslands (Lindholm et al 2008, Ramberg et al 2006b, Ramberg et al. 2011). Although no quantitative data on the relationships between water-dependent birds and hydroperiod exist, it may be expected that numbers of water birds show a similar response to discharge as do fish – that is that increased flooding initially supports larger numbers. It is not as simple as it seems, however: an increase in inflows will increase fish (or bird) production by a disproportionate amount in the initial few years, because of the increased primary productivity resulting from high nutrient inputs from recently flooded drylands. These inputs, and the nutrients from decaying terrestrial plants, will not persist under continuous flooding. The variation between the dry terrestrial phase and the flooded phase is necessary to sustain the high production which characterises the Delta. Seasonally flooded grassland is also a critical resource for the large mammalian herbivores, in particular herd-forming animals such as Cape buffalo (*Syncerus caffer*). Bonyongo et al. (2000) noted that grazing by large mammals was most prevalent in the *Paspalidium obtusifolium-Panicum repens* population in the floodplain they studied, in both wet and dry seasons, as *P. repens* responds to the advent of both local (summer) rain and winter floods.

Climate Change in a Human Development Context

If climate change were to result in drier conditions (Tyson et al. 2002, Arnell et al. 2003) under a scenario of increasing anthropogenic upstream water withdrawals (Andersson et al. 2003, Mukumbira 2003), a diminished water supply may cause significant 'shrinking' of the Delta (Andersson et al. 2003). The findings of Murray-Hudson et al. (2006) extend these predictions to ecological change. They found that under drier conditions, the area and proportions of permanent swamp, areas covered by sedge and grass vegetation (seasonal floodplains) and floodplain grasslands (intermittently flooded areas) would decline, and woody species would encroach onto the new dry land areas. These spatial changes will influence ecological processes in the Delta, in particular vegetative succession, primary production, and relationships of floodplains with the surrounding woodland and savannah as described above. Additional ripple effects up trophic levels are also expected. For example, viable habitat size and contiguity may fall below the minimum for some wetland and grazing species (and consequently for their predators), which will then be lost from the system. The available habitat may be considerably reduced, and thus populations will decline. Here, in particular, wattled crane, cheetah, and wild dog are at great risk, the Okavango populations being among the largest globally of these species that are already on the IUCN Red Data list.

Murray-Hudson et al. (2006) further suggest that tourist facilities presently on the fringes of the seasonal swamps could experience negative impacts as a result of reduced or increased flooding. Indeed, in an ecosystem valuation of the Okavango Delta from the perspective of the tourism sector, Mladenov et al. (2007) found that biodiversity positively affected the preservation value of the Delta, with more opportunities for wildlife viewing resulting in a greater recreational value for the Delta. Declines or losses of charismatic mammal species as a result of habitat loss clearly will affect this recreational value. Other potential negative impacts include changing hydrological status of viewing areas and facilities such as runways for aircraft in the Delta. Negative impacts from reduced water supply to wildlife-viewing opportunities and tourism revenue would thus impact Botswana's tourism industry, which contributes 4.5 per cent to Botswana's GDP (Mbaiwa 2003).

Human populations will also be affected by hydrologic changes. Kgathi et al. (2006) found that under conditions of natural shifts in the flow of water, which resulted in desiccation of river channels, local communities already have been adversely affected. The impacts manifested themselves as diminished access to water resources and impaired molapo farming, fishing, and harvesting of veld products (Kgathi et al. 2006).

Perspectives

It is evident that the social-ecological system of the Delta will be altered by climate change. Which direction that change may take is still highly uncertain, but with time and concerted effort our understanding will improve. The primary effect of climate change will be on the hydrology of the system, and this change will drive adjustments in the rest of the present largely water-dependent ecology. The relative contributions of different ecosystem services driving the local economy are, however, likely to change and this will perturb the human component of the system. This will require greater flexibility in land use regulation and planning than even the current highly variable system supports: innovative development and management approaches are needed to ensure improved human livelihoods and sustainability.

Management of the Okavango Resources

The Okavango Delta Management plan (ODMP)

In 1997 the Okavango Delta area was identified by Botswana's national Vision 2016 as requiring planning in order to promote prudent use of its resources for the benefit of the local communities, the nation and the international community. In 2003 the National Development Plan (NDP 9) set sustainability as one of the development principles and tourism as the main driver of economic diversification. The Okavango Delta was expected to play a significant role as the country's major tourist destination.

The ODMP was developed in fulfilment of Article 3.1 of the 1971 Ramsar Convention (where the Okavango Delta is listed from 1997) which states that 'The Contracting Parties shall formulate and implement their planning so as to promote the conservation of the wetlands included in the List, and as far as possible the wise use of wetlands in their territory'. Additionally, such a management plan was desirable at a national level, where the Delta was elevated to be part of the country's conservation and economic development strategy. The ODMP therefore has as much national significance as it does international.

Locally, however, the ODMP is a challenge to the existing sectoral management system of Botswana, which is characterized by limited integration of management activities between individual ministries. Sectoral management poses a threat to the ecosystems and the services they render, as well as to the livelihood strategies of the local communities. For example, local people complain about large land allocations for tourism or other commercial purposes that cut them off from their sources of livelihood. There is also fear that water in river channels may be polluted by mismanaged sewage and solid waste resulting from the establishment of tourist camps and villages that serve those camps. There are also management decisions, such as government-led channel clearing activities, which do not take into consideration the environmental impacts of such actions. The ODMP offers a way to integrate resource management for the Okavango Delta that will ensure its long-term conservation. The ODMP also aims to provide benefits for the present and future well-being of the people through sustainable use of its natural resources and collective and transparent decision making.

Challenges faced by the ODMP

The system of sector integration worked well during the planning phase when the project was funded and had dedicated staff within a project secretariat. The existence of an ODMP secretariat and planning team promoted the important horizontal lines of communication (Wolski et al. 2010). But during the current implementation phase integration between different government departments is more problematic. This is because the institutional landscape of Botswana is fundamentally sectoral and this was not permanently changed by the ODMP process. Although the integrated nature of the ODMP was supported by the prevailing culture of integration at district level, it was not at national level. The project secretariat was disbanded and replaced by a less resourced and less powerful Department of Environmental Affairs (DEA). This Department is the overall environmental coordinating agency in the district and it is one of its duties to 'encourage' other departments to implement ODMP activities from their recurrent budgets. However, such departments do not necessarily prioritise ODMP related activities. Implementation of the ODMP therefore has been a slow and frustrating process. Apart from a dedicated budget, a proper implementing agency is required for the ODMP to deliver on its objectives and the ideals of integrated water resource management.

A positive point is that the plan has been successfully used as the Botswana contribution to the basin-wide water resources management framework developed under the auspices of OKACOM (Permanent Okavango River Basin Commission). Given the right mandate by the riparian states, OKACOM is better placed to implement a sustainable development strategy for the Okavango basin, including the Delta.

Transboundary Water Resources Management Issues

The transboundary nature of the Okavango basin with three very different countries sharing its resources adds its own complications to future hydrological management. The current situation where waters of the system are little utilized and the Delta is left in an unregulated state is not the result of conscious management efforts. Rather, it results from natural, socio-economic and geo-political conditions not having favored development as well as the low potential of the catchment to support agricultural activities. Importantly also, Angola was engulfed in a protracted civil war from 1975 until 2002 during which no developments took place in the basin. In addition, the Okavango Delta in Botswana was infested with tsetse flies until spraying in 2002 removed them. This effectively kept livestock out and provided an exclusive environment for African wildlife. Furthermore, local and international political environmental concerns led to abandonment of several water development plans and efforts at flow regulation in the Delta.

It needs to be noted that from 1992 the whole of Southern Angola, including the Okavango River Basin, was fundamentally a war zone and a Union for the Total Independence of Angola (UNITA) rebel movement strong hold. This made it difficult for the riparian states of the Okavango River Basin to develop and implement a transboundary river basin management system. Nevertheless, the Permanent Okavango River Basin Water Commission (OKACOM), established on 15 September 1994, brought together the three riparian states of Angola, Botswana and Namibia to take on matters related to the conservation, development and utilization of water resources of the Okavango River Basin (OKACOM 1994). This agreement commits the member states to promote coordinated and environmentally sustainable regional water resources development, while addressing the legitimate social and economic needs of each of the riparian states. The three countries recognize the implications that developments upstream of the river can have on the resource downstream (OKACOM 1994).

For at least a decade OKACOM was active in both Namibia and Botswana with the intent to generate economic, social and ecological benefits from the very important perennial river system adjacent to their borders. Involvement by Angola was limited until the war in Angola ended in April 2004. The end of the war became a defining moment for OKACOM in Angola, and OKACOM became part of the rebuilding agenda.

These underlying geopolitical, historical, socio-economic and environmental circumstances have motivated the OKACOM riparian states' determination and commitment to cooperation and setting up of a joint management system. OKACOM's approach to transboundary management of the Okavango River Basin resources entails promoting coordinated and sustainable water and land resources management of the basin, while at the same time balancing the legitimate social and economic needs of the riparian states and communities. It is important to note, however, that joint management is fundamentally dependent on perceived mutual benefits and the understanding that the countries stand to gain more from working together than apart.

OKACOM's Approach to Basin Planning and Management

The three riparian Countries have different socio-economic priorities driven by country specific circumstances. Botswana and Namibia are characterized by high levels of water scarcity due to their semi-arid climatic conditions. As a result, the two countries are dependent on groundwater with economies based mostly on the mining, livestock and tourism sectors with marine fisheries important for Namibia. Angola, on the other hand, is more secure in fresh water availability due to high levels of precipitation. The country contributes more than 90 per cent of the annual inflow into the Okavango River Basin (Table 9.1). The economy is dominated by vibrant oil and diamond industries and there is high potential for agriculture and other economic activities such as fisheries and tourism.

While the economic activities and potentials of the three countries appear to have more similarities than differences, issues of prioritization, development planning and implementation strategies in the basin present specific challenges for OKACOM. Central to these challenges is the need to balance possibly diverging country-based development initiatives such as upstream commercial agriculture or hydropower generation with the maintenance of ecosystems integrity for the eco-tourism industry downstream.

It was against this backdrop that OKACOM, at a very early stage, proposed the development of a basin-wide environmental assessment to inform development priorities and strategies in the context of a basin-wide master plan. The aim of the basin master plan is to maintain the ecosystem functions and services of the basin and to maximize benefits for the basin as a whole. To meet this goal, OKACOM has most recently undertaken a United Nations Development Programme Global Environment Facility (GEF) funded Transboundary Diagnostic Analysis (TDA)/ Strategic Action Plan (SAP) process as a first step. The TDA is an objective assessment of the sources, causes, and impacts of transboundary water problems. The SAP compliments the TDA assessment as a negotiated document that identifies policy, legal, and institutional reforms and investments needed to address priority transboundary problems.

Ten to fifteen years ago, Angola's military conflict imposed impossible conditions for assessment of the status of land and water resources in the basin, a prerequisite for development planning and management. Today, however, hydrological gauging stations are being rehabilitated and upgraded in the country making it possible to collect necessary data. To ensure a catchment based TDA process, the researchers and policy makers adopted the 'environmental flows' assessment methodology (King et al. 2003) to minimize risks of generating country-specific assessments that might not connect to each other. Three research teams were formed, one per country, to bring in expert knowledge from selected sites along the basin. Different development scenarios were considered to guide OKACOM to define and negotiate an acceptable development space in the basin that can sustain fundamental development needs without undermining the stability and functionality of the river system (King 2009). Development space may be defined as the difference between current conditions in the basin and the furthest level of development found acceptable to stakeholders through consideration of relevant scenarios.

The TDA process has created a solid opportunity to develop a science-based approach for planning, management and decision making, a working framework that OKACOM believes should be the basis for its operations. The tense geopolitical landscape dominated by civil war that prevailed over decades brought challenges to the TDA process. The greatest of these was lack of data on the Angola part of the basin. But today the riparian states are building trans-border communication bridges that transcend individual government institutions and allow for smooth data flow. Government officers, researchers, academics, and the various segments of civil society are joining forces to develop a common shared vision and action plan for the joint management of the Okavango River Basin.

Management Challenges Faced by OKACOM

It has become evident that a number of challenges are still to be overcome to improve organizational and delivery performance of OKACOM, particularly in the areas of governance, enforcement, institutional capacity development, sustainable funding, and appropriate models for benefit sharing. In terms of institutional challenges, a number of grey areas still prevail. These include: a) the mandate and powers of country delegations to the Commission, b) the extent to which (and mechanisms through which) decisions taken at Commission level filter down to national institutions across all critical sectors to adequately internalize and act upon, and c) effectiveness in securing institutional memory to minimize dependency on individual champions. Responses to such challenges vary significantly from country to country and, in most cases, are highly dependent on such champions with technocratic passion for the cause.

Although there is a growing recognition that riparian states should take responsibility for sustainably financing OKACOM, the reality is that OKACOM still depends largely on international donors. New conceptual funding frameworks are emerging, such as establishing water tariffs from different water users, in order to strengthen the sense of ownership and long term sustainability of transboundary river initiatives.

Cost and benefit sharing within the Okavango River Basin offers real possibilities for sustainable development. It is expected that among other things the TDA/SAP processes will help the riparian countries define a suitable strategy for cost and benefit sharing. Contemporary debates on the approaches to benefit sharing in transboundary water management center around either volumetric sharing of water in the basin or the sharing of beneficial uses of the water. The first is rooted in the principle of portioning an amount of river-related goods and services to each riparian state, while the second is driven by the notion of investing in the most beneficial use of the available water and other natural resources to obtain maximum return. In other words, it is about maximizing benefits from the river and to the river (Sadoff and Grey 2005). It seems that the best development path is one in which the best return on investment is identified and agreed upon by the riparian states. The riparian states would subsequently embark on joint planning and management to maximize returns on investment, no matter where the development investment takes place (thus they think and act along transboundary lines). It is here that the previously discussed concept of acceptable development space recommended by the TDA process comes into play as the unifying common denominator of basin development planning. A cost and benefit sharing formula may be developed to secure equity, fairness and social justice across the riparian states. The process of negotiating this formula would need to take into account issues of legitimacy, technical and financial capacity as well as shared values and principles.

Put into perspective, it can be argued that the establishment of OKACOM through the 1994 agreement was a strong and bold political statement of willingness to cooperate and respond to the challenges imposed by the complexity of transboundary management of river systems. It demonstrated the vision of sustainable natural resource management prevailing in the basin irrespective of the constraints imposed by protracted conflict and the complexity of managing endorheic and ephemeral river systems (Seely et al. 2003). However, it has also emerged that a lot remains to be done to achieve the ideals of that cooperation. Much still needs to be done to develop the basin in a sustainable manner and put in place a robust and equitable cost and benefit sharing model. Furthermore, improved conservation of the natural resources of the Okavango Delta over the past few decades remains to be demonstrated.

Conclusions

The future integrity of the Okavango basin and Delta is at a crossroads. This wetland is still highly pristine and exhibits substantial hydrological integrity. One potential threat to the integrity of the water-dependent ecology of the Okavango Delta is climate change, especially if the scenario of increased drying (which currently appears as likely as continued wetness) becomes reality. Other critical challenges to sustainable management of the Okavango , such as population growth, the expanding tourism industry, poverty and hydrologic developments, are under some degree of control within the Delta Ramsar site but less so in the larger Okavango basin. The Okavango Delta Management Plan formulates the limits for sustainable development in the Botswana portion of the basin. For the larger basin, the good news is that initial steps have been taken toward improved management through OKACOM. Implementation in this case is primarily challenged by the very different roles that the Okavango/Cubango plays in the three countries. For Angola it is a remote (from Luanda) river basin where many formerly displaced people wish to return to a life of farming. For Namibia, the river holds the promise of water for irrigated crop-farming and abstraction for dry inland regions, while for Botswana a high degree of natural integrity (including that of water entering from the other two countries) is a prerequisite for a thriving tourism industry.

The paramount challenge remains to ensure ecosystem integrity of the wetland and at the same time reduce poverty and achieve improved living standards for the resident population. It is important to recognize that many wetland communities depend on agricultural production (crops and livestock) and fishing, hunting and gathering. Conservation efforts in the basin should not only concentrate on making these production systems sustainable, but should also promote ways to involve residents in community-based and commercial eco- tourism activities. This requires broadening of transboundary management and advancing the idea of benefits (and costs) sharing across the basin and within the riparian states within the bounds of acceptable and agreed upon development space.

Acknowledgement

This material supported in part by the National Science Foundation (NSF) Grant No. 0717451.

Any opinions, findings and recommendations expressed are those of the authors and do not necessarily reflect the views of the NSF.

Acronyms

AFESG - African Elephant Specialist Group
BWP - Botswana Pula
CBNRM - Community-Based Natural Resource Management
CITES - Convention on International Trade in Endangered Species
CMIP - Climate Model Intercomparison Project
CSO - Central Statistics Office (Botswana)
DWNP - Dept Wildlife and National Parks Botswana
GCMs - general circulation models
GDP - gross domestic product
GEF - Global Environment Facility programme
GHG - greenhouse gasses
HADCM - Hadley Centre Circulation Model
HEC - Human Elephant Conflict
IUCN - International Union for the Conservation of Nature
IWRM - integrated water resource management
ODMP - Okavango Delta Management Plan
OKACOM - Permanent Okavango River Basin Commission
PATTEC - Pan African Tsetse and Trypanosomiasis Eradication Campaign
SAP - Strategic Action Plan
TDA - Transboundary Diagnostic Analysis
UNITA - Union for the Total Independence of Angola
WCRP - World Climate Research Programme

References

African Elephant Specialist Group (AFESG). 1999. Review of the African elephant priorities, Second edition. *Working document of IUCN/SSG African Elephant Specialist Group.*

Andersson et al. 2006. Impact of climate change and development scenarios on flow patterns in the Okavango River. *J. Hydrology*, 331, 43-57.

Andersson, L., Gumbricht, T., Hughes, D., Kniveton, D., Ringrose, S., Savenije, H., Todd, M., Wilk, J. and Wolski, P. 2003. Water flow dynamics in the Okavango River Basin and Delta: a prerequisite for the ecosystems of the Delta. *Physics and Chemistry of the Earth*, 28, 1165-1172.

Aqualogic (Pty) Ltd. 2009. Liquid Waste Systems of Tourism Establishments in the Okavango Delta and Transportation, Handling and Storage of Hazardous Substances in the Okavango Delta. Final Report. Biokavango Project and HOORC, University of Botswana, Maun.

Arnell, N.W., Hudson, D.A. and Jones, R.G. 2003. Climate change scenarios from a regional climate model: estimating change in runoff in southern Africa. *Journal of Geophysical Research: Atmospheres*, 108, (D16), 4519, Available at: doi:10.1029/2002JD002782.

Arntzen, J., Buzwani, B., Setlhogile, T., Kgathi, D.L. and Motsholapheko, M.K. 2007. *Community-Based Resource Management, Rural Livelihoods and Environmental Sustainability*. Gaborone: Centre for Applied Research.

Arntzen, J., Molokomme, K., Tshosa, O., Moleele, N., Mazambani, D. and Terry, B. 2003. *Review of CBNRM in Botswana*. Gaborone: Centre for Applied Research.

Ashton, P. and Neal, M. 2003. An overview of key strategic issues in the Okavango Basin, in *Transboundary Rivers, Sovereignty and Development: Hydropolitical Drivers in the Okavango River Basin,* edited by Turton, A.R., Ashton, P., Cloete, T.E. Pretoria and Geneva: AWIRU and Green Cross International, 31-63.

Aylward, B. and Rajapakse, C. 2010. Benefit Sharing and Water Resources: A Case Study from the Okavango River (Unpublished Draft Paper).

Barnes, R. F. W. 1996. The conflict between humans and elephants in the central African forests. *Mamm. Rev*, 26, 67-80.

Blaikie, P. 2006. Is small really beautiful? Community-based natural resource management in Malawi and Botswana. *World Development*, 34(11), 1942-1957.

Bonyongo, M. C., Bredenkamp., G. J. and Veenendaal, E. 2000. Floodplain vegetation in the Nxaraga Lagoon area, Okavango Delta, Botswana. *South African Journal of Botany,* 66, 15-21.

Bonyongo, M. C. and Mazvimavi, D. 2007. Environmental monitoring of May-August 2006 aerial spraying of deltamethrin for tsetse fly eradication in the Kwando-Linyanti and Caprivi region. (Eds). HOORC/DAHP.

Calef, G.W. 1990. Elephant numbers and distribution in Botswana and north-western Zimbabwe, Management of the Hwange Ecosystem, Workshop sponsored by USAID.

Craig G.C. 1989. Population dynamics of elephants, in *Elephant Management in Zimbabwe, Dept National Parks and Wildlife Management, Zimbabwe*, edited by R.B. Martin, G.C. Craig and V.R. Booth. Harare: Department of National Parks and Wildlife Management, 67-72.

Craig G.C. 1991. *1991 Dry Season Survey of Northern Botswana*. Report to Dept. Wildlife and National Parks. Gaborone: Botswana.

Central Statistics Office (CSO), Botswana. 2002. *Human Population Census Report*. Gaborone: Botswana Government Printer.

Cumming D.H.M. 1981. The management of elephant and other large mammals in Zimbabwe, in *Problems in Management of Locally Abundant Wild Mammals*, edited by P.A. Jewell, S. Holt. New York: Academic Press, 91-118.

DWNP. 2002. *Internal Report on Elephant Population Census*. Gaborone: Botswana Government Printers.

Edwards, D., and Thomas P.L. 1977. The Salvinia molesta problem in the northern Botswana and Eastern Caprivi area, in *Proceedings of the 2nd National Weeds Conference of South Africa*. Stellenbosch, 2-4 February 1977, 221-237

Ellery, K. and Ellery, W. 1997. *Plants of the Okavango Delta, A Field Guide*. Durban: Tsaro Publishers.

Forno, I.W and Harley K.L.S. 1979. The occurrence of Salvinia molesta in Brazil. *Aquatic Botany*, 6, 185-187.

Forno, I.W. and Smith P.A. 1999. Management of the alien weed, Salvinia molesta, in the wetlands of the Okavango, Botswana, in *An International Perspective on Wetland Rehabilitation*, edited by W. Streever. Dordrecht: Kluver Academic Publishers, 159-166.

Gibson, D.S.T.C., Craig, G.C. and Masogo, R.M. 1998. Trends of the elephant populations in Northern Botswana from aerial survey data. *Pachyderm*, 25, 14-27.

Heinl, M., Sliva, J., Tacheba, B., and Murray-Hudson, M. 2008. The relevance of fire frequency for the floodplain vegetation of the Okavango Delta, Botswana. *African Journal of Ecology*, 46, 350-358.

Hewitson, B.C. and Crane, R.G. 2005. Gridded area-averaged daily precipitation via conditional interpolation. *Journal of Climate*, 18, 41-57.

Julien, M.H., Hill, M.P. and Tipping, P.H. 2009. Salvinia molesta D.S. Mitchell (Salviniaceae), in *Biological Control of Tropical Weeds using Arthropods*, edited by R. Muniappan, G.V.P. Reddy and A, Raman. Cambridge: Camridge University Press, 378-407.

Junk, W., Brown, M., Campell, I., Finlayson, M., Gopal, B., Ramberg, L., and Warner, B. 2006. The comparative biodiversity of seven globally important wetlands: a synthesis. *Aquatic Sciences*, 68, 400-414.

Kgathi, D.L., Kniveton, D., Ringrose, S., Turton, A.R., Vanderpost, C.H.M., Lundquist, J., and Seely, M. 2006. The Okavango; a river supporting its people, environment and economic development. *J. Hydrology*, 331, 3-17.

King, J.M., Brown, C.A. and Sabet, H. 2003. A scenario-based holistic approach to environmental flow assessments for regulated rivers. *Rivers Research and Applications* 19 (5-6). 619-640.

King, J. and Brown, C. 2009. Integrated basin flow assessments: concepts and method development in Africa and South-east Asia. *Freshwater Biology*. Available at: doi:10.1111/j.1365-2427.2009.02316.x.

Kurugundla, C.N. 2003. *Aquatic Vegetation Control Unit Annual Report*, September 1999-March 2003. Unpublished Report. Department of Water Affairs. Maun: Botswana.

Kurugundla, C.N., Khwarare, G., and Moleele, N. 2010. Addressing biodiversity loss through biocontrol and monitoring of alien aquatic invasive species (*Salvinia molesta*) in the Okavango Delta. In: Biodiversity and Climatic Chane: Achieving the 2020 Targets. *CBD Technical Series*, 1, 10-21. Nairobe, Kenya.

Lindholm, M., Hessen, D., Mosepele, K., Wolski, P. 2007. Food webs and energy fluxes on a seasonal floodplain: the influence of flood size. *Wetlands*, 27, 775-784.

Magole, L.I. and Magole L. 2005. *Ecotourism and socio-economic development of Basarwa: the case of Gudigwa village.* University of Botswana and University of Tromso Basarwa Collaborative Programme, Gaborone, Botswana.

Mazvimavi, D. and Wolski, P. 2006. Long-term variations of annual flows of the Okavango and Zambezi Rivers. *Physics and Chemistry of the Earth*, Parts A/B/C 31,944-951.

Mbaiwa, J.E. 2002. *The Socio-Economic and Environmental Impacts of Tourism Development in the Okavango Delta, Botswana: A Baseline Study.* Maun: Harry Oppenhiemer Okavango Research Centre.

Mbaiwa, J.E. 2003. The socio-economic and environmental impacts of tourism development on the Okavango Delta, north-western Botswana. *Journal of Arid Environments,* 54, 447-467.

Mbaiwa, J.E. 2005. The problems and prospects of sustainable tourism development in the Okavango Delta, Botswana. *Journal of Sustainable Tourism*, 13(3), 203-227.

Mbaiwa, J.E. 2010. *Tourism, Livelihoods and Conservation: The Case of the Okavango Delta, Botswana.* Saarbrucken: Lambert Academic Publishers.

Mbaiwa, J.E. and Darkoh, M.B.K. 2006. *Tourism and the Environment in the Okavango Delta, Botswana.* Gaborone: Pula Press.

Mbaiwa, J.E. and Sakuze, L.K. 2009. Cultural Tourism and Livelihood Diversification: The Case of Gcwihaba Caves and XaiXai Village in the Okavango Delta, Botswana. *Journal of Tourism and Cultural Change*, 7(1), 61-75.

Mbaiwa, J.E. and Stronza, A.L. 2010. The effects of tourism development on rural livelihoods in the Okavango Delta, Botswana. *Journal of Sustainable Tourism*, 18(5), 635-656.

McCarthy, T.S., Cooper, G.R.J., Tyson, P.D. and Ellery, W.N. 2000. Seasonal flooding in the Okavango Delta, Botswana: recent history and future prospects. *South African Journal of Science*, 96, 25-33.

McCarthy, T.S., Ellery, W.N. and I. G. Stanistreet. 1992. Avulsion mechanisms on the Okavango fan, Botswana: the control of a fluvial system by vegetation. *Sedimentology*, 39, 779-795.

McCarthy, T.S., Ellery, W.N. and Gieske, A. 1994. Possible ground water pollution by sewage effluent at camps in the Okavango Delta: suggestions for its prevention. *Botswana Notes and Records,* 26, 129-138.

Mendelsohn, J.M., VanderPost, C., Ramberg, L. Murray-Hudson, M., Wolski, P. and Mosepele, K. 2010. Okavango Delta: Floods of Life. Gland: IUCN.

Milzow, C., Kgotlhang, L. Bauer-Gottwein, P., Meier, P. and Kinzelbach, W. 2009. Regional review: the hydrology of the Okavango Delta, Botswana - processes, data and modelling. *Hydrogeology Journal*, 17(6), 1297-1328, Available at: doi:10.1007/s10040-009-0436-0.

Mladenov, N., Gardner, J.R., Flores, N.E., Mbaiwa, J.E., Mmopelwa, G. and Strzepek, K.M. 2007. The value of wildlife-viewing tourism as an incentive for conservation of biodiversity in the Okavango Delta, Botswana, *Development Southern Africa*, 24(3), 409-423.

Mmopelwa, G. and Blignaut, J.N. 2006. The Okavango Delta: the value of tourism. *South African Journal of Economic and Management Sciences,* 9(1), 113-127.

Mosepele, K., Moyle, P.B., Merron, G.S., Purkey, D. and Mosepele, B. 2009. Fish, floods and ecosystem engineers; interactions and conservation in the Okavango Delta, Botswana. *Bioscience*, 59, 53-64.

Mosojane, S. 2004. Human-elephant conflict along the Okavango Panhandle in northern Botswana. MSc Dissertation, Conservation Ecology and Planning, Department of Zoology and Entomology, University of Pretoria.

Mukumbira, R. 2003. Power plans face wall of objections: Namibia's plans to generate electricity using the waters of the Okavango river have led to a massive row with environmentalists and tour operators in Botswana. *African Business*, 1 April, 58.

Murray-Hudson, M. 2009. Floodplain Vegetation Responses to Flood Regime in the Seasonal Okavango Delta, Botswana. MSc Dissertation. University of Florida, Gainesville.

Murray-Hudson, M., Wolski, P. and Ringrose, S. 2006. Scenarios of the impact of local and upstream changes in climate and water use on hydro-ecology in the Okavango Delta, Botswana. *Journal of Hydrology*, 331, 73-84.

Naidu, K.C., Muzila, L.,Tyolo, I., and Katorah, G. 2000. Biological control of *Salvinia molesta* in some areas of Moremi Game Reserve, Botswana. *African Journal of Aquatic Science*, 25, 152-155

National Conservation Strategy Agency. 2008. *Okavango Delta Management Plan.* Gaborone: Ministry of Environment, Wildlife and Tourism, Government of Botswana.

NRP. 2000. Management Plan for the Okavango River Panhandle: Progress Report 3. Maun: NRP.

OKACOM. 1994. The Permanent Okavango River Basin Water Commission (OKACOM) Agreement. Agreement between the Governments of the Republic of Angola, the Republic of Botswana, and the Republic of Namibia.

Pinheiro, I., Gabaake, G. and P.Heyns. 2003. Cooperation in the Okavango River Basin: the OKACOM Perspective, in *Transboundary Rivers, Sovereignty and Development: Hydropolitical Drivers in the Okavango River Basin,* edited by Turton, A., Ashton, P. and T.E. Cloete. Pretoria and Geneva: AWIRU and Green Cross International, 105-118.

Ramberg, L. 1993. African communities in conservation: a humanistic perspective. *African Journal of Zoology*, 107, 5-18.

Ramberg, L. 2004. An overview of environmental effects caused by deltamethrine spraying of the Okavango Delta 2001,2002 and recovery monitoring 2003, in *Environmental recovery monitoring of tsetse fly spraying impacts in the Okavango Delta*-2003, edited by Perkins, J. and L. Ramberg. Okavango Report Series 3, 1-30. Maun: University of Botswana.

Ramberg, L., Wolski, P. and Krah, M. 2006a. Water balance and infiltration in a seasonal floodplain in the Okavango Delta, Botswana. *Wetlands,* 26(3), 677-690.

Ramberg, L. Hancock, P., Lindholm, M., Meyer, T., Ringrose, S., Sliva, J., Van As, J., and Vanderpost, C. 2006b. *Species diversity in the Okavango Delta, Botswana.* Aquatic Sciences 68, 310-337.

Ramberg, L., and Wolski, P. 2008. Growing islands and sinking solutes. Processes maintaining the endhoreic Okavango Delta as a freshwater system. Plant Ecology 196: 215-231.

Ramberg, L., Lindholm, M., Bonyongo, C., Hessen, D.O., Heinl, M., Masamba, W., Murray-Hudson, M., VanderPost, C. and Wolski, P. 2011. Aquatic ecosystem responses to fire and flood size in the Okavango Delta: natural experiments on seasonal floodplains. *Wetland Ecology and Management*, in review.

Roodt, V. 1998. Ecological Impact of Tourism on the Xakanaxa Area of Moremi Game Reserve. Unpublished Report, Maun.

Sadoff, C.W. and Grey, D. 2005. Cooperation on International Rivers: A Continuum for Securing and Sharing Benefits. *Water International*, 30(4), 1-8.

Schlettwein, C.H.G. 1985. The biological control of Salvinia molesta. Department of Water Affairs, Namibia, *Research Report* W85/5.

Seely, M. et al. 2003. Ephemeral and Endoreic River Systems: Relevance and Management Challenges, in *Transboundary Rivers, Sovereignty and Development: Hydropolitical Drivers in the Okavango River Basin,* edited by Turton, A., Ashton, P. and T.E Cloete. Pretoria and Geneva: AWIRU and Green Cross International, 187-212.

Smith, P.A. 1993. Control of Floating Weeds in Botswana, in *Proceedings of Workshop "Control of Africa's Floating Water Weeds",* edited by A. Greathead and P.A. de Grout. CSC and IIBC London, UK and BuN, Harare, Zimbabwe, 31-39.

Swatuk, L.A. 2005. From "Project" To "Context": community based natural resource management in Botswana. *Global Environmental Politics,* 5(3), 95-124.

Thakadu, O.T. 2005. Success factors in community based natural resources management in northern Botswana: lessons from practice. *Natural Resources Forum*, 29, 199-212.

Tchamba, M.N. 1996. The problem elephants of Kaele: a challenge for elephant conservation in northern Cameroon. *Pachyderm*, 19, 26-32.

Thouless, C.R. 1994. Conflicts between humans and elephants on private lands in northern Kenya. *Oryx*, 28, 119-127.

Thouless, C.R. and Sakwa, J. 1995. Shocking elephants: fences and crop raiders in Laikipia District, Kenya. Biol. *Conserve*, 72, 97-107.

Turpie, J., Barnes, J., Arntzen, A., Nherera, B., Lange, G-M. and Buzwane, B. 2006. Economic value of the Okavango Delta, Botswana and implications for management Republic of Botswana. Report for the Republic of Botswana.

Twyman, C. 2000. Participatory conservation? Community-based natural resource management in Botswana, *The Geographical Journal*, 166 (4), 323-335.

Wolski P. 2009. Assessment of hydrological effects of climate change in the Okavango Basin. Unpublished report, OKACOM, Maun, Botswana, October.

Wolski, P., Gumbricht, T. and McCarthy, T.S. 2002. Assessing future change in the Okavango Delta: the use of a regression model of the maximum annual flood in a Monte Carlo simulation, in *Proceedings of the Conference on Environmental Monitoring of Tropical and Subtropical Wetlands*, 4-8 December 2002, Maun, Botswana.

Wolski, P., Murray-Hudson, M.A., Mosimanyana, E. 2010. Carbon and water fluxes in the Okavango Delta, Botswana, under present conditions and under those of changing climate. *Proceedings of the Annual Meeting of the Society of Wetland Scientists*, June 27 - July 2, 2010, Salt Lake City, Utah.

Wolski, P., Ramberg, L., Magole, L., and Mazvimawi, D. 2010. Evolution of River Basin Management in the Okavango System, Southern Africa, in *Handbook of Catchment Management*, edited by Ferrier, R.C. and Jenkins, A. Oxford: Blackwell, 457-475.

Chapter 10

Wetlands and the Water Environment in Europe in the First Decade of the Water Framework Directive: Are Expectations Being Matched by Delivery?

Tom Ball

Introduction

In this chapter, the experience of regulation and restoration of the water environment in Europe is analysed as it relates to the management of tropical wetlands. Guidance is then offered, based on this experience, on its application in other countries in less economically developed regions. The objective is not to offer a prescriptive 'toolkit' but rather to identify areas in which particular problems have emerged in Europe and possible ways of dealing with them in other areas where a similar approach to ecological betterment is pursued.

The EU region provides an informative case study, since it arguably offers the most complex legislation implemented anywhere in relation to the water environment. Some countries outside the EU with developing statute books of environmental law have taken aspects of the approach and developed regulatory frameworks based on the centrepiece of the legislation – notably the Water Framework Directive (WFD) and its various precursor and daughter directives – for protection of the water environment. This trend partly reflects the growing international acceptance of the need for integrated water resource management (GWP 2002). On its own, this trend would justify a careful analysis of the strengths and weaknesses of the EU regime. But at this stage especially, given the application of the method to other regions, a look to the future is needed in order to see where problems that have emerged in implementation may be remedied in applying such legislation and policy elsewhere.

The Water Framework Directive: Origins and Approach to Implementation

The origins of the Water Framework Directive can be traced to a desire in the pre-enlargement EU to bring the *ecosystem approach* into environmental policy more explicitly. This approach combines the natural and social sciences to address environmental problems (Aptiz et al. 2006). On the one hand, it proposed a more integrated ecological definition of 'water'. On the other, it introduces the notion of public participation for policy implementation (Steyart and Ollivier 2007). The enshrinement of the catchment as the unit for assessment, and an attempt to balance economic interests against the status of ecosystems and the interests of the public, were all novel aspects of the legislation in Europe.[1]

The WFD required Member States to assess and strive to improve rivers, lakes, groundwater bodies and transitional/ estuarine and coastal waters, aiming to achieve 'good ecological status' – as described in Annex V of the WFD – by the year 2015.[2] Complexities in various aspects, notably how to define good status, were noted even in initial transposition of the framework (Chave 2001). However, compliance with the planning aspects of the WFD progressed smoothly in most member states. Characterisation of the status of river basins and transitional waters, the first step, was completed by 2005. The characterisation first required a surface water body typology based on the definition '...*a discrete and significant element of surface water such as a lake, a reservoir, a stream, river or canal, part of a stream, river or canal, a transitional water or a stretch of coastal water* (Article 2.10). Waters were then assessed for *pressures* and *impacts*, and subdivided into smaller units, based on this typology, in River Basin Management Plans (RMBP). These, with their associated Programmes of Measures, were completed in 2009, applying to the first six year planning cycle to 2015. Of the 27 States, 15 had, as of June 2010, adopted the River Basin Management plans required by Article 9, with 110 basin districts designated. Eight states were lagging at that stage, and it was noted with concern by the European Commission that most of these were in the Mediterranean belt where water environment governance issues were most pressing. Of the pre-2004 member states, most have imported the Directive's formal quantitative objectives on top of existing domestic regulations on surface water and groundwater quality.

Wetlands were implicated in the WFD in two broad senses: wetlands as part of the hydro-ecological system, impacted by hydrological modifications such as drainage, and wetlands as agents to deliver improvements in the water environment.

Some Articles had direct application to wetland management:

- Article 1, which *inter alia* established a wetland protection framework
- Article 4a, introducing *ecological quality* as an objective in addition to

---

1   Although as Hering et al. (2010) note, the recognition of multiple-uses within the Directive had precedents in US Clean Water Acts of the 1970s.

2   At the time of writing, many member states have acknowledged that the status targets will not be met by a substantial number of water bodies by 2015, and are instead working to the extended timescale given in the Directive, to 2027. However, Hering et al. (2010) note that even this revised timetable may be overambitious.

water quality (this complex area of double targeting is analysed in more detail below)

- Article 5, requiring wetlands to be characterised as part of the characterisation of river basin status (also Article 13, which implicitly required a characterisation in River Basin Management Plans)
- Article 8, which required assessment of the waters moving in and out of wetlands according to flow rate, chemical and ecological potential.

One of the innovative aspects of the WFD was its integrated conceptualisation of the water environment. This led to 'more integrated ecological definitions' (Steyaert and Ollvier 2007) of wetlands in the planning process, since they could be implicitly recognised as connected, in an ecological sense, to the water bodies that were being assessed. Conversely, the lack of development of a clear wetland typology within the WFD left much to member states, under the subsidiarity principle, to decide how to incorporate them. The approach was necessarily pragmatic. Wetlands as part of the hydro-ecological system vary in their morphological and ecological characteristics markedly across the EU. Notable in general are the large floodplain wetlands, riparian zones that have been heavily impacted by drainage and groundwater fed wetlands – fens – and coastal marshes. Due to their lack of dependence on a water body, the large number of rainwater fed wetlands of low trophic status were not implicated, yet many are also under high environmental pressure; conservation and restoration of these sites must fall back on other legislation, since they are not directly associated with surface water or groundwater bodies.

Although no explicit objectives were set in the WFD for wetland (as opposed to water body) status improvement, there are several implicit goals that reflect those for the water environment on which the wetlands depend. What is unclear from the Directive text alone is the extent to which wetlands themselves need to be conserved due to their interaction with surface water or groundwater bodies. The complexities of surface water-groundwater interactions were also not dealt with explicitly. In most assessment methods, groundwater classification has an intimate connection with the quality of water stored in wetlands. Indeed, the health of groundwater fed wetlands is frequently one of the indications used for general groundwater status (UKTAG 2003).

Much has, therefore, been left to national technical assessment methods to devise a workable methodology that is tied to policies compatible with conservation legislation. In establishing these, difficulties were encountered in both baseline assessment and the identification of improvement measures. Ecological objectives are hard to express, compared with, for example, 'reduce level of nitrates from x to y' – is this to be done by at-source prevention, or via a riparian 'wet zone' buffer strip with all its attendant ecological benefits, that may take the form of a constructed wetland? How should a catchment management body, with limited funding, choose between two alternative measures? Much may then depend on the provision of ecosystem services by the wetland and the ways these are valued. All these questions have been left up to implementing states to decide.[3] Maltby (2009) noted that there was limited evidence that European member states were developing formal worked strategies for wetlands as part of Directive implementation. This was the case in spite of much exhortation by the EU and other sources that clearly encourages wetland creation and restoration in order to further the Directive's goals (European Union 2003).

Most wetland conservation and restoration objectives that have been incorporated to RMBPs fell within the qualitative category, requiring tools to be developed to assess the ecological status of water bodies on which the wetlands depend. Interpretation has therefore been varied among member states as to how far the Directive mandates a *betterment* objective for surface waters and groundwaters (as compared with merely *conservation* objectives of preceding EU Directives such as the Habitats Directive) and indeed to which wetlands the objective should apply.

General issues relating to these are considered below, but in a very broad sense the framework approach arguably fails to develop a 'new paradigm' of wetland science as advocated by Maltby (2009). This on its own is not necessarily a drawback if, on top of the bare text, the practical implementation of the Directive is capable of implementing the approach in a manner that preserves and enhances ecosystems closely linked to the surface water environment.

Specific Technical Difficulties in WFD Implementation

Implementation of the WFD in general has been very challenging in technical terms, While the basic planning and district designation procedure proved generally straightforward, what has proven more problematic is delivery of the qualitative objectives that go to make up the concept of 'good ecological status'. While technical planning and assessment mechanisms and methodologies were largely left to member states, a Common Implementation Strategy (CIS) was also introduced which helped to secure consistency and integration in implementation across member states (European Commission 2001, Dworak et al. 2009). Stakeholder groups, including scientists, regulators and other governmental and non-governmental officials, came together in discussions on the CIS resulting in convergence of methodologies and techniques.

Technical guidance on implementation was produced early on in most member states, although the subsidiarity principle meant that the individual strategies took precedence over common guidelines. The CIS proved a success against its stated

---

3   There is already a wealth of literature on the regulation of the water environment in Europe and also its impact on wetlands. Given the scope of this chapter, which is to look at the potential for application of the European approach to wetland management in other areas, the reader is referred to these works for a more detailed synthesis.

aims of developing a 'common understanding and approach' for some sectors[4] with 24 common guidance documents produced, including the Wetlands Horizontal Guidance (European Union 2003). However, local specifics, even within member states, remained too marked for many common strategies to be agreed across all water body types and all states. The core strategy guidance document was silent on the extent to which wetlands were to be regarded as part of the surface water environment. A variety of approaches therefore developed. Most 'Article 5' typologies embraced wetlands only implicitly, as a constituent of a water body in a certain zone (e.g. a salt marsh in the estuarine transitional zone, or a groundwater-fed fen). Another difficulty emerged early on in the WFD implementation process, as member states were first delineating river basin districts. While the catchment was chosen as the unit of assessment, a key question was how catchment areas were to be chosen in areas that were similar enough from a water environment perspective that they could be regarded as *functionally* similar. This is a problem ultimately of classification and lack of good data on which to base it.

The interaction with conservation legislation as it relates to wetlands is complex. The existing EC 1979 Birds and 1992 Habitats Directives already provided, to quote the European Commission, the 'cornerstone of Europe's nature conservation policy'.[5] The Directives required management regimes to be set up for a series of sites in the 'Natura 2000' network. Many of those managed under formal compliance with both Directives are wetlands. The Birds Directive recognised the need to protect areas of importance for migratory species, especially wetlands, and of the 200 habitat types and 700 species named in the Habitats Directive, a large number were wetland types (European Commission 2002). Important measures set up under Natura 2000 by the various member state environmental agencies included payments for sound agricultural management in the vicinity of Natura 2000 sites. Such measures are also being pursued as a means of delivering the measures required under WFD River Basin Management Plans. In both cases no clear financial framework was established at an EU level. Compliance was, therefore, primarily aspirational and based on plan documents.

Closely linked is the general aspiration in the WFD for the water user to pay for water services, essentially a valuation of water uses that may impact on the water environment that allows for cost recovery of mitigation and restoration measures. The aspiration when the WFD was initially drawn up was to enshrine the principle that the water users should help meet costs by making an 'adequate contribution', but left it up to member states to decide what these costs and contributions should be. Difficulties have emerged from the lack of prescription of what constitutes '*adequate contributions*' by such users. '*...having regard to the social, environmental and economic effects of the recovery, as well as geographical and climatic conditions of the region*' (Article 9). The measure

---

4   See page 2 of the Common Implementation Strategy Document (European Commission 2001).

5   http://ec.europa.eu/environment/nature/legislation/habitatsdirective/index_en.htm

allows flexibility to member states to determine what constitutes 'adequate' and thus to avoid imposing disproportionate costs, but the danger is that an absence of clear valuation methods for environmental impact would lead to resistance among water users, who are often the very same individuals and bodies who are relied on to implement environmental improvements to secure good ecological status,

Borja et al. (2005) noted several further difficulties in relation to the protection of the water environment by the Directive not long after the implementing legislation for the WFD come into force. Many of these have been subsequently borne out by experience (Hering et al. 2010). Only a brief summary is offered here; a review of 170 River Basin Management Plans is currently under way by the Commission, to assess performance against the objectives (Falkenberg 2010).

Quantification, Including Establishing Appropriate Monitoring Networks

A pervasive problem underlying the delivery of programmes of measures in the Directive is scientific: it has lain in the difficulty of developing a workable system of measuring ecosystem status in terms of quality. A baseline has to be established first, quantifying the degree of degradation where present and the nature of that degradation (in terms of pressures and impacts) then a means must be found of quantifying the improvement that can be obtained from a series of steps. None of these has proven easy for the water environment, let along the wetlands linked to it in complex ways. The WFD distinguishes three types of monitoring: surveillance (for long term changes), operational, to establish the risk of a water body failing to meet its quality objectives, and investigative, which as its name suggests seeks to establish the underlying cause of any failure to meet objectives (Borja et al. 2008).

Existing monitoring networks in European countries, while extensive in some, tend to be oriented to the surveillance objective and focussed on water quality and hydrometry. It was not surprising, therefore, that initial characterisations of ecological pressures and impacts started from the catchments in which there was the best data. They focussed on deriving the water quality- ecology links, using data from monitoring schemes such as the EU Framework 6 programme project REBECCA ('Relationships Between Ecological and Chemical Status of Surface Waters'). Particular success has been noted for quantifying trends in acidification, excess metal loading and organic loading and changes in riparian and catchment land uses (Friberg 2010). However, while many catchments started to emerge from these quantification exercises as having water bodies with unambiguously poor status, baseline data were frequently inadequate to draw firm judgments on exactly how far the water environment had been impacted. For wetlands, the key remaining questions remained the links between land use change and catchment level water quality and between river habitat modifications and overall ecological status.

A further issue remains a lack of data on groundwater, a particular problem for wetlands outside the riparian zone. Groundwater tended to be monitored most actively in areas where it is used for water supply, with an emphasis on chemical water quality. Biological quality indicators tended to focus on the taxonomic rather than the functional, yet much of the scientific basis for linking indicator species to specific environmental quality categories remains in its infancy (Hering et al. 2010). Substantial differences have developed between member states in both choice of biological quality indicators and overall method.

The 'One out, All out' Problem

A severe problem of the assessment methods prescribed in the WFD is the possibility of false classifications of water body status. The goal of 'good ecological status' includes the ecosystems that have an ecology dependant on these waters. Such status is multifaceted, being based on physiochemical, and ecological characteristics. The latter, in turn, comprise biological and hydromorphological quality characteristics, which are frequently, but not always, interrelated. It was a novel aspect of the Directive that it attempted to deliver on both, by use of the Driver, Pressure State, Impact, Response approach (DPSIR), cf. EUWI (2002) and Borja et al. (2005). Analysis of pressures and impacts needed also to be forward thinking, considering how these pressures would be likely to develop, up to 2015, in ways that would place surface water bodies at risk of failing to achieve good ecological status if appropriate programmes of measures were *not* designed and implemented.

Targets for physiochemical status, expressed quantitatively, could be approached relatively easily in member states where robust regulatory and permitting regimes already existed. The links between achievement of such targets and measureable improvements have been drawn in a quantitative sense, at various scales from local to catchment (Freiberg 2010). But both biology and hydromorphology, partly because they are so intimately linked, have proven difficult to define on the simplified scale that the Directive used (from high, through good, moderate, poor, to bad), yet it has a vital bearing on wetland health where the wetland depends on the surface water, groundwater or transitional water in question. Since the status of surface water bodies is defined in the directive as the poorer of ecological and chemical status, even chemically perfect or near perfect water bodies can fail on hydro-ecological grounds.

It is axiomatic that hydromorphology is intimately tied to biological quality for dependant wetlands, as it is for surface waters. Development of workable biological quality indices therefore depends closely on an understanding of the link between such indices and general ecological health. If they were deemed to have poor status in just one category, the site could have ecological status below target level in the assessment and therefore merit intervention where this may not fairly be justified.

The appropriate spatial scale for classification of a habitat quality assessment also needed to be established, and benchmarking of quality indicators agreed. Once this had been carried out (usually by some form of survey, e.g. a river habitat survey; Raven et al. 2010), the indicators could be assessed on these spatial sub-units and a decision made on how the assessments of these sub-units impacted on overall water body quality. Hering et al. (2010) argue that errors in classification could result in a form of 'type 1 error', that is falsely classifying an ecosystem in a lower status category than fairly merited. The 'type 1 errors' could result in a larger number of ecosystems failing status targets than would otherwise have occurred.

A further interesting difficulty relating to wetlands can be phrased thus – 'even if the linked water body is defined as good status, is this level sufficient to conserve all wetland ecosystem services?' This uncertainty has led to a perceived need to focus on high status sites and the links with biodiversity (Hering et al. 2010). The 'one out, all out' structure of the Directive tends toward this type of error rather than a false good or high status ratings, something that was designed into the legislation for both safety and, to a certain extent, simplicity. The extent to which this type of error has occurred in practice is difficult to gauge, even for surface water bodies. Hering et al. (2010) made a timely call for more consideration of risk and uncertainty in practice, including the significant risk of such errors, a call that implementing organisations would do well to heed.

How to Value Wetland Ecosystem Services and Gain Agreement on Programmes and Payment Systems

The need to devise a workable assessment of the value of ecosystem services has informed much academic debate, and led to broad scale valuations (Costanza et al. 1997). However, devising an operational method to value wetland ecosystem services remains elusive.

Multiple environmental benefits are provided by wetlands (Maltby 2009). Particularly for diffuse pollution, which has proven a stubborn remediation problem under the WFD (Wakeham 2006), wetlands offer an economically efficient and effective mitigation measure, and if managed correctly an abundant supply of ancillary ecosystem services (Trepel 2010, Mannino et al. 2008, Turner et al. 2008). However, restoration or creation of wetlands at a large enough scale requires an approach that incentivises the use of wetlands by land managers alongside a suite of other measures that promote ecosystem health. There are barriers to this process. It is difficult to regulate for change and policies imposed in a 'top-down' fashion may encounter resistance (Blackstock et al. 2007). The ecosystem is often starting from a low base in terms of environmental quality, particularly where historical drainage has been widespread. Consequently, there may be a strong emphasis on the functionality (as opposed to the purely ecological aspects) that can, at least temporarily, override considerations of habitat provision. It may be that restoration has to be phased if more close attention is to be given to the functional aspects and their impact on the water environment. In the DSPIR framework, this essentially means drawing a feedback loop so that knowledge of the ecosystem response feeds back to correct modification of the driving processes in the water environment.

Alongside these problems is that of deciding how much public funding should be spent on the restoration of wetlands, even if these are restricted in definition to those linked to water bodies. Restoration does not come cheaply; centuries of neglect, drainage and invasive species introduction leads to questions about the optimal way of assessing both a baseline from which to work and a means of assessing success by a robust measurement of 'good (hydro)ecological status'.

As a functional zone in which wetlands and surface water bodies interact, the ecotones provided by floodplains, lake margins and intertidal zone are ones in which programmes of measures for restoration could be particularly targeted for both water quality and biochemical/ hydro-ecological benefit. At present, most of these goals are necessarily aspirational, partly because the benefits of ecosystem services provided by wetlands in these zones are not fully quantified. Approaches based on decision support, have been developed to integrate wetlands into river basin management, such as EVALUWET (Janssen et al. 2005). For restoration initiatives, there are some high profile examples that have been subject to research scrutiny. Trepel (2010) notes that a large-scale rewetting programme of peatlands in Schleswig-Holstein, Northern Germany, could mitigate diffuse N pollution more cost effectively than measures in a wastewater action plan. Lack of effective valuation system for these pollution-reducing ecosystem services is, however, acting as a constraint to delivery of diffuse pollution targets. The success stories have proven a scientific principle that could be effectively scaled up, based on accurate costing of such measures. This is considered in more detail below in relation to large tropical wetlands such as those of the Pantanal region (see section 'Application of a WFD approach beyond Europe').

Realising Synergies with other Policies and Involving Landowners and Stakeholders in the Planning Process

The requirement to set up a planning process that engaged the public and other stakeholders was present in the WFD. The importance of linking 'top down' to 'bottom up' approaches was noted early on in its life (WWF 2001). In many member states this has taken the guise of fora that bring together landowners, business sector representatives, public bodies, voluntary environmental groups and recreational groups. In some member states these have been praised, while in others they been criticised for some if not all of their roles (Blackstock et al. 2007).

It has been noted by the EU Directorate on the Environment that there is a need for more data on the extent to which the WFD has realised benefits through synergies with other policies (Falkenberg 2010). Arguably the most important area as it relates to wetlands is the synergy with agricultural policy. Currently under review, The EU system of financial support for agriculture, the Common Agricultural Policy (CAP) was long ago decoupled from production goals. The current structure separates general farm payments (Pillar I) from specific agricultural schemes and targeted rural development programmes directed at environmental betterment (Pillar II).

The difficulty of implementing a programme of measures to deal with such aspects of the water environment as diffuse pollution makes a strong case for Common Agricultural Policy reform led to a great extent by the needs of the WFD. Evidence from the first round of RBMPs suggests that, without such reform, it will prove very difficult to uprate the status of water bodies affected by diffuse pollution even by the time of the extended compliance deadline for good status, 2027. Quoting one Government's strategy for implementation of measures to reduce diffuse pollution, a key question is '*will implementation options that have been or are about to be put in place and/or potential land-use change between now and 2015 be sufficient to address diffuse pollution, or will additional options be required?*' (Scottish Government 2009).

Key to addressing this question is the use of robust implementation tools, such as GIS grid-based approaches that can identify and compare the extent and sources of diffuse pollutants under the current baseline, and to look at the likelihood of real progress in this arena by 2015. Such tools have emerged in a number of WFD implementing member states. The use of Pillar II measures, such as Agri-Environment schemes, in specific member states is advocated in the context of both assessments of natural mitigation potential and for their establishment to mitigate diffuse pollution. A common problem is that funding mechanisms do not prioritise effectively the measures, including wetlands that may provide many additional services, e.g. for biodiversity and conservation. Compensation for the provision of these services at the farm level can only be provided at the level of income forgone. This has tended to limit their uptake in farms that do not generate substantial profits. Yet these are often, by their nature, low intensity farms that have high potential for such ecosystems to be established.

One particular example that pointed to the potential for use of wetlands, the wise use of floodplains project (WUF) was an analysis of examples and pointers to good practice in using floodplains (and coastal transitional water such as salt marsh) in sustainable management of river basins. Based on case studies in the UK, Ireland, and France, a key driver of this project was the desire to appraise how far the policy-based implementation of the WFD needs to adapt in order to further restoration objectives more effectively. In setting objectives for wetland conservation and restoration, a need was identified in this project to agree aims with stakeholders from the outset, and at this stage focus on the appropriate scale, which combined the local level (a time consuming and onerous task), and the catchment level. The resulting policy analysis called, pointedly, for complementary funding and more co-ordinated approaches. Specific barriers to change were identified in relation to floodplains that linked closely with the stakeholder liaison aspects of the WFD. At a highest level, policy barriers were identified which ranged from reform of the farm payments system under the Common Agricultural Policy (CAP) to national strategies that set objectives in water environment management plans. Agri-environment schemes (under 'Pillar II' of CAP) are the only land management schemes that currently further this purpose. No funding system was established under the WFD to support restoration projects (although reference is made in it to cost recovery, cost effectiveness and the polluter pays principle).

Therefore, such schemes tended to remain oriented to conservation objectives for a wide range of habitats and ecosystems, not specifically focussed on wetlands. Scope was identified for targeted reform, setting up simplified objective–led approaches for riparian 'buffer strips', and marshland and washland payment systems, as well as introducing wider education programmes for landowners. Lack of agreement on what constituted problems was a key obstacle. The hardest case in point was gaining agreement with landowners that floodplain drainage was a problem, since so many farmers were used to thinking in terms of maximising productivity and had grown used to the status quo, never perhaps having seen the land in its pre-drained state (Oates 2008). Yet in some case studies agreements had been reached, often through the constructive intervention of intermediary bodies. Integration of catchment land use planning with water environment planning was held to be essential in the future. The need for a neutral planning organisation to play this intermediary role was pointed out, and ideally this should take a strategic role as an integrated catchment management forum made up of community and other organisations' representatives with an interest in the management of the water environment and its resources.

The UK and Scotland: A Case Study in Devolved Implementation

The transposition of the WFD in the UK is, in its scientific sense, typical of all implementing countries: a study in the difficulties of assessing hydromorphology as an intrinsic aspect of water environment, and wetland health. In a socio-political context, it reveals regional contrasts in approach owing to the devolution of government relating to environmental law and policy, with separate administration in England and Wales, Scotland and Northern Ireland. These regional contrasts shed light on the flexibility of the mechanism to cope with more local hydro-ecological and socio-political factors.

Due to the moderate to high rainfall, land drainage has been a vital aspect of ensuring agricultural productivity in the UK, as with much of northern Europe. In the uplands, pastoral productivity boomed following the application of improved drainage techniques in the 19th century with the use of clay tile, and further increased from the 1960s onwards with the advent of plastic pipe drains. In the lowlands of the South and East, in common with many European nations, the enhanced drainage was associated with increased fertiliser and pesticide input. The water environment has been affected in terms of both quality and quantity. In some cases, peak river flows have increased, although there has been an increase in precipitation as well, notably in winter and this is projected to continue through the 21st century (Werritty et al. 2002, UKCP 2009). The water chemical environment has been detrimentally impacted, mainly in the arable zones of the South and East. The result is 76 per cent of the water bodies in England and Wales at risk of failing good status targets by 2015.

Considering the management of wetlands, disconnection between river and floodplain, river channel works, and drainage of the riparian zone are reasons for failure of status in the UK that have a direct bearing on wetland health. However, full restoration of floodplains to their pre-modification condition is neither economically viable nor, in an engineering sense, cost effective. The proposed measures in River Basin Management plans over the 2009-2015 cycle sheds light on the compromises that are being sought in the UK.

The situation for groundwater dependent wetlands in the UK does not match this mixed picture for rivers. The conditions of wetlands dependent on groundwater were classified as universally 'good' in Scotland, although several areas with extensive wetlands in England and Wales have areas that are at risk, particularly from groundwater with high nitrate content. Yet this assessment depends on the extent and quality of monitoring of surface water/ groundwater interactions in wetlands, a science that is still hampered by the limited monitoring taking place on the interface between surface water and groundwater in such ecosystems.

Public participation in the UK has progressed by a variety of methods. In England and Wales, it is secured by the Catchment Management Planning process that seeks to integrate land use, pollution control and flood risk management. In Scotland, Area Advisory Groups in 8 river basin sub-districts perform a similar role, with a greater emphasis on influencing landowners to undertake measures rather than exercising control over the water bodies.

In Scotland, the first round of river basin management plans for the two main districts (Scotland, Solway-Tweed) identified 43 per cent of water bodies 'adversely affected by human activities' (including heavily modified water bodies). Of these bodies, it was notable that the status failures of water bodies in the wetter North and West were generally on hydromorphological grounds, whereas those in the South and East failed in general more due to water quality. Alteration of water flows and levels of rivers and lakes, modification of bed, banks and shores and barriers to river continuity for fish migration collectively accounted for 48% of failures to reach good status by surface waters in the Scotland district (SEPA 2009). The impact of hydroelectric power is major in some catchments, sufficient to force several rivers into the 'heavily modified' category (SEPA 2009: 15). For rivers, channel realignment, particularly straightening, and removal or degradation of bank vegetation, were the main adverse impacts, accounting for 577 water bodies at less than good status, with realignment causing most impact.

It is noteworthy from the Scottish RMBPs that water bodies with few water quality issues are classified as 'poor' or in some cases 'bad' status. The question arises, based on the 'one out all out' problem noted above, how far were these failures really a reflection of the overall quality of the water body, and how far are they due to the assessment method that causes failures on the basis of poor hydro-ecological status? The question is relevant since it is important to appraise whether those water bodies that fail for these reasons, and their dependant wetlands, fail justifiably and in a manner that prioritises restoration of the water body/ wetland system in a holistic manner, not piecemeal efforts on small sections of water body. The question is also relevant to judge the effectiveness of the regulatory structure that has been implemented in order to prevent deterioration of status for those bodies at good status or better. Dangers of such regulatory structures arise if they are not perceived as closely related to the environmental objective by those affected. Also, if that objective is not clear itself, there is a risk that those affected by the regulatory structure will not respect its overall goals, leading to a breakdown in implementation.

In Scotland, the principal regulatory tool for implementation has been the Controlled Activities (Scotland) Regulations 2005. This imposed a charging scheme linked to a licensing regime for abstractions and activities, in turn coupled to an assessment mechanism based on river hydro-ecological sensitivity of the watercourse subunit. The same assessment method is used to assess the potential of a water body for ecological improvement. The regulatory structure, with its focus on regulation with local effect, runs the risk of creating a 'sticking plaster', as opposed to strategic, approach to restoration, prioritising areas that fail status over those that, in spite of being in good status, have similar issues (e.g. drainage) in formerly wetland areas.

High environmental value wetlands have been encouraged by general environmental stewardship policies that frequently have as one of their goals improvement of the water environment. These are now given further backing by the programmes of measures in the RBMPs. Yet the lack of an explicit link between programmes of measures and funding streams has inhibited a co-ordinated approach. Landowners are often not made aware of the possible funding streams, do not have the time or knowledge to complete applications, or are disinclined to do so since their only compensation is frequently income forgone. It falls to nongovernmental co-ordinating bodies to arrange a holistic planning approach and to play the role of go-between. Where present, these are, in every case, non-statutory (as a successful example, the Tweed Forum has played such a role in the Scottish Borders for 20 years). The lead planning organisation for the RBMP has, by contrast, little executive control over the funding streams that could be used to spur landowners into action for wetland provision.

The most pervasive proposed measure in the UK's RMBPs is reduction of point source and diffuse pollution. Point source efforts have proven reasonably successful, continuing a trajectory that began before the WFD was implemented in the UK. In Scotland, the process has arguably been smoothed by the continuing state influence over Scottish Water, contrasting with the privatised systems in England and Wales. It is backed by legislation to control pollution and over-abstraction. A significant secondary objective is to '*restore the banks, beds and shores of surface waters to a more natural condition*' (SEPA 2009: 4). The proposed measures for dealing with degraded areas partly rely on encouragement and public and landowner agreement, backed by regulations that restrict abstraction, impoundment and other channel works. The main emphases are awareness-raising, including training and guidance for land managers. Yet the installation of '...*buffer zones, including woodland planting and wetlands; capture polluted runoff from steadings (e.g. in constructed farm wetlands)*' also directly encourages land use change involving wetland creation, an example of the second role of wetlands in the WFD referred to above. Constructed wetlands are also named as remedial measures for environmental damage due to mining and quarrying activity, although there is only limited appraisal of the link between wetland form and function. There is hope that the construction of wetlands will be stimulated by changes in emphasis in the system for agricultural support, particularly under Pillar II of the Common Agricultural Policy.

A recent (2010) major thrust in Scottish policy, coming on the back of the 2009 RMBPs, is the funding of restoration projects to address the impacts of the past activities identified above, principally channel straightening, loss of bed and bank vegetation, the increase in invasive species, and the removal of barriers to fish migration. In terms of planned improvements to bank and shore vegetation, the measures are partly legislative but also involve funding of a limited number of restoration projects (for Scotland, the targets are 6 water bodies by the 2015 round rising rapidly to 40 by the 2021 round).

Funding implications of these initiatives are highly uncertain. A problem that may emerge in securing beneficial changes is that metrics used to assess degradation do not always translate well to judging the benefits of future restoration. The mechanics of the process involve the use of assessment tools, an example being the MIMAS tool developed in the UK (UKTAG 2003) for application to surface water bodies. Degradation is assessed as ecosystem capacity absorption, expressed generally as a percentage absorption figure. This figure, in turn, depends for its accuracy on assessment of the baseline state of the ecosystem, which must frequently be inferred. However, the same metrics are also used to assess the potential response of the water environment and dependant wetlands to channel impacts such as impoundments, for the purpose of the CAR regulatory regime described above. Because the regime is designed to prevent back-sliding of impact (from good to moderate, moderate to poor status and so on), the focus may not prove sufficiently holistic enough to act as a comprehensive benefit assessment method that can cover both prevention of harm and positive restoration initiatives. Weight to this argument as it relates to riparian wetlands is given by the fact that the assessment stops at the riverbank, not considering in detail the wetlands that may depend on periodic flooding. However, since embanked rivers do have a certain amount of ecological capacity absorbed, there is an incentive provided to breach embankments and allow periodic floodplain inundation that might lead to wetland re-establishment. What are the relative benefits of such breaching versus bank and floodplain vegetation? How far should straightened channels be re-aligned? These are complex questions for which no quick answer is available. The response must depend on local circumstances, but the lack of a uniform prioritisation approach causes concerns for land managers.

In view of these problems, the RSPB, a wildlife conservation organisation in the UK, concluded that the following specific barriers remained relating to implementation (RSPB 2010):

- Despite being strategic plans, it was argued that the UK RBMPs failed to quantify the main pressures. In turn, it may be inferred that this led to problems in identifying impacts.

- Failure to identify actions and remedial measures. With little insight into pressures affecting water bodies, this led to plans that contain few new actions.
- Failure to engage with key sectors blamed for diffuse pollution. Farms, urban areas and roads were particularly to blame, causing widespread damage to freshwater ecosystems and costing water customers hundreds of millions of pounds. Yet remedial measures for this sector were criticised as being limited in the first planning round.

Application of a 'WFD Approach' beyond Europe

A number of countries outside the EU have imported aspects of the EU approach to conservation of the water environment. Is this approach likely to be of benefit? It is worth recalling that legal instruments such as the Ramsar Convention, though widely lauded now, were, as Maltby (2009) noted, initially perceived by poorer nations as indulgences by the richer nations that could afford them. The key to changing this mindset was the enshrinement of the 'wise use' concept at the core of the conservation method. In turn, this was connected with the addition of further, linked benefits of wetlands as suppliers of ecosystem services such as flood prevention. Will the utility of the framework become clear once implementation commences? Can 'wise use' be compatible with a framework approach to the water environment that, as we have seen, elevated flexibility above all other criteria?

Taking tropical wetland management in general and the Pantanal in particular, we can identify particular aspects of the WFD approach that have potential benefit and those where a different focus might be appropriate. With three bordering countries, there are strengths, at least in theory in a supranational legislative framework that carries with it enforcement potential. Legislation that prescribes steps rather than limiting values has the advantage of high adaptability to the situation in a particular jurisdiction. We have seen that the WFD supplies both, but specific limiting values may not be easy to negotiate and enforce where monitoring has not been in progress for long periods.

In the Pantanal, one superficial similarity with the situation in Europe is that intensive agriculture is the principal agent oversupplying nutrients (notably from grazing) and sediments by soil erosion. The WFD approach, in its emphasis on anticipatory planning and water body target setting, has some advantages as a framework, but the continuing difficulties of delivering on targets for diffuse pollution in the EU show the extent of the challenge. As a first stage, a co-ordinated effort to assess pressures and impacts, along the lines of the initial assessments completed for the WFD in the mid-2000s, would help to identify what measures would be appropriate. But such work is non-trivial in that requires careful assessment of the response of ecological indicator species to the hydrological and nutrient pulses. A systematic categorisation of the extent of modification of water bodies is necessary in order to identify areas where there are constraints to action governed by, for example, hydroelectric power dams, which are numerous in the Pantanal region. Given the extent of modification of the lower order tributaries of the rivers feeding the Pantanal, an accurate quantification of the baseline condition of ecological status of these rivers and the water bodies they feed may never be achievable. Nonetheless, a programme of monitoring the effects of discharge variations on the rivers, coupled to conceptual modelling of pollutant loads, ecological impacts of these loads, is an achievable goal.

A consequence of broad based legislation can be the lack of specificity; while the flexibility of the WFD legislation permits a local focus, it does not necessarily require it. In relation particularly to application in other areas, particularly tropical and other highly biodiverse ecoregions, an area in which this poses potential difficulties is in defining zones that have specific management needs, where fine scale ecological variation can demand a specific focus that integrates conservation and water environment legislation at an ecosystem level. This is particularly true in the Pantanal. Already in the more biodiverse ecoregions of Europe, the importance of developing a locally specific approach alongside larger scale approaches has been noted (see, e.g., Hughes 2005, in relation to Macronesian archipelagos of S. Europe). This arguably applies to an ever greater extent the more diverse the region is, with grounds for local assessment methods and programmes of measures rather than the more blanket regulatory approach. Thus, in large tropical wetlands such as in the Pantanal, as with these highly biodiverse regions of the EU, we can envisage a finer grained approach to management as having great potential alongside a broad based target-setting approach. The approach should take account of local socioeconomic criteria as well as environmental variation.

The above examples of success in implementing the WFD in a holistic way show that taking account of local characteristics is helped by a management body that ideally is not representative of a state regulator. This may be an agency with the knowledge of the land and water use characteristics of its area that can approach landowners, water users and statutory agencies in a neutral capacity. Such agencies are not prescribed by the WFD but experience seems to bear out that they are crucial to delivery of the ecosystem approach.

Cost recovery is an issue that pervades implementation of water environment legislation. In developing countries, such issues become ever more pressing since water is a good whose use has traditionally be neither regulated nor costed. Securing the right balance between state subsidy for benefits that are public goods and a private contribution that is both manageable and reflects the real ecological impact of the activity, is a significant obstacle to implementation of the approach in the less economically developed nations.

A further difficulty is securing the monitoring necessary to judge success on the ecological front. North-West Europe has generally a high density of hydrometric and water quality monitoring networks. Elsewhere, these rarely reach the same level of density seen in Europe. However, this is not a difficulty if the focus is right, and it has been seen how the emphasis on physiochemical and hydrometric monitoring in the WFD can to an extent distract away from the question of defining ecological status for the purpose of measuring it. Such definitions, though partly hydrological in the case of wetlands, demand a more nuanced approach that does not necessarily require a high density of centrally managed data, but takes advantage more of local knowledge through engagement with communities that depend on, and in some cases manage, the wetland environment.

Conclusions

There is little room for doubt that the WFD, taken together with existing conservation legislation, and IWRM, has begun to shift the balance in Europe away from narrowly economic criteria in relation to water, and toward those of the environment and public good associated with it. Whatever the objection to the shortfalls of the process identified here, the river basin planning procedures have certainly helped to reveal the true state of the water environment. This process has allowed prioritisation of areas that have experienced most impact, which ultimately should prove to the benefit of wetlands dependant on surface water and groundwater. A key remaining challenge is in delivering on objectives that require proper valuation of ecosystem services, an approach that is still in its infancy. Application of the approach to wetland tropical wetland ecosystems should learn lessons from these challenges and how they have been met.

Acronyms

CAP - Common Agricultural Policy
CIS - Common Implementation Strategy
DPSIR - Driver, Pressure State, Impact, Response approach
EUWI - European Union Water Initiative
GWP - Global Water Partnership
IWRM - Integrated Water Resource Management
RMBP - River Basin Management Plan
RSPB - Royal Society for the Protection of Birds
SEPA - Scottish Environment Protection Agency
UKCP - UK Climate Projections
UKTAG - United Kingdom Technical Advisory Group on the Water Framework
    Directive
WFD - Water Framework Directive (code 2000/60/EC)
WWF - World Wide Fund for Nature

References

Aptiz, S.E., Elliot, M., Fountain, M. and Galloway, T.S. 2006. European environmental management: moving to an ecosystem approach. *Integrated Environment Assessment Management*, 2(1), 80-85.

Blackstock, K.L., Kelly, G.J. and Horsey, B.L. 2007. Developing and applying a framework to evaluate participatory research for sustainability. *Ecological Economics*, 60, 726-742.

Borja, Á, Galparsoro, I, Solaun, O, Muxika, I, Tello, E-M, Uriarte, A. and Valencia, V. 2006. The European Water Framework Directive and the DPSIR, a methodological approach to assess the risk of failing to achieve good ecological status. *Estuarine Coastal and Shelf Science*, 66, 84-96.

Borja, Á., Tueros, I, Belzunce, M.J., Galparsoro, I., Garmendia, J.M., Revilla, M., Solaun, O. and Valencia, V. 2008. Investigative monitoring within the European Water Framework Directive: a coastal blast furnace slag disposal as an example. *Journal of Environmental Monitoring*, 10, 453-62.

Chave, P.A. 2001. *The EU Water Framework Directive: An Introduction*. London: IWA Publishing.

Costanza, R., d'Arge, R., De Groot, R., Farber, S., Grasso, M., Hannon, B., Limburg, K., Naeem, S., O'Neill, R.V.O., Paruelo, J.,Raskin., R.G., Sutton, P. and van den Belt, M. 1997. The value of the world's ecosystem services and natural capital. *Nature*, 387, 253-260.

Dworak, T., Falsch, M., Thaler, T., Grandmougin, B., Strosser, P. 2009. UK – TAG (Technical Advisory Group) WGD 21- Comparison of Draft River Basin Management Plans. Report to Scotland and Northern Ireland Forum for Environmental Research (SNIFFER). Vienna: Eco-logic Institute

European Commission. 2001. *A Common Implementation Strategy for the Water Framework Directive (2000/60 EC) Strategic Document – 2 May 2001*. Available at: http://ec.europa.eu/environment/water/water-framework/objectives/pdf/strategy3.pdf. [accessed: 1 October 2010].

European Commission. 2002. *Commission Working Document on Natura 2000. EU Commission, Brussels, December 2002*. Available at: http://ec.europa.eu/environment/nature/info/pubs/docs/nat2000/2002_faq_en.pdf [accessed: 1 October 2010].

European Union. 2003. *Wetlands Horizontal Guidance*. Horizontal guidance document on the role of wetlands in the Water Framework Directive. Brussels: Wetlands Working Group, Common Implementation Strategy for the Water Framework Directive, EU.

EUWI. 2002. *Directing the Flow: A New Approach to Integrated Water Resources Management. European Commission, Brussels*. Available at: http://ec.europa.eu/research/water-initiative/pdf/iwrm_060217_en.pdf [accessed: 1 September 2010].

Falkenberg, K. 2010. *The EU Water Framework Directive: Aspirations and Lessons Learned*. Directorate-General Environment, European Commission. Available at: http://www.siwi.org/documents/Resources/Water_Front_Articles/2010/ The_EU_Water_Framework_Directive.pdf [accessed: 1 September 2010].

Freiberg, N. 2010. Pressure-response relationships in stream ecology: introduction and synthesis. *Freshwater Biology*, 55(7), 1367-1381.

GWP. 2002. *ToolBox, Integrated Water Resources Management.* Available at: http://gwpforum.netmasters05.netmasters.nl/en/index.html [accessed: 1 September 2010].

Hering, D., Borja, A., Carstensen, J, Carvalho, L, Elliot, M., Feld, C.K., Heiskanen, A-S., Johnson, R.K., Moe, J. and Pont, D. 2010. The European Water Framework Directive at the age of 10: A critical review of the achievements with recommendations for the future. *Science of the Total Environment*, 408, 4007-4019.

Hughes, S.J. 2005. Application of the Water Framework Directive to Macaronesian freshwater systems. *Biology and Environment,* 105B(3), 185-193.

Janssen, R., Goosen, H. Verhoeven, M.L., Verhoeven, J.T.A., Omtzigt, A.Q.A. and Maltby, E. 2005. Decision support for integrated wetland management. *Environmental Modelling and Software*, 20, 215-229.

Maltby, E. 2009. The changing wetland paradigm, in *The Wetland Handbook*, edited by E. Maltby and T. Barker. Oxford: Wiley-Blackwell, 3-42.

Mannino, I, Franco, D., Picconi, E., Favero, L., Mattiuzzo, E. and Zanetto, G. 2008. A Cost-effectiveness analysis of seminatural wetlands and activated sludge waste water treatment systems. *Environmental Management*, 41, 118-129.

Oates, R. 2008. *Policy Analysis: Analysing Barriers to Change: a tool to assist river basin planning. Technical Report: Wise Use of Floodplains Project (EU-LIFE Environment Project)* Available at: http://www.floodplains.org/pdf/guidance_ notes/Guidance%20Note%20No%203.pdf [accessed 1 November 2010].

Raven, P.J., Holmes, N.T.H., Vaughan, I.P., Dawson, F.H. and Scarlett, P. 2010. Benchmarking habitat quality: observations using River Habitat Survey on near-natural streams and rivers in northern and western Europe. *Aquatic Conservation: Marine and Freshwater Ecosystems*, 20, S13-S20.

SEPA. 2009. *The River Basin Management Plan for the Scotland River Basin District.* Available at: http://www.sepa.org.uk/water/idoc.ashx?docid=fbcdf339-4d78-4ccb...1 *[accessed: 1 August 2010].*

RSPB. 2010. *Water and Wetlands: The Water Framework Directive.* Available at: http:// www.rspb.org.uk/ourwork/policy/water/policyissues/waterframeworkdirective. aspx . Amended August 2010 [accessed: 1 October 2010].

Scottish Government. 2009. *Initial Evaluation of Effectiveness of Measures to Mitigate Diffuse Rural Pollution, Part 2.* Scottish Government Publications, Edinburgh. Available at: http://www.scotland.gov.uk/ Publications/2009/01/08094303/4 [accessed: 1 September 2010].

Steyaert, P. and G. Ollivier. 2007. The European Water Framework Directive: how ecological assumptions frame technical and social change. *Ecology and Society*, 12(1), 25. Available at: http://www.ecologyandsociety.org/vol12/iss1/ art25 [accessed: 1 November 2010].

Trepel, M. 2010. Assessing the cost-effectiveness of the water purification function of wetlands for environmental planning. *Ecological Complexity*, 7(3), 320-326.

Turner, R.K., Georgiou, S.G. and Fisher, B. 2008. *Valuing Ecosystem Services: The Case of Multi-functional Wetlands*. London: Earthscan.

UKCP. 2009. *Key findings: Change in winter mean precipitation for the UK*. Available at: http://ukclimateprojections.defra.gov.uk/content/view/1272/499 [accessed 1 November 2010].

UKTAG. 2003. *Guidance on the identification of groundwater dependant terrestrial ecosystems*. Available at: http://www.wfduk.org/tag_guidance/Article_05/Folder.2004-02-16.5332/TAG2003%20WP%205a-b%20%2801%29 [accessed: 1 September 2010].

Wakeham, H. 2006. The Water Framework Directive and Farming: Tackling Diffuse Pollution. Paper presented at the HGCA conference, 25-26 January 2006. Home Grown Cereals Authority, Warwickshire, UK.

Werritty, A., Black, A., Duck, R., Finlinson, B., Thurston, N., Shackley, S. and Crichton, D. 2002. *Climate Change: Flooding Occurrences Review*. Scottish Government Publications, Edinburgh. Avabilable at: http://www.scotland.gov.uk/Resource/Doc/156664/0042098.pdf. [accessed: 1 November 2010].

WWF. 2001. *Elements of Good Practice in Integrated River basin Management. A Practical Resource for Implementing the Water Framework Directive*. Brussels: WWF.

Chapter 11

Managing the Everglades and Wetlands of North America

Victor C. Engel, Carol Mitchell, Bruce Boler, Joffre Castro,
Leonard Pearlstine, Dilip Shinde

Introduction

The Everglades is a 9,000 km² subtropical wetland ecosystem located in the southern portion of the state of Florida, in the United States. The Everglades are the only subtropical wetlands in North America and the only wetlands in the US which are also covered by the UNESCO and the Ramsar convention. The Convention on Wetlands of International Importance, signed in Ramsar, Iran, in 1971, is an intergovernmental treaty that provides the framework for national action and international cooperation for the conservation and wise use of wetlands and their resources (http://www.ramsar.org). The Everglades National Park (ENP) protects more than 565,000 ha of habitat and is the second largest national park in the United States. However it represents only a portion of the original Everglades ecosystem, encompassing the downstream terminus of the major freshwater drainage features in this system, as well as large expanses of coastal oligohaline marshes, mangrove swamps, and the offshore benthic communities of Florida Bay. Due to its location downstream of the other parts of the remaining ecosystem, the freshwater supply to ENP has been affected by all upstream land use changes and water resource allocations in south Florida. Its location along the coast also presents new risks associated with sea level rise due to global warming. The vast majority of ENP lies less than 2 m above sea level. ENP has been on the UN World Heritage List since 1979. After Hurricane Andrew in 1992, the United Nations placed ENP on the List of World Heritage in Danger in 1993. The Park was temporarily removed from the danger list in 2007, but was reinstated in 2010. ENP has also been designated as an International Biosphere Reserve, and a Wetland of International Significance.

The headwaters to the ecosystem begin in the lakes and wetlands near the city of Orlando which flow into the Kissimmee River and discharge into Lake Okeechobee. Prior to human intervention, seasonal high water overflowed the southern rim of Lake Okeechobee and formed a shallow, slow-moving river approximately 100 km wide which travelled over 160 km to Florida Bay and the Gulf of Mexico. Often termed "sheetflow" this wide expanse of slow moving water helped shaped the Everglades ecosystem as it is known today. The entire region is underlain by a limestone shelf of Pleistocene origin, and the typically high transmissivity of this formation affecting hydrology is a primary determinant of the Everglades ecosystem characteristics. Another primary determinant of ecosystem functioning is the small-scale topographic variation in the peat surfaces beneath the marsh communities. These "micro-topographic" differences in peat surface elevations lead to significant variability in hydroperiods and water depths across relatively small spatial scales (10^1 to 10^2 m) in the marsh. This type of habitat is characterized by parallel "ridges" dominated by sawgrass (*Cladium jamaicense*) occurring on the slightly elevated (10 to 30 cm) peat deposits, separated by deeper-water, long hydroperiod "sloughs" dominated by *Eleocharis* and *Nymphaea* spp. This habitat is collectively referred to as the "ridge-slough" habitat. Approximately two-thirds of the historical extent of the Everglades exhibited the parallel ridge-slough formations. Much of the original patterning in this system was lost or degraded with the impoundment and drainage activities that began in the late 1890s. Restoring this habitat is a primary objective of the large restoration project currently underway in the Everglades.

This chapter focuses first on the fundamental characteristics of the Everglades ecosystem and how it has changed as a result of human impacts over the last 120 years. A description of the institutional frameworks in the US which govern wetland management will then be discussed in relation to management initiatives and restoration activities in the Everglades. Finally, the results of new scientific investigations designed to guide restoration and improve management of the Everglades will be highlighted.

The remaining Everglades wetlands are divided into Water Conservation Areas (WCA) where water levels are managed independently using discharge control structures such as the S12s. The large metropolitan areas of West Palm Beach, Fort Lauderdale, and Miami are adjacent to this ecosystem.

The Freshwater Everglades

Climate

The average annual air temperature in south Florida is approximately 24°C and the average annual precipitation is roughly 1,320 mm (McPherson and Halley 1997) with slightly more rainfall occurring in the southern than the northern parts of the region. The region is characterized by distinct wet (June to November) and dry seasons. Rainfall amounts in the wet season are approximately 75% greater than the dry season rains, and the strong wet-dry seasonal cycles create hydrologic pulses which influence many ecosystem characteristics (Duever et al. 1994, Frederick and Ogden 2001, Harvey and McCormick 2009). Tropical storms impact the region on average 1 out of every 3 years. Annual to decadal-scale variability in Everglades rainfall has been correlated with the El-Nino Southern Oscillation (ENSO) cycles, and with the phase of the Atlantic Multi-Decadal Oscillation (Enfield et al. 2001,

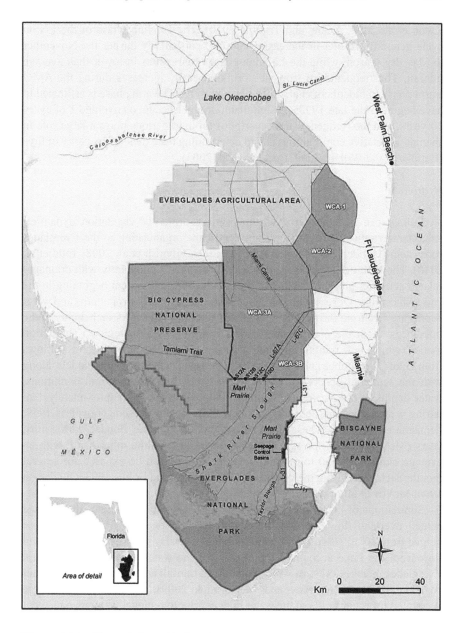

Figure 11.1 Everglades National Park (ENP) that Encompasses the Southernmost Portion of the Historic Everglades Ecosystem of South Florida

Kwon et al. 2006, Abtew and Trimble 2010). The El Nino phase of the ENSO cycle generally results in increasing rainfall, particularly during the November to May dry season, while the La Nina phase often results in lower than average rainfall. The frequency of tropical storms generally increases during the AMO warm phase (Goldenberg et al. 2001). The latest AMO warm phase is estimated to have begun in the late 1990s. A significant drought occurs on average 1 of every 10 years, and the frequency and severity of drought periods plays a large role in shaping vegetative communities and in determining the extent and severity of fires (Bernhardt and Willard 2009, Beckage et al. 2003).

Hydrology

Hydrology can be considered as the primary driver of vegetation dynamics, soil characteristics, nutrient status and landscape organization in the Everglades ecosystem (Gunderson 1994, Larsen et al. 2007, Givnish et al. 2008, Todd et al. 2010). The hydrology of the Everglades has changed significantly with drainage-related activities and the construction of the levee-canal system beginning in the late 19th century. Observations or direct measurements of hydropatterns and water depths are rare from the period before these activities began. As such, many of the inferences about the pre-drainage Everglades are based on regional simulation models which link rainfall and inflows from the Kissimmee River to outflows from Lake Okeechobee and water levels in the open marshes. These models, along with the scattered observational record, suggest that under natural conditions, water depths in the *Nymphaea*-dominated sloughs averaged approximately 1 m during wet seasons and less than 0.5 m during the dry seasons (Loveless 1959, Fennema et al. 1994). These slough communities remained flooded during most years (McVoy et al. *in press)*. The *Cladium* ridges, on the other hand, with an average elevation of 30 to 60 cm above the slough, can be exposed for several months each year during the dry seasons (Gunderson and Loftus 1993, Gunderson 1994, McVoy et al. *in press*).

Simulations of pre-drainage, regional hydrology models suggest the average annual overland flow entering the central Everglades from Lake Okeechobee ranged from 10^7 to 10^9 m^3.yr^{-1} (Fennema et al. 1994). By some estimates, overland flow from the Lake accounted for just 8% of the water present in the historic Everglades marshes, 81% of which came from rainfall and 11% came from other contributing basins (Harvey and McCormick 2009). Groundwater discharge accounted for roughly 1% of the water budget prior to regional water management operations. Today, rainfall accounts for only 66% of the Everglades water budget, and managed, surface water releases to the marshes from agricultural areas account for 18% (Harvey and McCormick 2009). The exchanges between surface water and the groundwater which characterize this karstic environment have also increased with development, and new data suggest the high ionic strength of the groundwater affects many of the biogeochemical characteristics of the current system, including peat decomposition rates (Harvey and McCormick 2009).

Groundwater discharge from the freshwater wetlands to coastal upwelling zones is also recognized as an important driver of the coastal wetland and offshore benthic communities (Price et al. 2006).

Topographic gradients in the Everglades basin are very shallow and on average range from 3 to 5 cm.km^{-1} from Lake Okeechobee to the Gulf of Mexico. These shallow gradients, coupled with the hydraulic resistance of the vegetation and relatively low water depths result in exceedingly slow water velocities in the ridge-slough wetlands. Modelling studies suggest the velocities in the deeper-water sloughs over most of the Everglades were greater prior to the construction of the levee-canal system (Larsen and Harvey 2010). In impounded zones of the Everglades today surface water velocities rarely exceed 1 cm.s^{-1} (Harvey et al. 2009, Ho et al. 2009, Variano et al. 2009*)*. Under natural conditions, periodic high flow events may also have been involved in forming the Everglades peatlands (Harvey et al. 2009, Larsen and Harvey 2010). The shallow hydraulic gradients and the distinct seasonal rainfall patterns which are present in this region also historically resulted in large-scale, dry season recession of surface waters across the Everglades. These annual recessions help to concentrate fish and other aquatic species into local depressions, and this recurrent, reliable food source helped to drive the formation of the wading bird "super-colonies" which once characterized this system (Ogden et al. 1994, Frederick and Ogden 2001). Disturbances to the timing and location of the recession patterns have been altered with development, and these disturbances have had important implications for modern-day wading bird foraging behaviour and nesting success (Gawlik 2002, Russell et al. 2002).

Vegetation

The freshwater wetlands of the Everglades are characterized by several different habitat types, including seasonally-flooded swamps dominated by *Persea*, *Salix*, and *Taxodium spp.*, short-hydroperiod prairies dominated by *Rhyncospora*, *Panicum*, and *Muhlenbergia* spp., and peat-based, longer hydroperiod marshes dominated by *Cladium jamaicense* interspersed with emergent or floating-leaved aquatic plants such as *Eleocharis spp. or Nymphaea spp.* Thick benthic and floating or attached algal communities dominated by blue-green species, collectively termed "periphyton" are nearly ubiquitous throughout the wetland. Periphyton photosynthesis results in the precipitation of low-density, calcium carbonate "floc" in the water column. These conditions suggest the majority of the freshwater wetlands present in the Everglades should be classified as a fen peatland or alkaline mire (Richardson, 2010) instead of a marsh.

Uplands, usually forested habitats interspersed throughout the marshes, important for many local species, are an integral part of the greater Everglades ecosystem. Some of the most important of these upland forested habitats, the tree islands, occupy only 3 per cent of the total area of the Everglades, and are dominated by temperate and tropical hardwood hammock species such as *Bursera simaruba*, *Eugenia axillaris*, and *Ficus spp.* (Armentano et al. 2002). Tree islands may flood from less than 1 to 6 months per year depending on their elevation above the surrounding marsh, and they are susceptible to burning during drought periods (Gunderson and Loftus 1993, Wetzel et al. 2005). Paleo-ecological studies suggest many of the tree islands in the southern Everglades formed during climatically dry periods between 400 and 1200 YBP (Years Before Present) (Willard et al. 2006) on small (0.5 to 1.5 m) limestone outcrops. Peat-based tree islands dominate in the northern parts of the system (Brandt et al. 2002). Recent archaeological findings suggest shell and bone deposited by early humans may have been the origin of the topographic high-points where many Everglades tree islands have developed (Graf et al. 2008). Water flowing around these islands has resulted in their characteristic "tear-drop" shape, with the widest part and highest elevations of the islands occurring on the upstream end (Sklar and van der Valk 2002, Hanan and Ross 2010).

Soils

The peat base of much of the Everglades are Holocene Histosols which historically reached approximately four metres in thickness downstream of the southern side of Lake Okeechobee, tapering to 1 m or less in the oligohaline transition zone near the Gulf of Mexico and Florida Bay (Gleason and Stone 1994). These soils formed in the wettest parts of the system in low lying areas, and were deposited beginning less than 6,000 YBP when rising sea levels stabilized and freshwater discharge rates slowed after the end of the last ice age. The other dominant soil types in this system are classified as Inceptisols (Gunderson and Loftus 1993). This soil is characterized as calcitic marl and is formed by the precipitation of calcium carbonate by cyanobacteria (Gleason et al. 1974, Browder et al. 1994). These soils are present in the shorter- hydroperiod wetlands which border the primary Everglades peatlands in the southernmost portions of the ecosystem. Layers of Inceptisols are also frequently found embedded within the layers of the Histosol peat deposits, indicating large-scale and persistent fluctuations in climate and water levels occurred during the development of this ecosystem (Rischardson 2010).

Nutrient Status

The Everglades is generally considered a phosphorus (P) limited ecosystem (Koch and Reddy 1992, Hagerthey et al. 2008). The background water column total phosphorus (TP) concentrations in the open marshes are extremely low (< 10 µg/L) and most of the native plant communities evolved under these very low nutrient conditions. The low TP levels in the marsh roughly match the concentrations found in rainfall (Ahn and James 2001) suggesting that in an undisturbed state, rainfall is the primary source of nutrients in this system. Nitrogen-fixing cyanobacteria in the water column further enhance the ecosystem P limitations (Craft and Richardson 2008).

One of the defining features of the Everglades is the presence of strongly anisotropic soil nutrient distributions, and some communities are more limited by nitrogen (N) than P. In general, soil TP increases with elevation above the marsh, and is therefore also inversely related to water depths (Ross et al. 2006, Wetzel et al. 2010). Soils occurring in the elevated tree islands exhibit the highest soil TP concentrations that can reach up to 5 per cent by mass. The foliar $^{15}N:^{14}N$ isotopic ratio has recently been shown to increase with soil P concentrations, and further confirm that N is more of a growth-limiting factor than P on these islands (Wang et al. 2010). As is shown below, the mechanisms by which nutrient enrichment occurs on these islands have been a subject of intense investigation in the last several years. Wading bird guano deposits are considered to be a primary mechanism of nutrient enrichment on some, if not all, of the Everglades tree islands (Orem et al. 2002, Frederick and Powell 1994). Groundwater discharge to the islands driven by root water uptake (i.e. biological pumping) has also been proposed as a primary mechanism by which nutrients accumulate in tree island soils (Wetzel et al. 2005). Similar mechanisms may also lead to the anisotropy in soil nutrients observed in the ridge-slough habitat (Ross et al. 2006). The breakdown of soil nutrient anisotropy is concurrent with the breakdown in the parallel *Cladium* ridge-slough formations which tends to follow hydrologic manipulation in this system (Watts et al. 2010). Van der Valk et al. (2009) note that the anisotropic distribution and "patchiness" of soil fertility in the Everglades mirrors the conditions often found in arid environments.

Landscape Patterning

The precise set of mechanisms leading to the peat formations beneath the parallel, highly productive *Cladium* ridges and the deeper-water sloughs in the central Everglades is still a matter of speculation. Recent model results suggest that periodic high water velocities in the marsh caused by extreme high water events such as those associated with the passage of a tropical storm result in the net transport of floc and organic matter from the sloughs to the ridges (Larsen et al. 2007, Larsen and Harvey 2010). Differential peat accretion and respiration rates in the permanently flooded sloughs compared to the periodically exposed ridges may also be a factor (Watts et al. 2010). The combined results of many of the recent investigations suggest that biological processes (e.g. productivity, root water uptake) and surface water hydrology must both play significant roles in the development of ridge-slough-tree island patterning, peat surface microtopography, and the anisotropy in soil nutrient distributions. The breakdown of this patterning and the loss of microtopographic variability in the Everglades marshes are now understood to represent one of the most detrimental effects of the hydrologic manipulations which have impacted the system over the last century.

The heterogeneity of habitats generated by intact ridge and slough patterning increases the number of niches available to support secondary and upper level consumers (Gawlik 2002). For example, the series of interconnected, permanently flooded sloughs in the historic system provided ecological connectivity and migration routes for many fish species (Trexler et al. 2002). The loss of topographic variability in the Everglades today contributes to a reduction in the occurrence and distribution of isolated pools of deep water refugia for fish during the dry season. This lack of refugia, in turn, reduces the availability of prey for nesting wading birds (Gawlik 2002). Alligator holes, which also serve as deep water refugia and preferential feeding zones for wading birds have also declined with hydrologic management and with the gradual migration of alligators from the marshes to the surrounding canals (Mazzotti and Brandt 1994). System-wide investigations into the complex linkages between hydrology, vegetation patterning, habitat quality, and trophic dynamics have defined much of the Everglades research over the past several years.

The History of Hydrologic Manipulations and Major Ecological Impacts

Early Efforts to "Reclaim" the Everglades

The early efforts to reclaim the Everglades in the 1800s focused first on channelizing and straightening the once-meandering Kissimmee River, which emptied into Lake Okeechobee (Godfrey 2006). A network of secondary canals connected surrounding agricultural and cattle ranching areas to the Kissimmee River. The runoff carried with it high nutrient concentrations, eventually resulting in the hyper-eutrophication of Lake Okeechobee observed today. For example, in 1973 water column P concentrations were measured at 40 mg.L^{-1} and had increased to over 100 mg.L^{-1} by 1999 (Richardson 2010). Subsequent drainage operations focused on the building of large canals from Lake Okeechobee to the Atlantic and Gulf of Mexico coasts, including the St. Lucie Canal, the Caloosahatchee Canal, the North New River Canal, and the Miami Canal.

The large canals in this region were excavated to transport water from Lake Okeechobee and the inland marshes to the coast to be discharged into tidal waters (Light and Dineen 1994). Discharging this water to tide, however, inevitably robbed the downstream portions of the ecosystem from a vital water source, and degradation quickly followed. As described below, building the infrastructure capable of storing the water discharged to tide and sending it to the southern parts of the ecosystem became a fundamental component of the Everglades restoration plans developed in the late 1990s. However, in the decades immediately following the completion of these canals, extensive agricultural areas developed south of Lake Okeechobee in the rich organic soils which had only recently been drained. These early efforts at reclamation for agriculture were only partially successful, and in 1926 a large hurricane resulted in flooding from Lake Okeechobee causing the deaths of more than two thousand people. In response to this disaster, the United States Army Corps of Engineers (hereafter referred to as the USACE) built the Herbert Hoover Dike around the Lake perimeter to a height of approximately

10 m and excavated additional canals in the surrounding areas (Godfrey 2006). The 3000 km² agricultural zone now called the Everglades Agricultural Area (EAA) which developed in this region now produces 50 per cent of the total US sugar output (Haman and Svedsen 2006). The high nutrient loads emanating from the EAA over the past several decades has caused numerous and potentially irreversible ecological changes in the historically oligotrophic downstream marshes (Richardson 2010).

While enabling large-scale human development for the first time in south Florida, the early drainage efforts also promoted soil subsidence through oxidation (Sklar et al. 2002) and increased the frequency of fires. Soil loss was estimated to have reached more than 2.5 cm yr⁻¹ in some areas. As much as 3 m of soil have been lost to oxidation in portions of the EAA. Over time, this widespread loss of soil has altered the regional topographic gradients which once regulated surface water movement and sheetflow patterns. These changes in the regional topographic gradients caused by soil loss are now understood to represent one of the largest challenges to restoring natural sheetflow patterns across the Everglades. The early drainage efforts also resulted in salt water intrusion along the Florida coast. Several early municipal drinking water wells for the city of Miami were abandoned in the late 1930s as new wells were installed further inland, closer to the Everglades marshes. This dependence on the Everglades marshes as a source of drinking water and as a barrier to salt water intrusion in south Florida continues to this day. The impacts of the drainage efforts from the late 1800s to the late 1930s quickly became apparent in the plant communities and animal populations of the Everglades. The large colonies of wading birds which once characterized the system were reduced by 90 per cent due to the effects of the altered hydrology on their food base (Ogden 1994, Frederick and Spalding 1994) and because the navigable canals created significantly more hunting opportunities. The impacts on the wading birds populations were so severe that the Florida state government outlawed hunting of these species in 1901. Alligator populations also declined dramatically with the lowered water levels and overhunting, prompting the Federal government to declare it an endangered species in 1967. Alligator populations have since recovered with conservation measures.

The Creation of Everglades National Park and the C&SF Project

In response to the growing public concern over the fate of the Everglades ecosystem, the US Congress authorized the creation of Everglades National Park (ENP) in 1934. ENP was the first National Park designated solely for its role in providing habitat for plants and animals.

Shortly after the establishment of ENP, it was recognized by the NPS and several environmental groups led by Marjory Stoneman Douglas that without a reliable water source from the northern portions of the ecosystem, the area set aside for ENP could not provide habitat suitable for the protection of the flora and fauna in the region (Godfrey 2006). The USACE during this time also recognized the need for an integrated, region-wide infrastructure and regulatory framework for managing the region's water resources established the Central and Southern Florida Flood Control Project (C&SF). The state-controlled operations of the C&SF system led to the current South Florida Water Management District (SFWMD).

The C&SF Project is intended to provide flood control, water supply for municipal, industrial and agricultural uses in south Florida, to prevent salt water intrusion, protect fish and wildlife resources and to provide water to ENP. Construction began in 1950 and was essentially complete in the 1970s although some work is still under way. The primary system includes more than 1,609 km of canals, 1,150 km of levees, 150 water control structures and 16 major pump stations. Today, the C&SF Project provides flood protection and water supply to over six million people and almost 405,000 ha of agricultural lands. It encompasses almost 730,000 ha of the Everglades habitat, including ENP and three Water Conservation Areas (WCA 1, 2, and 3). This water management infrastructure may control up to 3.8 million m^3 of water per day (Haman and Svedsen 2006).

In addition to fundamentally altering the natural hydrologic cycles of the Everglades, the canals associated with the C&SF project have impacted ecosystem function several other ways. Many of the larger canals are excavated through the limestone bedrock and subject to direct inputs from deep groundwater. This groundwater input raises the $CaCO_3$, magnesium, sodium, potassium, chloride, and sulphate concentrations in canal outfalls relative to surface water (McCormick 2010). The effects include increasing pH and altered periphyton communities. Under the naturally acidic surface water conditions which occurred in the northern Everglades prior to drainage, periphyton was dominated by green algae and diatom taxa. These areas are now dominated by blue-green, calcareous and mat-forming taxa (Davis et al. 2005, McCormick 2010). This increase in periphyton $CaCO_3$ encrustation (McCormick et al. 2009) could be contributing to the loss of slough habitat which has been observed in this area by inhibiting sediment re-suspension and transport during high flow events.

Canals are one of the primary vectors for non-native and invasive species to enter the Everglades. The majority of the invasive species are from tropical regions in Asia and South America. Many species, such as the Peacock Bass (*Cichla ocellaris*) were released intentionally into the canals to support game fisheries. Today, the canals serve as deep water refugia from periodic dry periods in the marsh and cold weather (Trexler et al. 2000, Shafland et al. 2007). Both native and non-native fish in canals move into adjacent marshes during the wet season.

Recognition of the importance of the WCAs to fish and wildlife led to the involvement in 1950 of both the US Fish and Wildlife Service and the Florida Game and Freshwater Fish Commission in their management. Despite the involvement of these conservation-based agencies, the USACE and the SFWMD have been frequently criticized by environmental groups and by the NPS for operating the C&SF system primarily for the purposes of flood control and water supply, with wildlife and habitat preservation as distant secondary goals (Godfrey 2006). In response, and following several years of extreme drought in the 1960s in which salt water intrusion and severe fires threatened many of the interior marshes of ENP, new components of the C&SF project were expanded southward to provide ENP with a permanent water supply. And in 1989, the Secretary of the Army was directed under the Everglades National Park Protection and Expansion Act "to take all measures feasible and consistent with the purposes of the C&SF Project to protect natural values associated with ENP" (Godfrey 2006). However, by the time these additional features were completed in the 1980s, there existed still very little information on the precise amounts of water needed to support the freshwater wetland and coastal ecosystems in ENP. Uncertainties regarding the proper timing, amount, and distribution of water deliveries needed to support the wildlife and ecosystems in ENP remain to this day. As is shown below, significant research continues to be directed towards addressing these uncertainties.

Delivering Water to ENP

Tamiami Trail (US Highway 41) is a two lane road built in 1929 that links the east and west coasts of Florida and crosses the central Everglades ridge and slough habitat. The road now forms the northern border of ENP, and is considered one of the primary obstructions to the north-south overland sheetflow which characterized the historic system.

In general, drought conditions in ENP during periods of low rainfall are exacerbated because water is held in the northern, upstream WCA impoundments for municipal water supplies. This condition has led to severe, peat burning fires in the southern parts of the system. For example, in the mid-1980s and in the early 1990s, two periods of very low rainfall, severe fires swept across much of ENP. And as recently as 2008, the Mustang Corner fire in northeast ENP burned roughly 16,000 ha, including countless tree islands (Ruiz et al. 2010). In contrast, during the 2008 dry season, water levels at a gauge in central WCA 2 were above the 30th percentile for this month (calculated over the period between 1991 and 2010). Thus, the pattern of prolonged hydroperiods and deep water conditions in the northern WCAs is concurrent with more frequent and more severe drought conditions in the southern part of the system. The deep water conditions in the northern Everglades have also led to unintended ecological consequences. For example, more than 70 per cent of the historic tree islands in WCA 2 have been lost due to high water stress (van der Valk et al. 2008).

Despite the efforts of many government agencies and environmental organizations over the past several decades, the regulation of water levels in the northern Everglades and the quality, timing, amount, and distribution of discharges to ENP are still widely considered to be inadequate to sustain ecosystem function. Perhaps the most striking indicator of degradation in the Everglades is the loss of the parallel ridge and slough patterning which once characterized large expanses of the interior marshes (SCT 2003, Givnish et al. 2008, Watts et al. 2010). The precise set of factors which have caused decline in this pattering are not fully understood. In overly drained areas the hydroperiods in the low-lying slough communities have decreased, and these once patterned areas are now colonized almost entirely by *Cladium* stands. In contrast, in impounded areas where water levels and hydroperiods on the emergent ridges have increased, *Cladium* cover has declined, and the once parallel ridges are now divided by the relatively more prevalent deep water sloughs. Only one large, contiguous area of well-preserved ridge and slough habitat remains in central WCA 3A. In this region, water levels and water quality are thought to more closely resemble historic conditions than anywhere else in the remaining Everglades. However, sheetflow directions in this region have been altered, and recent measurements (e.g. Variano et al. 2009) suggest the influence of nearby canals frequently results in sheetflow directions that are not aligned with the ridge and slough orientation. A misalignment between the ridge and slough patterning and sheetflow directions has also been observed in degraded northern ENP (He et al. 2010). Surface water flow velocities in both of these regions are thought to be reduced compared to historic conditions, and they may only infrequently exceed the critical velocity needed to suspend and redistribute sediments from the sloughs to the ridges (Harvey and McCormick 2009). Modelling results from Larsen et al. (2007) and Larsen and Harvey (2010) suggest periodic sediment transport and redistribution are necessary for the formation and maintenance of the ridge-slough habitat.

Since 1963, surface water discharges from WCA 3 into western ENP have been controlled primarily through the four large gated culverts called the S12 structures. Water enters eastern ENP through several free-flowing culverts connected to the L29 canal and which pass beneath a 17 km section of the Tamiami Trail roadbed. The amount of water that passes annually from WCA 3 to ENP is highly variable and has changed with the different C&SF operational schemes. Prior to the compartmentalization of the WCAs, 60 per cent or more of the annual north-south overland flow into ENP would have passed across the eastern half of Tamiami Trail into Shark Slough, the primary drainage feature of ENP (Fennema et al. 1994). However, with the construction of the C&SF levee-canal system, the vast majority of the water currently passing into ENP (approximately 70% annually) is now delivered through the S12 structures on the west side of Tamiami Trail. In addition to the changes in the east-west distribution of water entering ENP, the total amount of water has also decreased, and by some estimates the current water releases (approximately 10^8 $m^3.yr^{-1}$) represent only one-half of the historic flow volumes (Fennema et al. 1994). Decreasing freshwater flows are considered the leading cause of high salinity conditions now present in the Everglades coastal wetlands (Brewster-Wingard and Ishman 1999).

The S12 structures discharge directly onto short-hydroperiod marl prairies, and the increased hydroperiods and water depths now found in these prairies are driving changes in the vegetative communities (Nott et al. 1998, Ross et al. 2004, Sah et al. 2010). The impacts of the altered hydrology in the marl prairies are particularly damaging for the nesting cycle of a sub-population of the endangered and endemic Cape Sable Sea Side Sparrow (CSSS, *Ammodramus maritimus mirabilis* ;Lockwood et al. 2001, Pimm et al. 2002). Increasing the water deliveries to the eastern side of ENP and reducing the S12 discharges is considered by many agencies and stakeholder groups to represent a possible mitigation strategy for the impacts to the CSSS and the marl prairies, while simultaneously returning more water to Shark Slough and eastern ENP.

In 1989 the Everglades National Park Protection and Expansion Act (PL 101-229) was signed by President Bush. This act authorized acquisition of 43,500 ha of privately-owned and state lands located south of Tamiami Trail. The 1989 Act also directed the USACE to increase flows into ENP to the extent practicable.

The Modified Water Deliveries (MWD) to Everglades National Park Project Act (16 USC. § 410r-S) authorized the USACE in 1992 to restore natural hydrologic conditions in ENP. A major objective in the original plan was to deliver roughly 113 $m^3.s^{-1}$ to northeastern ENP. Progress to meet this target has been slow due to political, economic, and engineering constraints, and the original MWD plan has been subject to several revisions in 2003, 2005, and most recently in the 2008 Limited Re-evaluation Report (LRR). As of December 2010, a 1.6 km bridge authorized in the 2008 LRR is still under construction. While this plan may help to increase flows, ENP officials consider the limited extent of the 1.6 km bridge and the modest increases in water levels and flows associated with this project to be insufficient to fully restore the ecohydrologic characteristics of the system. Additional analyses on the ecological benefits of longer bridge spans described below have led to the potential expansion of the bridges to a total of approximately 9 km.

In the southern and eastern portions of ENP, flood control operations for the metropolitan areas and agricultural zones in south Miami-Dade County have led to chronic low water levels and inadequate flow volumes. These wetlands form the headwater of Taylor Slough, the second largest drainage feature in ENP after Shark Slough. Taylor Slough discharges to northeast Florida Bay, and the reductions in overland flow have led to hyper-saline conditions in the near shore embayments in this region. The altered hydrology and high salinity threaten the breeding grounds of the endangered American Crocodile (*Crocodylus acutus*) and a large population of the Roseate Spoonbill (*Platalea ajaja*, Lorenz et al. 2009). To address this ecological problem, a series of elevated detention basins (the S332 basins) were constructed in the late 1990s by the USACE. The basins are aligned north-south between ENP and the C111 and the L31N and were designed to control seepage from the Park into these canals. In this scheme, water from the L31N canal is pumped into these detention basins to create a hydrologic ridge that minimizes surface water and groundwater flow leaving the park. The degree to which these basins control seepage through the highly transmissive limestone substrate is difficult to quantify given the complex and time-varying hydraulic gradients which develop from intermittent rainfall and changing management operations. Recent modelling results and observations suggest much of the water pumped into the basins returns to the L31N and C111 canals. Construction on a new restoration project described later in this chapter was recently started in this area to further reduce seepage from this region of ENP.

Water Quality Impacts and Mitigation

Phosphorus (P) concentrations have increased with surface water runoff from the EAA and Lake Okeechobee across large areas of the northern Everglades (Craft and Richardson 1993, Richardson 2010). The combination of the increased P and other pollutants with the hydrologic effects of the canals draining this area has had profound and negative effects on natural Everglades wetlands. Nutrient inputs from the EAA have caused shifts from *Cladium* ridge and slough habitat to mono-specific stands of *Typha* over very large areas downstream of canal and structural outfalls in WCA 1 and 2 (Urban et al. 1993).

Sulphate contamination also originates from canals draining the EAA (Bates et al. 2002). At canal point source sites in WCA 2A there is a pattern of seasonally high dissolved sulphide content in sediment porewater and a high level of sulphur content with depth in the sediment (Bates et al. 2002). Sulphur is used in soil amendments in the agricultural zones to enhance P uptake in crop plants. In the drained, aerobic soils of the EAA, the sulphur is oxidized and is discharged with storm water runoff into the Everglades via the C&SF canals. In the marshes, sulphate-reducing bacteria mediate the production of methylmercury, a neurotoxin that is bioaccumulated, thus making sulphur from canal discharges a major factor in the production and distribution of methylmercury (Bates et al. 2002). Mercury contamination in the Everglades has been implicated in the decline of wading bird populations observed during the 20th century. The increase in the frequency of wet and dry-down periods in the Everglades that accompanies water management activities may promote the mobilization and bioaccumulation of mercury in the food web. Mercury and methylmercury contamination remain a significant challenge to managing the Everglades. According to the Florida Department of Environmental Protection, (http://www.dep.state.fl.us/labs/bars/sas/mercury/docs/flmercury.htm) more than 40,000 ha of Everglades marshes contain fish with tissue mercury levels that exceed health-based standards.

As a result of degradation in the ecosystem observed downstream of the EAA outfalls, in 1988 the US Government sued the State of Florida for violation of the Clean Water Act, citing P as the primary pollutant. The result of this lawsuit was a 1992 Federal decree that mandated the State to implement measures to clean the water. State legislation, called the Everglades Forever Act, followed in 1994 to support the development of quantitative criterion for P and a timeline for reaching a point of "no-imbalance" to natural systems. According to the ruling by the US Environmental Protection Agency, the legal limit on P concentrations in water discharged to the Everglades wetlands is now 10 mg.L^{-1}. Legal and policy action since the passage of the Everglades Forever Act has included the 2006 Florida Legislature's extension (by a decade) of the timeline required to reach the goals of the original Act, and a recent (September 2010) federal judicial order mandating that the State implement a more aggressive schedule for meeting the legal limit. In the case of water quality, at this time in 2010, the legal mandates are likely to drive many of the large-scale decisions facing wetland managers and restoration practitioners in the Everglades, and will constrain decision-makers' ability to manage the ecosystem based on other, equally important and fundamental ecological principles.

Since the mid-1990s, the SFWMD has constructed 16,564 ha of filter marshes in 6 separate storm water treatment areas (STAs) in an effort to comply with the Everglades Forever Act and to reduce P loads to the Everglades. The first of these STAs became operational in 1994–1995. Prior to the completion of the STAs, water flowing into WCA 1 and WCA 2 contained more than 70 mg.L^{-1} P, on average, between 1978 and 2000 (Richardson 2010). Maximum discharge concentrations were measured at higher than 1,400 mg.L^{-1}. Since the completion of the STAs, the P concentrations in waters entering the WCAs have generally declined, but frequently exceed the 10 mg.L^{-1} criterion. Heavy rainfall events associated with hurricane activity have been correlated with decreases in the P removal efficiency of the STAs and increases in P loads to the Everglades marshes (SFWMD 2006). In 2004 and 2005, the region was struck by several hurricanes, and the flow-weighted mean P concentrations in water leaving the STAs reached as high as 100 mg.L^{-1} (Richardson 2010). Continued releases of water with high P concentrations will further degrade the downstream Everglades marshes. More area must be dedicated to building STAs and treating runoff from the EAA to avoid further degradation of the system.

A recent initiative called the *River of Grass* and led by the State of Florida seeks to convert large tracts of former agricultural land into additional treatment marshes and storage reservoirs. If completed, River of Grass could lead to water quality improvements across the northern Everglades and a more reliable water source for the southern parts of the system. The initiative started in June 2008 when Florida Governor Charlie Crist announced that the SFWMD would begin negotiating an agreement to acquire as much as 75,700 ha of agricultural land privately owned by the US Sugar Corporation. In December of 2008 following extensive negotiations and public meetings among the regions stakeholder groups, the SFWMD voted to accept the proposal to acquire the land for the price of US$ 1.34 billion, contingent upon financing and affordability. This would have made it the largest and most expensive land purchase for restoration ever to occur in the continental US However, larger-scale economic problems driven by the global recession in 2008 and 2009 forced several sequential renegotiations of the land purchase. In May 2009, the SFWMD and US Sugar Corporation amended the agreement providing for an initial purchase of close to 30,000 ha for US$536 million, with options to purchase the remaining 43,300 ha during the next ten years when economic and financial conditions improve. Finally, in August of 2010 and in light of continued economic impacts, the decline in SFWMD tax revenues, and the need to address recent federal court orders related to the Everglades Forever Act, a second amendment was made to the agreement with US Sugar Corporation. On October 12, 2010 under the modified agreement, the SFWMD purchased 10,850 ha of strategically located land with high restoration potential for US$197 million while preserving the option to acquire 62,000 ha of additional lands, if future economic conditions allow. The potential benefits of these lands to improve the hydrology and water quality of the Everglades have not yet been assessed.

US Institutional Framework for Managing and Restoring Wetlands

Federal Environmental Legislation

The fundamental set of federal environmental laws that influences management of the Everglades and all wetland systems in the US includes the Clean Water Act (1972), the Endangered Species Act (1973), and the National Environmental Protection Act (1969). The Clean Water Act protects surface water quality for the purpose of supporting "the protection and propagation of fish, shellfish, and wildlife and recreation in and on the water." A lawsuit filed by the US Federal government against the State of Florida for violations of the Clean Water Act resulted in the Everglades Forever Act in 1994. This Act set the legal limit for P concentrations in surface waters at 10 mg.L^{-1}, and was the primary impetus behind the construction of the STA filtration marshes built to treat runoff from the EAA.

Under the Clean Water Act, ENP waters have an Outstanding Waters (Class III) designation, which requires that surface waters be free from acutely toxic chemicals. In a recent revision of the Federal Insecticide, Fungicide, and Rodenticide Act (FIFRA), this designation was cited in conjunction with the results from several ecotoxicity assays by the US Environmental Protection Agency (EPA) in a revision to the regulations governing the use of one particularly toxic pesticide called Endosulfan. On June 9, 2010, the EPA announced that all new uses of Endosulfan were being cancelled throughout the US, and that the pesticide would be phased out completely over a five year period.

The Endangered Species Act provides for the protection of species listed as Threatened and Endangered by the US Fish and Wildlife Service, and is a powerful legal tool requiring that water managers consult with the Service for projects involving changes to water operations or infrastructure. If a proposed action is determined to have potential negative impact on a listed species, the Service may require changes to the proposed action in order to protect the species. The Endangered Species Act, although one of the most powerful legal instruments for environmental protection, has had unexpected effects during its application to the Everglades. This Act protects individual species, particularly those which may be naturally rare, ecological specialists or geographically limited in the natural system, characteristics which tend to lead to endangered status. The Act in no way protects underlying fundamental characteristics of the ecosystem, and in fact in an artificially fractured and managed ecosystem like the Everglades, can lead to actions that actually increase the fractured nature and move illogically in a direction away from a functioning ecosystem towards single species management. For example, the timing of surface water discharges from WCA 3 into ENP is regulated by a finding under this Act that states water releases must be limited during certain times of the year to avoid interfering with the nesting cycle of the Cape Sable Sea Side Sparrow (CSSS), which lives downstream. The negative ecological consequences of extended periods of high water upstream in WCA 3 have been linked by some stakeholder groups to these restrictions on discharge volumes.

The purpose of the National Environmental Policy Act (enacted 1969 amended 1984) is "...to declare a national policy which will encourage productive and enjoyable harmony between man and his environment (...)". This legislation requires all Federal agencies to conduct an analysis of the environmental impact of any major action, and to strive to minimize the negative effects on the environment. In the context of Everglades management or restoration, this Act requires the development of some form of Environmental Impact Statement for each proposed restoration project, and requires consultation with the US Fish and Wildlife Service to determine impacts to endangered species. It is in the process of developing the Environmental Impact Statement that different alternative configurations of a given project are developed, their environmental impacts quantified and compared to one another, and an "environmentally preferred alternative" is identified. The policy does not, however, mandate the selection of the environmentally preferred alternative: cost, impact to social aspects of the project area and other considerations are taken into account in the final decision on a given project. For example, during the nearly 20-year process to plan and construct the bridging for the Tamiami Trail, several planning documents identified full bridging over the entire 17 km length of the roadway to be the "environmentally preferred alternative", as this design maximizes sheetflow and connectivity of the natural system. As described in more detail below, impacts to residents and businesses along the roadway, including tribal properties, were considered and taken into account in the 2010 recommended design to add 9 km of bridges, which optimized environmental benefits while considering social impacts and cost.

Formulating and Evaluating Wetland Management Plans

The USACE is a federal agency made up of roughly 34,000 civilian and military personnel, making it the world's largest public engineering, design and construction management agency. The Clean Water Act of 1972 granted the USACE authority to regulate the construction of water resource projects throughout the US Later, the Water Resources Development Acts of 1986, 1990, 1992, and 1996 all granted authority to the USACE to participate in ecosystem restoration projects. When undertaking a project, the USACE includes other Federal, State, and local agencies in the planning process with the intent of producing more widely-accepted strategies for dealing with water resources issues. Collaborators are defined as Federal, State, and local agencies and institutions that possess financial and other resources with which to actively participate in a study and that can commit those resources to the planning process and project implementation (Holloway 2007).

The USACE manages the large wetland restoration projects of national importance such as those occurring in the Florida Everglades, the coastal zones of the state of Louisiana, the Chesapeake Bay along the Atlantic Coast, and in the San Francisco Bay Area Wetland Restoration Program. The stated objective of the USACE in ecosystem restoration is to restore degraded ecosystem structure, function, and dynamic processes to a less degraded, more natural condition. Restored ecosystems should mimic, as closely as possible, conditions which would occur in the area in the absence of human changes to the landscape and hydrology (USACE 2000). For all ecosystem restoration projects, the USACE by law must have a local, State-level sponsor for cost-sharing purposes, and to ensure water resource projects meet the needs of the local population. The construction costs of projects such as those listed above are often estimated in the billions or tens of billions US dollars and they may take decades to complete. Thus strong local support is a fundamental requirement for Congressional authorization of these large projects. Surprisingly, only a very small percentage (generally less than 3%) of the USACE budget for water resource projects can be dedicated to monitoring and adaptive management. A complete description of the process by which water resource and wetland restoration projects are selected by the USACE and local collaborators is beyond the scope of this chapter. However, a brief account of the methods by which ecosystem benefits are calculated and used to compare different project alternatives will be provided since it is these aspects of the decision-making process which have the greatest impact on the wetlands of the United States.

Calculating Ecosystem "Benefits"

In the formulation and selection of alternative plans, the USACE utilizes a method referred to as the "future with- and without-project analysis" (USACE 2000). The future without-project condition is the most likely condition expected to exist in the future in the absence of a proposed water resources project, and it constitutes the benchmark against which alternative plans are evaluated. On the other hand, the with-project condition is the most likely condition expected to exist in the future with the project implementation. In the USACE planning process, comparisons of the with-project to the without-project conditions in the study areas are used to identify the beneficial and adverse effects of the proposed plans. These with- and without-project comparisons thus provide the framework for the evaluation of alternative ecosystem restoration or protection plans.

The forecasts of with- and without-project conditions are based on national and regional forecasts of socio-economic parameters such as income, trends in employment, land use, and population growth. Of interest in light of the current economic setting of the last several years is that the national projections used by the USACE are based on an assumption of full employment. According to the USACE process, the current trends in ecosystem functioning should also be included in the with- and without-project comparisons. A potential criticism of this requirement is that in this period of climate change and sea level rise, the hydro-climatic conditions in a study area may change over the lifetime of the project and this may independently affect ecosystem trends in as yet unknown ways. Thus, the current trends in ecosystem function may be different from those which may occur in the future. Adequate field data are also not often available to fully describe the current ecological trends, and this approach does not consider the impacts of stochastic factors like large-scale disturbances or invasive species. Factors such as these tend to increase the uncertainty regarding ecosystem trajectories and the potential ecological benefits that may be derived from planning large water resource projects. For these reasons, the with and without-project projections of ecosystem function are more appropriately considered as planning tools rather than actual predictions of ecological trajectories, and the benefits that are calculated under each alternative are meaningful primarily only in comparison to each other.

The USACE typically employs a benefit-cost analysis and incremental cost analysis to examine the costs of a project in comparison to the potential ecological benefits. Under this scheme, a plan is considered cost effective if it provides a desired level of output for the least cost. Incremental cost analysis is used to compare the additional costs associated with incremental increases in ecological benefits. This approach often requires generating non-monetary metrics of ecosystem performance for each alternative. According to the USACE policy, incremental analysis is intended to help to identify and display variations in costs among different increments of restoration measures and alternative plans.

The calculations of benefits in the USACE ecosystem restoration projects are often based on units that measure some aspect of ecosystem "value". Some examples of possible metrics include the area of increased spawning habitat for anadromous fish, number of stream kilometres restored to provide fish habitat, and increases in the population of target species. Many previous restoration efforts have used the US Fish and Wildlife Service's 1980 version of Habitat Evaluation Procedures (HEP-80) to quantify and evaluate potential project benefits. This methodology uses a Habitat Suitability Index (HSI) to rate various metrics of habitat quality (e.g. average water depths or hydroperiod) on a scale of 0 to 1 (1 being optimum) for target species or community types. HSI values are multiplied by the area of available habitat in the project domain to obtain Habitat Units (HUs). One HU is defined as 1 acre (0.4 ha) of optimum habitat. By comparing existing (i.e. without-project) HUs to the HUs expected to be gained with a proposed action (i.e. with-project), the benefits of each alternative can be quantified. Specific targets must be defined for each ecosystem metric used in the USACE benefits analysis. The set of metrics, targets and the methodology used to weigh each metric in the overall calculations of project benefits often reflects the interests of the various local and federal agencies involved in the planning process. For example, while some stakeholders may be more interested in the number of waterfowl a particular project might produce, others are more interested in maximizing landscape heterogeneity and resilience to disturbance. In many cases, the targets which are developed by the project collaborators and upon which the evaluations of alternatives are based cannot be achieved simultaneously under a single management scheme. In large, complex ecosystems where restoration projects may cover several jurisdictions, such as in the Everglades, the conflicting interests of participating agencies can lead to significant delays in enacting restoration or new management plans.

Stakeholder Participation in the Everglades

The Everglades ecosystem is managed within a complex institutional mix of federal, state, and locally managed jurisdictions, each with their own mission and management purpose. Furthermore, a suite of laws and policies, including the federal laws mentioned previously as well as state and local laws and policies, governs the overarching aspects of water use, flood protection, water quality and protection of the environment. The juxtaposition of numerous land areas with different management purposes is challenging when trying to manage and restore ecosystem functioning through the regulation of ecohydrologic drivers such as overland flow and nutrient transport.

The entire Everglades watershed is included within the boundaries of the SFWMD. The SFWMD is the local authority for the C&SF project and is responsible for water supply and flood control operations. Along with the Florida Department of Environmental Protection, the SFWMD is also responsible for ensuring that water deliveries to the natural systems in the region meet the water quality (e.g. P concentration) standards set by the US Environmental Protection Agency. Nature-based recreational activities such as hunting and fishing are managed by the Florida Fish and Wildlife Conservation Commission over most of the State-owned natural Everglades, including both freshwater areas and nearshore estuarine/marine areas. This authority includes not only direct regulation of people's activities, but also species and habitat management for recreational and conservation purposes and thus has strong influence on water management operations. On the Federal side, the USACE is the entity responsible for flood control and thus for modifications to the water management infrastructure for the purpose of flood control. The USACE and the SFWMD share responsibility of operating the C&SF water management system.

The US Department of the Interior (DOI) has a large role in the Everglades ecosystem in both land management (via three National Parks and eleven National Wildlife Refuges) and management of threatened and endangered species over the entire landscape. Land owned or managed by the DOI makes up approximately one-half of the remaining Everglades' wetland ecosystems, and thus the mission of Interior's land management agencies (specifically the US National Park Service and the US Fish and Wildlife Service), is an important factor in water and natural resources management decision-making on a regional level.

In addition, two Native American tribes, the Miccosukee Tribe of Indians of Florida, and the Seminole Tribe of Florida, have rights over resources and lands within the Everglades ecosystem. The Miccosukee Tribe lives within the Miccosukee Reserved Area, a 2.8 km^2 sector of Everglades National Park, and possess a reservation in the central Everglades with a total area of 329 km^2. Furthermore, the Miccosukees have traditional use rights to a large sector of Water

Conservation Area 3, an area of 76,000 ha. The Seminole Tribe has an Indian Reservation north of the Miccosukee Indian reservation, with a total area of 212 km².

The diversity of stakeholders in the Everglades, and the scale and the scope of the wetland management effort mean that water and land use conflicts arise as individual restoration projects are designed and implemented. To address issues of conflict and balance the natural system and human interests, several regional groups have been formed, beginning in the mid-1990s, with the purpose of providing a forum for coordination and conflict resolution. These include the Governor's Commission for a Sustainable South Florida and the South Florida Ecosystem Restoration Task Force. The formation of these groups represents a tendency toward managing on the level of the ecosystem; however, the jurisdictional, policy and legal tools that they have to work with focus on smaller geographic scales, or on individual components of the ecosystem.

New Management and Restoration Initiatives

Bridging Tamiami Trail and Removing Barriers to Flow

New plans developed in 2008 as part of the Modified Water Deliveries project called for increasing the amount of water flowing into eastern ENP by elevating and bridging a 1.6 km section of eastern Tamiami Trail. While this plan would help to increase flows into ENP, many considered the limited extent of the 1.6 km bridge and the modest increases in water levels and flows to be insufficient to fully restore the ecohydrologic characteristics of the system.

The Omnibus Appropriations Act of 2009 (H.R. 1105; P.L. 111-008, March 11, 2009) directed the Department of the Interior and the National Park Service (NPS) to evaluate the feasibility of additional bridging of the Tamiami Trail beyond the current 1.6 km being constructed. The additional bridging is considered necessary to improve the ecological connectivity within the remaining natural Everglades, including ENP and WCA 3. The Omnibus Act further directed that more natural water flow and habitat restoration within the Everglades be achieved. The Final Environmental Impact Statement (FEIS) prepared by ENP in technical collaboration with the USACE represents partial fulfilment of the requirements of the Act and is scheduled to be released to the public on December 17, 2010. During the preparation of the FEIS, the NPS worked with an interagency team consisting of other federal, state, local government representatives to develop the methodology for evaluating the ecological costs and benefits of several alternative construction plans for the new Tamiami Trail bridges. Representatives from the Miccosukee Tribe of Indians of Florida also participated. The key finding in the FEIS is that an additional 9 km of bridging on the Tamiami Trail is necessary to maximize the ecological benefits. The projected costs of constructing these bridges in expected to reach approximately US$310 million. However, when coupled with other projects being planned as part of the Comprehensive Everglades Restoration Plan (CERP), the additional bridging will be capable of providing unconstrained flows to northeast ENP and Shark Slough that closely match historic volumes.

Restoring Hydroperiods and Water Quality in Eastern ENP

The C111 canal in south Miami-Dade County is the southernmost canal of the C&SF Project. The canal was constructed under the authority of the Flood Control Act of 1962 to extend flood protection for a roughly 260 km² basin of agricultural and suburban land use located to the east and southeast of ENP. Excavation of the C111 was completed in the mid-1980s. Within a few short years, the detrimental impact this canal had on hydrologic conditions and ecosystem function in ENP and other nearby wetlands became apparent. Water levels dropped significantly in Taylor Slough, a major drainage feature of ENP, and increased salinity levels in the downstream embayments of Florida Bay harmed local fish and bird populations. The public outcry over the massive sea grass die-off that occurred in this region of Florida Bay in the early 1990s provided the political pressure needed to address the tremendous ecological problems facing the Everglades. In 1994, a re-evaluation of the C111 project was undertaken by the USACE with participation from several agencies, including the US Department of Interior and the National Park Service.

Two restoration projects have resulted from the analysis of impacts of the C111 canal: the C111 South Dade project and the C111 Spreader Canal Western project. These two projects use similar technologies that are expected to improve conditions in Taylor Slough by retaining water in the slough and establishing more natural water flows in this region. This, in turn, will improve the timing, distribution and quantity of water in Florida Bay. The C111 Spreader Canal Western project also contains features that are expected to lead to restoration of the large wetland areas in this region of Miami-Dade County. It is estimated that about 102,000 ha of wetlands and coastal habitat may be affected by the proposed project.

To accomplish the goals of restoring Taylor Slough and Florida Bay, a 14 km hydraulic ridge adjacent to Taylor Slough will be created by integrating several existing and new water control features, including an unregulated, secondary drainage canal parallel to the C111 and next to ENP, a new 240 ha above-ground detention basin, the installation of two 6 $m^3.s^{-1}$ pump stations, reinforced earthen plugs in connecting canals, and operational changes at the downstream control structures. Continuous water quality monitoring for nutrients and salinity initiated several years before construction will be used to gauge the effectiveness of this project.

The impacts of pesticides in agricultural runoff are increasingly recognized as a threat to the health of Taylor Slough and downstream Florida Bay. For example, more than 76% of the pesticide known as Endosulfan used in the southeastern United States has been reported to be released into Florida Bay's watershed (Scott et al. 2002). Endosulfan has been found to exceed established standards repeatedly in surface waters draining directly into ENP, and in high concentrations in arthropod and fish tissue at several locations within the eastern boundary of the Park (Carriger et al. 2008a, Carriger et al. 2008b, Rand et al. 2010). These organisms are the common prey of endangered and threatened birds which feed in these areas, such as the Roseate Spoonbill (*Ajaja ajaja*). Citing these results, a petition filed by officials at ENP to the US Environmental Protection Agency has contributed to a permanent ban of Endosulfan in the agricultural lands across the US While this is a positive step towards restoration in the long term, the slow release of Endosulfan bound in the soils and groundwater in this region are likely to continue for several years.

The Comprehensive Everglades Restoration Plan

In the mid-1990s, unusually high rainfall led to significant ecological problems in the ecosystems of south Florida, including Lake Okeechobee, the central Everglades, and Florida Bay. In the opinion of many stakeholders at the time, these high rainfall amounts, when coupled with the increasingly difficult task of providing adequate flood protection in the now densely populated urban areas of south Florida, exceeded the capacity of the original C&SF system to maintain ecologically-favourable water levels in many parts of the Everglades. As a result of the political pressures that developed, the USACE was directed in the Water Resources Development Act (WRDA) of 1996 to develop a comprehensive plan for Everglades Restoration. And in 2000, President Bill Clinton signed a bill authorizing the Comprehensive Everglades Restoration Plan (CERP).

The fundamental objectives of the CERP are to divert the water being released to tide from Lake Okeechobee and the northern Everglades into large reservoirs, improve the quality of this water by creating thousands of new treatment marshes, and deliver this water to the remaining Everglades freshwater marshes. The total cost of the original plan is now estimated to be US$11.9 billion, making it one of the most expensive restoration projects in history. CERP provides a framework and guide to restore, protect and preserve the water resources of central and southern Florida, including the Everglades. It covers 16 counties over a 47,000 km² area. However, CERP was approved only on the condition that the water supply and flood protection for the current and future residents of south Florida (close to 5 million at the time CERP was signed into law) was maintained at the present standards. The re-engineering of the public water supply and flood control infrastructure needed to accomplish these goals relies on a complex set of very large engineering projects including wastewater reuse, seepage management, and aquifer storage and recovery. Many of the technologies upon which CERP relies have not been demonstrated at the large scales needed to ensure the overall success of the programme. Pilot projects are currently underway to address some of the uncertainties associated with the effectiveness of several component of the original Plan. Ten projects and the adaptive assessment programme, totalling US$1.1 billion have already been recommended for initial authorization.

As it was originally envisioned, CERP was made up of 68 individual projects and was intended to create approximately 88,000 ha of reservoirs and wetland-based water treatment areas. CERP is expected to result in new regulations to govern Lake Okeechobee water levels and recommend possible dredging of nutrient-enriched lake sediments to help achieve water quality targets. As part of the new management scheme for Lake Okeechobee the discharge of excess water to tide through the St. Lucie and Caloosahatchee canals in the wet seasons will be greatly reduced. To accomplish this, CERP relies heavily on new underground storage wells that will need to hold up to six million m³ per day. The injected freshwater is to be recovered by pumping during dry periods and used as water supply for the natural and man-made systems. To improve water quality, CERP seeks to create approximately 14,400 ha of stormwater treatment areas (STAs), built to treat urban and agricultural runoff water before it is discharged to the natural areas throughout the system. These are in addition to the roughly 18,000 ha of STAs already being constructed as a result of the Everglades Forever Act. Another primary objective of CERP is to improve the volume, timing and quality of water delivered to the ecosystems, including a 25 per cent increase in the amount of water delivered to ENP. This translates into an average of nearly 620 million m³ of water reaching ENP each year.

The primary ecological benefit of CERP is expected to be derived from the removal of more than 380 km of canals and internal levees within the Everglades to re-establish the natural sheetflow of water through the system. Plans have been developed to fill in portions of the Miami Canal in WCA 3 and to rebuild as much as 32 km of the Tamiami Trail with bridges and culverts to allow water to flow more naturally into ENP. Collectively called "decompartmentalization", the plan to back-fill so many kilometres of canals and the removal of levees and other barriers to flow in the central WCAs has proven to be exceptionally contentious among the CERP stakeholders. The canals are widely used for recreational activities, and many consider these deep water habitats to be important refugia for aquatic species during low rainfall periods. Removing levees and back-filling canals is also very expensive, and in many cases, there is insufficient material remaining in the levees to completely fill the adjacent canals. Thus, in order to completely restore the natural landscape and to fill all the canals as originally envisioned in CERP, massive amounts of material will have to be transported into remote regions of the marsh, adding substantially to the estimated cost of this project.

Within the CERP planning process, the USACE leads interagency teams whose objective is to develop alternative plans for each component project of the overall restoration. The ecological benefits and monetary costs of each alternative must be evaluated as part of this process. In the initial planning phases for the Decompartmentalization project, there existed no consensus among stakeholders on how to quantify the benefits of sheetflow to the Everglades ecosystem. The loss of the peat base to oxidation and the resulting changes in topographic gradients in many areas of the Everglades suggests that simple removal of levees and canals may not fully restore the historic overland flow conditions. Introducing sheetflow to the historic flow paths is expected to lead to improvements in water quality and to increase ecological connectivity across large areas, but there remains significant scientific uncertainty regarding the precise timing, amount, and velocity of sheetflow needed to restore ecological processes in degraded areas, such as formation of the ridge and slough patterning. New research and a pilot project have been formulated to address these uncertainties.

Adaptive Management and the DPM

The Decompartmentalization Physical Model (DPM) is a US$10 million pilot project designed to periodically introduce overland flow into a degraded portion of the ridge and slough, and to monitor the biological and physical responses. A primary objective of the DPM is to collect enough information on the relationships between overland flow and ridge-slough processes to develop ecological metrics for quantifying the benefits of full sheetflow restoration. A large, gated culvert is being constructed along the L67A levee to allow overland flow across approximately two km of currently impounded marshes. Ridge and slough processes, including sediment and nutrient transport in the study area will be monitored continuously downstream of the culvert. Segments of the L67C canal will be completely filled, partially filled or left open and the levee downstream of the L67C canal will also be removed. Fish population dynamics in each of the different canal treatments will be monitored separately, and the impacts of each canal treatment on sediment transport and on the hydrologic characteristics of overland flow will also be monitored. The investigations in the canal are designed to inform decision-makers about the ecological benefits versus the costs of each canal alternative (i.e. complete, partial, or no fill) including the costs associated with the loss of recreational opportunities. The DPM experiment follows a Before-After-Control-Impact (BACI) statistical design (Smith et al. 1993) whereby the full set of ecological measurements will be conducted at both the control (i.e. no flow) and impacted (i.e. flowing) sites before, during, and after the opening of the culvert on L67A. The before-impact, baseline studies began in November 2010, and the results of the five year study will be used to help select alternative plans for the full Decompartmentalization project. This project is considered by many to represent a fundamental realization of the principles of adaptive management in wetland restoration. The current science plan for this project developed by Drs. Scot Hagerthey, Sue Newman, and Fred Sklar at the SFWMD can be found at (http://www.evergladesplan.org/pm/projects/docs_12_wca3_dpm_ea.aspx).

Emerging Science

A primary research initiative seeks to improve the understanding of the natural and man-made processes that lead to both the formation, and the breakdown, of the parallel ridge, slough and tree island patterns which characterized much of the historic Everglades peatlands. A second, related initiative deals with combining synoptic, system-wide measurements of water depths with spatially-explicit information on species composition to better understand the hydrologic optima and tolerances of the various community types. This information, in turn, is being used in the development of gridded, dynamic vegetation and community succession models for evaluating potential impacts of future hydrologic management and restoration plans.

Landscape Patterning and the Hydro-ecology of the Everglades

The important role that shallow, overland flow (called sheetflow) plays in shaping the Everglades ridge-slough habitat has been recognized for some time (SCT 2003, Ogden 2005). However, until recently, many of the physical and biogeochemical mechanisms which have been proposed to lead to ridge and slough formation and which are mediated by flow had not been measured. Early model simulations of the potential processes involved in the formation of this habitat were also inconclusive (Ross et al. 2006, Larsen et al. 2007). Hydrodynamics, differential peat accretion rates, nutrient dynamics, plant physiological properties, and sediment transport are now understood to play a role in the formation of the parallel vegetation bands (Larsen et al. 2010).

Some of the more recent advancements in understanding ridge-slough processes developed out of an improved understanding of the physical characteristics of sheetflow in and around emergent vegetation (e.g. Nepf 2004). These studies have shown that vegetative resistance to sheetflow changes with water depths, and this is largely because of the differences in vegetative frontal area and stem density with height above the substrate (Riscassi and Schfrannek 2004, Lee et al. 2004, Harvey et al. 2009, He et al. 2010). Investigations of large-scale sheetflow patterns have recently been made for the first time using methods of deliberate surface water tracer release and recovery. The goal of the Everglades Tracer Release Experiments (EverTREx) is to develop a method to measure sheetflow patterns in the ridge and slough habitat, and allow improved understanding of how large-scale advection and dispersion patterns in the Everglades reflect controlling factors such as water depths, vegetation, and hydrologic management. EverTREx represents the first application of sulphur hexafluoride (SF_6) in a shallow-water vegetated environment. SF_6, a gas tracer that has been applied previously in investigations of transport processes in estuarine and coastal waters (e.g. Ho et al. 2002, Caplow et al. 2003), has advantages over fluorescent dyes in that its concentration can be measured over a greater dynamic range, is considerably less expensive, and does not suffer from photodegradation, thus enabling experiments to be conducted over larger areas and over a longer time period (Ho et al. 2006). In these experiments, a SF_6 injection into the water column can be used to track advection and dispersion for as long as 10 days and over areas up to several square kilometres. Since 2006, six EverTREx campaigns have been carried out in the central Everglades marshes (Ho et al. 2009, Variano et al. 2009). The measured flow directions suggest that basin-scale forcing from water management structures and operations can override the effects of local landscape features such as the ridge and slough patterns in guiding the flow. Management effects were particularly evident in two regions where the historic, natural landscape patterning has degraded. In these degraded regions, the direction of sheetflow was significantly different from the historic orientation of the parallel ridge and slough patterns.

The improved understanding of the factors controlling flow velocities and flow patterns in the open marsh has led to a greater appreciation of the roles particulate and nutrient transport play in the formation and maintenance of the ridge and slough habitat. The most-widely accepted conceptual model of ridge and slough formation contains interacting feedback cycles that regulate local rates of peat accretion and vegetation growth that are regulated by water depths, stages, flow velocities, and the supply of P, a limiting nutrient (Davis 1994, Noe et al. 2001). Recent modelling results from Larsen and Harvey (2010) suggest that a specific set of relationships between water surface slope, water velocity, bed shear stress and sediment entrainment are required to form the ridge-slough habitat. These results also indicate that because of changes in vegetative hydraulic resistance that follow the expansion of *Cladium* or other emergent species into sloughs, the water surface slope and water depth conditions needed to restore degraded ridge-slough habitats to their original patterning will not match the conditions that gave rise to these formations. This suggests that a more active form of hydro-ecological engineering may be necessary to restore this habitat and that simply recreating the historic hydrologic conditions which gave rise to these formations may not be sufficient to meet restoration goals. This poses a unique challenge to restoration planners.

Regional Hydrologic Monitoring

Synoptic measurements of water surface elevations and water surface slopes throughout the Everglades have significantly improved the understanding of water movement and hydrology in the open marsh. These measurements are made possible by the recent completion of the Everglades Depth Estimation Network (EDEN) by the United States Geologic Survey (Palaseanu and Pearlstine 2008). EDEN is comprised of 253 automated water level recorders, precisely referenced to the same vertical datum and distributed throughout the central Everglades. Daily average water level data from these stations are routinely processed and are available on a public website (http://sofia.usgs.gov/eden). Interpolation of water levels between stations on a 400 m^2 grid is used to generate contour maps of geo-referenced water surface elevations (Palaseanu and Pearlstine 2008). These water surface elevation maps have provided remarkable insight into how the water surface changes in the Everglades with management operations and rainfall. Because all of the stations are georeferenced to the same datum, the EDEN water surface projections can also be used to calculate depths and hydroperiods at any location and at any height in the marsh. For example, these projections are now routinely used to estimate the water depths and hydroperiods of hundreds of tree islands with known elevations (Liu et al. 2009). Real-time EDEN data can be used to estimate daily the cumulative flooding or drought stress that these islands may be experiencing due to management, climate or both.

Nutrient Focusing and Everglades Tree Islands

Tree islands of the freshwater Everglades are rare upland habitats that support numerous species (including wading birds) and which hold tremendous cultural value for several stakeholder groups, including the local Native American tribes. Roughly half of the original tree island areal extent in WCA 2, WCA 3, and ENP has been lost due to stressors associated with hydrologic management, including fire (Patterson and Finck 1999, Sklar et al. 2004, Hofmockel et al. 2008). The loss of islands is not equally distributed. Some areas have lost more than 90 per cent of the original tree island areal extent, while other areas exhibit significantly less decline. Tree islands located in the ridge-slough habitat are elongated, teardrop-shaped features with the narrow end pointing in the downstream direction. This characteristic shape suggests flow is an important factor in their formation and restoring flow will be necessary to maintain these unique habitats.

Tree islands are characterized by very high soil P concentrations in the otherwise oligotrophic Everglades. Recently, Wetzel et al. (2010) proposed the Focused Nutrient Redistribution (FNR) model of tree island biogeochemistry, which suggests that patches of woody vegetation alter local biogeochemical cycling. According to the FNR model, tree islands concentrate resources (especially P) through a combination of evapotranspirational pumping of surface and groundwater (McCarthy and Ellery 1994, Ross et al. 2006), dry deposition (Weathers et al. 2001, Krah et al. 2004), and deposition of animal bones and feces under trees (Frederick and Powell 1994, Coultas et al. 2008).

The hypothesis of evapotranspirational pumping in the FNR model suggests that the high transpiration rates of tree island communities cause them to take up water and nutrients from the surrounding marsh and groundwater, especially during the dry season, and that nutrients gradually accumulate through this process (Wetzel et al. 2005). This hypothesis is supported by two previous studies (Saha et al. 2009, Saha et al. 2010) using [18]O isotopic data which show trees and ferns located on the head or near the tail use shallow soil water or marsh surface water if it is available, regardless of season. However, if the shallow soil and marsh surface water are not available, either because the marsh is dry, or if the trees are significantly higher than the surrounding marshes, then the island plants obtain water from the deeper groundwater sources.

The observations of fluctuations in deep and shallow groundwater on a few of these islands are considered as support of the hypothesis of evapotranspirational pumping (Wetzel et al. 2005, Ross et al. 2006). These fluctuations are presumably caused by a daytime drawdown of soil water from transpiring vegetation which then recovers at night with recharge from deep groundwater sources or from the surrounding marshes. The groundwater beneath these islands also exhibits a remarkable build-up of cations such as Na+. This increase in ionic strength is indicative of groundwater discharge beneath the islands driven by root water uptake and exclusion of dissolved salts. These results suggest diurnal groundwater fluctuations and the build-up of cations may provide a metric of "healthy" tree island function. However, it is not known if groundwater beneath degraded islands exhibits the same characteristics. Research is now being funded by the USACE to clarify the degree of hydrologic connectivity between marsh surface water, soil water, and groundwater in both disturbed and well-preserved tree islands.

Dynamic Vegetation Models and Community Succession

Knowledge of how plant species composition changes with the prevailing hydrologic conditions across the landscape provides a means to estimate the optimum water depths and hydroperiod (and tolerances) of the various Everglades plant communities (Ross et al. 2003). Richards and Gann (2009) for example, identified water depth, length of draw down periods and variability of mean annual water depth among the critical drivers of vegetation dynamics. Ross et al. (2003) suggested that a narrow threshold of 5–10 cm change in water depth or a 10–60 day hydroperiod change can alter the dominance of vegetation types. Change in fresh water marsh species abundance may occur rather quickly, but with a lag period of within 3–4 years (Zweig and Kitchens 2009, Armentano et al. 2006) following hydrological alterations.

The Everglades Landscape Vegetation Succession model (ELVeS), currently under development is a spatially-explicit simulation of vegetation community change over time in response to environmental drivers, including hydrology. The model uses empirically-based probabilistic functions of vegetation community niche space, disturbance responses, and temporal lags in expected community response on an annual basis. The model has been parameterized for WCAs 1, 2, 3 and ENP.

ELVeS has been developed to provide scientists, planners and decision-makers a simulation tool useful in landscape-scale analysis and restoration planning. The model is also intended for integration with wildlife models to provide a temporally-dynamic vegetation input layer. ELVeS will simulate succession and disturbance-related changes between 27 distinct community types. These communities span a wide suite of environmental conditions including freshwater marsh systems and wet prairies, tropical and temperate hammocks and upland pine forests, saline seagrass communities, mangroves, and coastal salt marshes. Six variables are input as drivers of community response: mean annual depth, standard deviation of mean annual depth, seventeen day moving average water depth maximum and minimum, soil loss on ignition, and soil total phosphorus. Fire and storm impacts and temporal lag effects are planned, but not yet implemented. Model rules defining niche spaces for vegetation communities are derived primarily from the literature, expert opinions of plant community responses to ecological drivers, and GIS statistical analysis of vegetation community relationships to mapped hydrological and soil metrics. Hydrological metrics were derived from daily continuous mapped surfaces of water depth provided by the EDEN (Palaseanu and Pearlstine 2008). Transitions between community states are defined with conditional probabilities. Future versions will weigh the probabilities by spatial neighbourhood community abundance and temporal lag periods specific to each respective plant community. ELVeS is written in Java as open source, freely distributed programme executables and source code to facilitate sharing, expansion and modification by a wide and diverse end user clientele. The ELVeS modelling framework has the capacity to integrate modules for climate change, hurricanes, and fire scenarios providing opportunity to explore potential habitat modifications for marine, freshwater, coastal vegetation and their dependent wildlife communities.

Conclusions

In many ways, the ecosystem services provided by the Everglades (e.g. water supply, flood control, rich soils) have supported the rapid growth of the south Florida economy. However, the ecosystem continues to degrade as a result of 150 years of the altered hydrology and poor water quality which have accompanied development in the region. Hydrologic management has also led to a 'flattening' of the landscape and widespread loss of soil due to oxidation. This altered topography presents one of the primary challenges to restoring more natural surface flow patterns in the Everglades. Recent modelling and field observations suggest pulsed, high flow events which transport and redistribute sediments may be necessary to both maintain and restore the ridge-slough patterning which characterized the historic Everglades freshwater marshes.

The restoration initiatives currently underway in the Everglades hold promise for improving ecosystem function over the short and long terms. However, the pace of restoration has been slowed by violations of the Clean Water Act and the requirements to reduce the nutrient content of water before it is discharged to the oligotrophic Everglades. Restoration has also been slowed by the difficulties in obtaining funding for the very large, very expensive construction projects which are needed to improve water quality across the landscape and to re-engineer the flood control and water supply system for the metropolitan areas of south Florida. These public-works projects are necessary before the hydrology and water quality of the Everglades can be managed separately from the needs of the adjacent municipal and agricultural areas. Because of the costs associated with such large projects, the regulatory framework in the United States requires that ecosystem benefits of restoring a more natural hydrology to the Everglades must be quantified. This presents a challenge to the ecologists in the region who are often asked to project the potential ecological outcomes of restoration alternatives with limited data on historic trends and imperfect knowledge of current ecosystem functioning and trophic interactions. New modelling suggests the trajectories the ecosystem may take from a degraded condition to a more "natural" or restored state will not simply be a reversal of the trajectory taken during the degradation. Several recent research and modelling initiatives are intended to further improve the ability to forecast the potential responses of the Everglades ecosystem to hydrologic manipulations. It is anticipated the products of these investigations will be incorporated into an iterative and adaptive management scheme for Everglades restoration, whereby the initial results will be used to inform basic engineering designs while subsequent monitoring of ecosystem parameters will be used to help refine operations and improve additional construction. This approach, if successful, holds promise for wetland management and restoration schemes throughout the US However, the slow pace by which restoration projects are authorized and funded present a significant barrier to preserving this ecosystem. Over the past two decades, sea level rise and new invasive plants and animals have been recognized as stressors likely to create widespread negative effects on the Everglades ecosystem. Non-stationary climate modes represent an additional stressor which must be considered in all future management schemes. Continued monitoring of ecosystem parameters and research into biogeochemical processes are necessary to distinguish the effects of such stressors from the potential benefits of upcoming restoration projects.

Acronyms

AMO - Atlantic Multi-Decadal Oscillation
C111, L29, L67A, L67C - levee-canal water control structures
C&SF - Central and Southern Florida Flood Control Project
CERP - Comprehensive Everglades Restoration Plan
DOI - US Department of the Interior
DPM - Decompartmentalization Physical Model
EAA - Everglades Agricultural Area
EDEN - Everglades Depth Estimation Network
ELVeS - Everglades Landscape Vegetation Succession model
ENP - Everglades National Park
ENSO - El-Nino Southern Oscillation
EPA - US Environmental Protection Agency
EverTREx - Everglades Tracer Release Experiments
FIFRA - Federal Insecticide, Fungicide, and Rodenticide Act
HEP-80 - Habitat Evaluation Procedures
HSI - Habitat Suitability Index
HUs - Habitat Units
LRR - Tamiami Trail Limited Re-evaluation Report
MWD - Modified Water Deliveries
NPS - US National Park Service
SFWMD - South Florida Water Management District

STAs - Storm water treatment areas
Ramsar - The Convention on Wetlands of International Importance, Ramsar, Iran, 1971
UN - United Nations
UNESCO - United Nations Educational, Scientific and Cultural Organization
USACE - United States Army Corps of Engineers
WCA - Water Conservation Areas
WRDA - Water Resources Development Act

References

Abtew, W. and Trimble, P.J. 2010. El Niño–Southern Oscillation link to South Florida hydrology and water management Applications. *Water Resources Management,* 24(15), 4255-4271, Available at: doi:10.1007/s11269-010-9656-2.

Ahn, H. and James, R.T. 2001. Variability, uncertainty, and sensitivity of phosphorus deposition load estimates in south Florida. Water, Air, and Soil Pollution, 126, 37-51.

Armentano, T.V., Jones, D.T., Ross, M.S. and Gamble, B.W. 2002. *Vegetation pattern and process in tree islands of the southern Everglades and adjacent area, in Tree Islands of the Everglades,* edited by F.H. Sklar and A. van der Valk. Dordrecht: Kluwer Academic Publishers, 225-281.

Armentano, T.V., Sah, J.P., Ross, M., Jones, D.T., Cooley, H.C. and Smith, C.S. 2006. Rapid response of vegetation to hydrological changes in Taylor Slough, Everglades National Park, Florida, USA. *Hydrobiologia,* 569, 293-309.

Bates, A.L., Orem, W.H., Harvey, J.W. and Spiker, E.C. 2002. Tracing sources of sulfur in the Florida Everglades. *Journal of Environmental Quality,* 31, 287-299.

Beckage, B., Platt, W.J., Slocum, M.G. and Panko, B. 2003. Influence of the El Niño Southern Oscillation on fire regimes in the Florida Everglades. *Ecology,* 84(12), 3124-3130.

Bernhardt, C.E. and Willard, D.A. 2009. Response of the Everglades' ridge and slough landscape to climate variability and 20th century water-management. *Ecological Applications,* 19, 1723-1738.

Bernhardt, C.E., Willard, D.A., Marot, M. and Holmes, C.W. 2004. Anthropogenic and natural variation in ridge and slough pollen assemblages. USGS Open-File Report 2004-1448. Reston, VA: US Geological Survey.

Brandt, L.A., Silveira, J.E. and Kitchens, W.M. 2002. Tree islands of the Arthur R. Marshall Loxahatchee National Wildlife Refuge, in *Tree Islands of the Everglades,* edited by F.H. Sklar and A.van der Valk. Dordrecht: Kluwer Academic Publishers, 311-335.

Brewster-Wingard, G.L. and Ishman, S.E. 1999. Historical trends in salinity and substrate in central Florida Bay: A paleoecological reconstruction using modern analogue data. *Estuaries and Coasts,* 22(2), 369-383, Available at: doi: 10.2307/1353205.

Browder, J.A., Gleason, P.J. and Swift, D.R. 1994. Periphyton in the Everglades: spatial variation, environmental correlates and ecological implications, in *Everglades: The Ecosystem and its Restoration,* edited by S.M. Davis and J.C. Ogden. Delray Beach: St. Lucie Press, 379-418.

Caplow, T., Schlosser, P., Ho, D.T. and Santella, N. 2003. Transport dynamics in a sheltered estuary and connecting tidal straits: SF6 tracer study in New York Harbor. *Environmental Science and Technology,* 37, 5116-5126.

Carriger, J.F. and Rand, G.M. 2008a. Aquatic Risk Assessment of Pesticides in Surface Waters in and Adjacent to the Everglades and Biscayne National Parks: I. Hazard Assessment and Problem Formulation. *Ecotoxicology,* 17(7), 660-679.

Carriger, J.F. and Rand, G.M. 2008b. Aquatic Risk Assessment of Pesticides in Surface Waters in and Adjacent to the Everglades and Biscayne National Parks: II. Probabilistic Analyses. *Ecotoxicology,* 17(7), 680-696.

Coultas, C.L., Schwandron, M. and Galbraith, J.M. 2008. Petrocalcic horizon formation and prehistoric people's effect on Everglades Tree Island Soils, Florida. *Soil Survey Horizons,* 49, 16-21.

Craft, C.B. and Richardson, C.J. 1993. Peat accretion and nitrogen, phosphorus and organic carbon accumulation in nutrient enriched and unenriched Everglades peatlands. *Ecological Applications,* 3, 446-458.

Craft, C.B. and Richardson, C.J. 2008. Soil characteristics of the Everglades peatland, in *The Everglades Experiments: Lessons for Ecosystem Restoration,* edited by C.J. Richardson. New York: Springer, 59-72.

Duever, M.J., Meeder, J.F., Meeder, L.C. and McCollom, J.M. 1994. The climate of south Florida and its role in shaping the Everglades ecosystem, in *Everglades: The Ecosystem and its Restoration,* edited by S.M. Davis and J.C. Ogden, St. Lucie Press, Delray Beach, 225-248.

Enfield, D.B., Mestas-Nunez, A.M. and Trimble, P.J. 2001. The Atlantic Multidecadal Oscillation and its relationship to rainfall and river flows in the continental. *US Geophysical Research Letters,* 28, 2077-2080.

Fennema, R.L., Neidrauer, C.J., Johnson, R.J., MacVicar, T.K. and Perkins, W.A. 1994. A Computer Model to Simulate Natural Everglades Hydrology, in *Everglades: The Ecosystem and its Restoration,* edited by S.M. Davis and J.C. Ogden. Delray Beach: St. Lucie Press, 533-570.

Frederick, P.C. and Ogden, J.C. 2001. Pulsed breeding of long–legged wading birds and the importance of infrequent severe drought conditions in the Florida Everglades. *Wetlands,* 21(4), 484-491.

Frederick, P.C. and Powell, G.V.N. 1994. Nutrient transport by wading birds in the Everglades, in *Everglades: The Ecosystem and its Restoration,* edited by S.M. Davis and J.C. Ogden. Delray Beach: St. Lucie Press, 571-584.

Frederick, P.C. and Spalding, M.G. 1994. Factors affecting reproductive success of wading birds (Ciconiiformes), in *Everglades: The Ecosystem and its Restoration* edited by S.M. Davis and J.C. Ogden. Delray Beach: St. Lucie Press, 659-691.

Gaiser, E.E., Childers, D.L., Jones, R.D., Richards, J.H., Scinto, L.J. and Trexler, J.C. 2006. Periphyton responses to eutrophication in the Florida Everglades: Cross-system patterns of structural and compositional change. *Limnology and Oceanography*, 51, 617-630.

Gawlik, D.E. 2002. The effects of prey availability on the numerical response of wading birds. *Ecological Monographs*, 72, 329-346.

Givnish, T.J., Volin, J.C., Owen, V.D., Volin, V.C., Muss, J.D. and Glaser, P.H. 2008. Vegetation differentiation in the patterned landscape of the central Everglades: importance of local and landscape drivers. *Global Ecology and Biogeography*, 17, 384-402.

Gleason, P.J., Cohen, A.D., Brooks, H.K., Stone, P.A., Goodrick, R., Smith, W.G. and Spackman, W. Jr. 1974. The Environmental Significance of Holocene Sediments from the Everglades and Saline Tidal Plain, in *Environments of South Florida: Past and Present*, edited by P. J. Gleason. Miami: Miami Geological Society, 146-181.

Gleason, P.J. and Stone, P.A. 1994. Age, origin and landscape evolution of the Everglades peatlands, in *Everglades: The Ecosystem and its Restoration*, edited by S.M. Davis and J.C. Ogden. Delray Beach: St. Lucie Press, 149-198.

Godfrey, M.C. 2006. *River of Interest: Water Management in South Florida and the Everglades, 1948-2000*. Belfast: Historical Research Associates.

Goldenberg, S.B., Landsea, C.W., Mestas-Nuñez, A.M. and Gray, W.M. 2001. The recent increase in Atlantic hurricane activity: causes and implications. Science 293, 474-479, Available at: doi:10.1126/science.1060040.

Graf, M.T., Schwadron, M., Stone, P.A., Ross, M. and Chmura, G.L. 2008. An enigmatic carbonate layer in Everglades tree island peats. *Eos Trans. AGU*, 89(12), Available at: doi:10.1029/2008EO120001.

Gunderson, L.H. 1994. Vegetation of the everglades: Determinants of community composition, in *Everglades: The Ecosystem and its Restoration*, edited by S.M. Davis and J.C. Ogden. Delray Beach: St. Lucie Press, 323-340

Gunderson, L.H. and Loftus, W.T. 1993. The Everglades, in *Biodiversity of the Southeastern United States: Lowland Terrestrial Communities*, edited by W.H. Martin, S.G. Boyce, and A.C.E. Echtemacht. New York: Wiley, 199-255.

Hagerthey, S.E., Newman, S., Rutchey, K., Smith, E.P. and Godin, J. 2008. Multiple regime shifts in a subtropical peatland: community-specific thresholds to eutrophication. *Ecological Monographs*, 78, 547-565.

Haman, D.Z. and Svedsen, M. 2006. Managing the Florida Everglades: changing values, changing policies. *Irrigation and Drainage Systems*, 20, 283-302.

Harvey, J.W. and McCormick, P.V. 2009. Groundwater's significance to changing hydrology, water chemistry, and biological communities of a floodplain ecosystem, Everglades, South Florida, USA. *Hydrogeology*, 17, 185-201.

Harvey, J.W., Newlin, J.T. and Krupa, S.L. 2006. Modeling decadal timescale interactions between surface water and ground water in the central Everglades, Florida, USA. *Journal of Hydrology*, 320, 400-420.

Harvey, J.W., Schaffranek, R.W., Noe, G.B., Larsen, L.G., Nowacki, D. and O'Connor, B.L. 2009. Hydroecological factors governing surface-water flow on a low-gradient floodplain. *Water Resources Research* 45, W03421, Available at: doi:10.1029/2008WR007129.

He, G., Engel, V., Leonard, L., Croft, A., Childers, D.L., Laas, M., Deng, Y. and Solo-Gabriele, H. M. 2010. Factors controlling surface water flow in a low-gradient subtropical wetland. *Wetlands*, in press.

Ho, D.T., Schlosser, P. and Caplow, T. 2002. Determination of longitudinal dispersion coefficient and net advection in the tidal Hudson River with a large-scale, high resolution SF6 tracer release experiment, *Environmental Science and Technology,* 36, 3234 -3241.

Ho, D. T., Schlosser, P., Houghton, R. and Caplow, T. 2006. Comparison of SF6 and Fluorescein as tracers for measuring transport processes in a large tidal river, *Journal of Environmental Engineering*, 132, 1664-1669.

Ho, D., Engel, V., Variano, E., Schmieder, P. and Condon M.E., 2009. Tracer studies of sheetflow in the Florida Everglades. *Geophysical Research Letters* 36, L09401, Available at: doi:10.1029/2009GL037355.

Hofmockel, K., Richardson, C.J. and Halpin, P.N. 2008. Effects of hydrologic management decisions on Everglades tree Islands, in *The Everglades experiments: lessons for ecosystem restoration*, edited by C.J. Richardson. New York: Springer, 191-214.

Holloway, C. 2007. *Project Planning in Collaboration with Government Entities*: *Practical Approaches, United States Army Corps of Engineers* Report 07-R-2. Alexandria: US Army Corps of Engineers, Institute for Water Resources.

Huang, Y.H., Saiers, J.E., Harvey, J.W., Noe, G.B. and Mylon, S. 2008. Advection, dispersion, and filtration of fine particles within emergent vegetation of the Florida Everglades. *Water Resources Research,* 44, W04408, Available at: doi:10.1029/2007WR006290.

Kadlec, R.H. 1994. Phosphorus uptake in Florida marshes. *Water Science and Technology*, 30, 225-234.

Koch, M.S. and Reddy, K.R. 1992. Distribution of soil and plant nutrients along a trophic gradient in the Florida Everglades. *Soil Science Society of America* 56, 1492-1499.

Krah, M., McCarthy, T.S., Annegarn, H. and Ramberg, L. 2004. Airborne dust deposition in the Okavango delta, Botswana, and its impact on landforms. *Earth Surface Processes and Landforms* 29, 565–577.

Kwon, H.H., Lall, U., Moon, Y.I., Khalil, A.F. and Ahn, H. 2006. Episodic interannual climate oscillations and their influence on seasonal rainfall in the Everglades National Park. *Water Resources Research*, 42, W11404, Available at: doi: 10.1029/2006WR0050.

Larsen, L.G., Harvey, J.W. and Crimaldi, J.P. 2007. A delicate balance: ecohydrological feedbacks governing landscape morphology in a lotic peatland. *Ecological Monographs*, 77(4), 591-614.

Lee, J.K., Roig, L.C., Jenter, H.L. and Visser, H.M. 2004. Drag coeffcients for modeling flow through emergent vegetation in the Florida Everglades, *Ecological Engineering,* 22, 237-248

Leonard, L., Croft, A., Childers, D., Mitchell-Bruker, S., Solo-Gabriele, H. and Ross, M. 2006. Characteristics of surface-water flows in the ridge and slough landscape of Everglades National Park: implications for particulate transport, *Hydrobiologia,* 569, 5-22

Light, S. and Dineen, J.W. 1994. Water Control in the Everglades: A Historical Perspective, in *Everglades: The Ecosystem and its Restoration,* edited by S.M. Davis and J.C. Ogden, St. Lucie Press, Delray Beach, 47-84.

Liu, Z., Volin, J.C., Owen, V.D., Pearlstine, L.G., Allen, J.R., Mazzotti, F.J. and Higer, A.L. 2009. Validation and ecosystem applications of the EDEN water-surface model for the Florida Everglades. *Ecohydrology,* 2, 182-194.

Lockwood, J.L., Fenn, K.H., Caudill, J.M., Okines, D., Bass, O.L., Duncan, J.R. and Pimm, S. L. 2001. The implications of Cape Sable Seaside Sparrow demography for Everglades restoration. *Animal Conservation,* 4, 275-281.

Lorenz, J.J., Langan-Mulrooney, B., Frezza, P.E., Harvey, R.G. and Mazzotti, F.J. 2009. Roseate spoonbill reproduction as an indicator for restoration of the Everglades and the Everglades estuaries. *Ecological Indicators,* 9(6, S1), S96-S107.

Mazzotti, F.J. and Brandt, L.A. 1994. Ecology of the American alligator in a seasonally fluctuating environment, in *Everglades: The Ecosystem and its Restoration,* edited by S.M. Davis and J.C. Ogden. Delray Beach: St. Lucie Press, 485-506.

McCarthy, T.S. and Ellery, W.N. 1994. The effect of vegetation on soil and ground water chemistry and hydrology of islands in the seasonal swamps of the Okavango Fan, Botswana. *Journal of Hydrology,* 154,169-193.

McCormick, P.V., Newman, S. and Vilchek, L.W. 2009. Landscape responses to wetland eutrophication: loss of slough habitat in the Florida Everglades, USA. *Hydrobiologia,* 621, 105-114.

McCormick, P.V. 2010. Soil and periphyton indicators of anthropogenic water-quality changes in a rainfall-driven wetland. *Wetlands Ecology and Management,* Available at: doi:10.1007/s11273-010-9196-9.

Mcpherson, B. F. and Halley, R. 1997. *The South Florida Environment: A Region under Stress.* US Geological Survey Circular 1134. Reston: US Geological Survey.

McVoy, C.W., Said, W.P., Obeysekera, J., VanArman, J. and Dreschel, T. *Landscapes and Hydrology of the Pre-drainage Everglades.* University Press of Florida, Gainesville, FL. *In press.*

Nepf, H.M. 2004. Vegetated flow dynamics, in *The Ecogeomorphology of Tidal Marshes,* edited by S. Fagherazzi, M. Marani and L.K. Blum. Washington, DC: American Geophysical Union, 137-163.

Noe, G.B., Childers, D.L. and Jones, R.D. 2001. Phosphorus biogeochemistry and the impact of phosphorus enrichment: Why is the Everglades so unique? *Ecosystems,* 4, 603-624.

Noe, G.B., Harvey, J.W., Schaffranek, R.W. and Larsen, L.G. 2010. Controls of suspended sediment concentration, nutrient content, and transport in a subtropical wetland. *Wetlands,* 30, 39-54.

Nott, M.P., Bass, O. L. Jr., Fleming, D. M., Killefer, S. E., Fraley, N., Manne, L., Curnutt, J. L., Brooks, T. M., Powell, R. and Pimm, S. L. 1998. Water levels, rapid vegetational change, and the endangered Cape Sable Seaside-Sparrow. *Animal Conservation,* 1, 23-32.

Ogden, J.C. 1994. A comparison of wading bird nesting colony dynamics (1931-1946 and 1974-1989) as an indication of ecosystem conditions in the southern Everglades, in *Everglades: The Ecosystem and its Restoration,* edited by S.M. Davis and J.C. Ogden. Delray Beach: St. Lucie Press, 533-570.

Ogden, J.C., 2005. Everglades ridge and slough conceptual ecological model. *Wetlands,* 25, 810-820.

Orem, W.H., Willard, D.A., Lerch, H.E., Bates, A.L., Boylan, A. and Corum, M. 2002. Nutrient geochemistry of sediments from two tree islands in Water Conservation 3B, the Everglades, Florida, in *Tree islands of the Everglades,* edited by F.H. Sklar and A. van der Valk. Dordrecht: Kluwer Academic Publishers, 153-186.

Orem, W.H., 2007. *Sulfur Contamination in the Florida Everglades: Initial Examination of Mitigation Strategies.* USGS Open-File Report 2007-1374. Reston: US Geological Survey.

Palaseanu, M. and Pearlstine, L. 2008. Estimation of water surface elevations for the Everglades, Florida, *Computers and Geosciences,* 34, 815-826.

Pimm, S.L., Lockwood, J.L., Jenkins, C.N., Curnutt, J.L., Nott, P., Powell, R.D. and Bass, O.L. Jr. 2002. *Sparrow in the Grass: A Report on the First Ten Years of Research on the Cape Sable Seaside Sparrow* (*Ammodramus maritimus mirabilis*). Report to Everglades National Park, Homestead, FL.

Patterson, K. and Fink, R. 1999. *Tree islands of the WCA-3 Aerial Photointerpretation and Trend Analysis Project Summary Report.* Report to the South Florida Water Management District by Geonex Corporation, St Petersburg, FL.

Price, R.M., Swart, P.K. and Fourqrean, J.W. 2006. Coastal groundwater discharge-an additional source of phosphorus for the oligotrophic wetlands of the Everglades, *Hydrobiologia,* 569, 23-36.

Rand, G.M. et al., 2010. Endosulfan and its Metabolite, Endosulfan Sulfate, in Freshwater Ecosystems of South Florida: A Probabilistic Aquatic Ecological Risk Assessment; *Ecotoxicology,* 19, 879-900.

Richards, J.H., Childers, D.L., Ross, M., Lee, D. and Scinto, L. 2009. *Hydrological Restoration Requirements of Aquatic Slough Vegetation.* Final Report to Everglades National Park, Homestead, FL.

      Tropical Wetland Management

Richardson, C.J. 2010. The Everglades: North America's subtropical wetland. *Wetlands Ecology and Management,* Available at: doi: 10.1007/s11273-009-9156-4.

Riscassi, A.L. and Schaffranek, R.W., 2004. *Flow Velocity, Water Temperature, and Conductivity in Shark River Slough, Everglades National Park, Florida: June 2002-July 2003.* USGS Open-File Report 03-348. Reston: US Geological Survey.

Ross, M.S., Sah, J., Ruiz, P.L. and Lewin, M.T. 2003. Vegetation: environment relationships and water management in Shark Slough, Everglades National Park, *Wetlands Ecological and Management,* 11, 291-303.

Ross, M.S., Sah, J.P., Ruiz, P.L., Jones, D.T., Cooley, H.C., Travieso, R., Snyder, J.R. and Robinson, S. 2004. *Effect of Hydrology Restoration on the Habitat of the Cape Sable Seaside Sparrow.* Report to Everglades National Park, Homestead, FL.

Ross, M.S., Mitchell–Bruker, S. Sah, J.P., Stothoff, S., Ruiz, P.L., Reed, D.L., Jayachandran, K. and Coultas, C.L. 2006. Interaction of hydrology and nutrient limitation in the ridge and slough landscape of the southern Everglades. *Hydrobiologia,* 569, 37–59.

Ross, M. S., J. P. Sah, P. L. Ruiz, D. T. Jones, H. Cooley, R. Travieso, F. Tobias, J.R. Snyder and D. Hagyari. 2006. *Effect of Hydrologic Restoration on the Habitat of the Cape Sable Seaside Sparrow. Annual Report of 2004-05.* Homestead: South Florida Natural Resource Center.

Russell, G. J., Bass, O.L. and Pimm, S.L. 2002. The effect of hydrological patterns and breeding-season flooding on the numbers and distribution of wading birds in Everglades National Park, *Animal Conservation,* 5, 185-199.

Sah, J. P., Ross, M.S., Snyder, J.R., Ruiz, P.L., Stoffella, S., Colbert, N., Hanan, E., Lopez, L., and Camp, M. 2010. *Cape Sable Seaside Sparrow Habitat - Vegetation Monitoring.* FY 2009 Final Report submitted to U. S. Army Corps of Engineers, Jacksonville, FL.

Saha, A.K., Sternberg, L.S.L. and Miralles-Wilhelm, F. 2009. Linking water sources with foliar nutrient status in upland plant communities in the Everglades National Park, USA. *Ecohydrology,* 2, 42-54.

Saha, A.K., Sternberg, L.S.L., Ross, M.S. and Miralles-Wilhelm. F. 2010. Water source utilization and foliar nutrient status differs between upland and flooded plant communities in wetland tree islands. *Wetlands Ecology and Management,* 18(3), 343-355.

Scott, G.I. et al. 2002. Toxicological Studies in Tropical Ecosystems: An Ecotoxicological Risk Assessment of Pesticides Runoff in South Florida Estuarine Ecosystems, *Journal of Agricultural and Food Chemistry* 50, 4400-4408.

SCT, Science Coordination Team, 2003. *The Role of Flow in the Everglades Ridge and Slough Landscape, South Florida.* Ecosystem Restoration Working Group. National Research Council. Available at: http://sofia.usgs.gov/publications/papers/sct_flows [accessed: 10 October 2010]

Shafland, P.L., Gestring, K.B. and Stanford, M.S. 2008. Florida's exotic freshwater fishes. *Florida Scientist,* 7(3), 220-245.

Sklar, F. H. and Van Der Valk, A. G. 2002. Tree islands of the Everglades: an overview, in *Tree Islands of the Everglades*, edited by F.H. Sklar and A.van der Valk. Dordrecht: Kluwer Academic Publishers, 1-18.

SFWMD. 2006. *South Florida Environmental Report for 2005*. West Palm Beach: South Florida Water Management District.

Smith, E.P., Orvos, D.R. and Cairns, J. Jr. 1993. Impact assessment using the before-after control- impact (BACI) model: Concerns and comments. *Canadian Journal of Fisheries and Aquatic Sciences,* 50, 627-637.

Todd, M.J., Muneepeerakul, R., Pumo, D., Azaele, S., Miralles-Wilhelm, F., Rinaldo A. and Rodriguez-Iturbe, I. 2010. Hydrological drivers of wetland vegetation community distribution within Everglades National Park, Florida. *Advances in Water Resources,* Available at: doi:10.1016/j.advwatres.2010.04.003.

Trexler, J.C., Loftus, W.F. Jordan, F., Lorenz, J.J., Chick, J.H. and Kobza, R.M. 2000. Empirical assessment of fish introductions in a subtropical wetland: an evaluation of contrasting views. *Biological Invasions,* 2, 265-277.

Trexler, J. C., Loftus, W.F. Jordan, F., Chick, J.H., Kandl, K.L., McElroy, T.C. and Bass, O.L. Jr. 2002. Ecological scale and its implications for freshwater fishes in the Florida Everglades, in *The Florida Everglades, Florida Bay, and Coral Reefs of the Florida Keys: an Ecosystem Sourcebook*, edited by J. W. Porter and K. G. Porter. New York: CRC Press, 153-181.

Urban, N.H., Davis, S.M. and Aumen, N.G. 1993. Fluctuations in sawgrass and cattail density in Everglades Water Conservation Area 2A under varying nutrient, hydrologic and fire regimes. *Aquatic Botany* 46, 203-223.

USACE. 2000. *Planning Guidance Notebook.* Engineering Regulations ER 1105-2-100 and ER 1165-2-501. Washington DC: USACE.

Van Der Valk, A. G., Wetzel, P., Cline, E. and Sklar, F. H. 2008. Restoring Tree Islands in the Everglades: experimental studies of tree seedling survival and growth. *Restoration Ecology,* 16, 281-289.

Variano, E.A., Ho, D.T., Engel, V.C., Schmieder, P.J. and Reid, M.C., 2009. Flow and mixing dynamics in a patterned wetland: kilometer-scale tracer releases in the Everglades. *Water Resources Research* 45, W08422, Available at: doi:10.1029/2008WR007216.

Wang, X., Sternberg, L.S.L., Ross, M. and Engel, V.C. 2010. Linking water use and nutrient accumulation in tree island upland hammock plant communities in the Everglades National Park, USA. *Ecosystems. In press*

Watts, D.L., Cohen, M.J., Heffernan, J.B. and Osborne, T.Z. 2010. Hydrologic modification and the loss of self-organized patterning in the ridge–slough mosaic of the Everglades. *Ecosystems,* 13(6), 813-827.

Weathers, K.C., Cadenasso, M.L. and Pickett, S.T.A. 2001. Forest edges as nutrient and pollutant concentrators: potential synergisms between fragmentation, forest canopies, and the atmosphere. *Conservation Biology* 15, 1506-1514.

Wetzel P.R., van der Valk A.G. and Newman S. et al. 2005. Maintaining tree islands in the Florida Everglades: nutrient redistribution is the key. *Frontiers in Ecology and Environment,* 3(7), 370-376.

Wetzel, P.R., van der Valk, A.G., Newman, S., Coronado, S., Troxler-Gann, T.G., Childers, D. L., Orem, W.H. and Sklar, F.H. 2009. Heterogeneity of phosphorus distribution in a patterned landscape, the Florida Everglades. *Plant Ecology,* 200, 83-90.

Wetzel P.R., Sklar, F.H., Coronado, C.A., Troxler, T.G., Krupa, S.L., Sullivan, P.L., Ewe, S., Price, R.M., Newman, S. and Orem, W.H. 2010. Biogeochemical processes on tree islands in the Greater Everglades: initiating a new paradigm. *Critical Reviews in Environmental Science and Technology. In press*

Willard, D.A., Bernhardt, C.E., Holmes, C.W., Landacre, B. and Marot, M. 2006. Response of Everglades tree islands to environmental change. *Ecological Monographs,* 76, 565-583.

Willard, D.A., Holmes, C.W., Korvela, M.S., Mason, D., Murray, J.B., Orem, W.H. and Towles, D.T. 2002. Paleoecological insights on fixed tree island development in the Florida Everglades: Environmental controls, in *Tree Islands of the Everglades,* edited by F.H. Sklar and A. van der Valk. Dordrecth: Kluwer Academic Publishers, 117-152.

Wu, Y.G., Wang, N.M. and Rutchey, K. 2006. An analysis of spatial complexity of ridge and slough patterns in the Everglades ecosystem. *Ecological Complexity,* 3, 183-192

Zweig, C.L. and Kitchens, W.M. 2008. Effects of landscape gradients on wetland vegetation communities: Information for large-scale restoration. *Wetlands,* 28, 1086-1096.

Zweig, C.L. and Kitchens, W.M. 2009. Multi-state succession in wetlands: a novel use of state and transition models. *Ecology,* 90, 1900-1909.

Chapter 12

Wetland Management Challenges in the South-American Pantanal and the International Experience

Wolfgang J Junk and Catia N da Cunha

Introduction

The Pantanal plays a major role in the worldwide discussion of sustainable management and protection of wetlands. The Brazilian constitution has declared the Pantanal as National Patrimony. The government has fostered and supported the creation and operation of the Pantanal Research Center (CPP) as well as it has financed the construction of the Pantanal Research Institute at the Federal University of Mato Grosso (UFMT) in Cuiabá. This institute, together with CPP and the university's recently established research network, the National Institute for Science and Technology in Wetlands (INCT-INAU), will provide additional momentum to current research and management activities.

All such effort requires a thorough "on-the-ground" situation analysis in order to fine tune scientific and technological planning, for an efficient and effective wetlands management strategy. This book has summarized the basic data on the social, hydrological, and ecological aspects of the Pantanal (Chapters 2, 3, 4, 5), present a model for the establishment of protected areas (Chapter 6), provide a detailed analysis of the stakeholders and their values (Chapters 7, 8), and discuss, for purposes of comparison, the management of the Okavango Delta, European wetlands, and the Everglades (Chapters 9, 10, 11). The aim of this chapter (Chapter 12) is to provide a synthesis of the information presented in these previous chapters, with an analysis that draws upon additional information (Junk et al. 2011) and offers some perspectives on the future of the Pantanal.

Geological and Geographical Peculiarities

The geological and geographical peculiarities of the Pantanal were addressed by Calheiros et al., Figueiredo et al., and Ide et al. in Chapters 3, 4, and 5, respectively. The Pantanal occupies the central part of the Cuiabá Depression and can be considered as a sediment deposition basin of the Paraguai River and its major tributaries. These form internal deltas in the Pantanal, with the largest being that of the Taquari River. The individual flood pattern and the dynamics of the internal deltas are reflected in the distinct sub-regions of the Pantanal (Brasil 1983, Hamilton et al. 1996, Silva and Abdon 1998) and its large geomorphological diversity.

Total precipitation in the Pantanal decreases from about 1,150 mm yr[1] in the north to 850 mm yr[1] in the south. The explicit seasonality of the precipitation leads to dry and wet periods of about four and six months, respectively, with brief transition periods. The natural vegetation cover is typical savannah (*cerrado*) in the upland and "hyperseasonal savannah," which is a savannah vegetation with many wetland-specific elements (Eiten 1983).

On the Brazilian side, the Cuiabá Depression is surrounded by tablelands and mountains of different ages and chemical composition that provide much of the water and sediments to the Pantanal. The volume and chemical composition of the dissolved and solid components of the waters of the tributaries reflect the mineralogy of their catchment areas, thereby increasing the diversity in environmental parameters at a regional scale, as shown in Chapter 4. However, these parameters are increasingly affected by human impacts, e.g., the use of upland areas for large-scale agriculture and cattle ranching as well as gold and diamond mining which dramatically increase the sediment load of most of the tributaries.

Hydrology, Water and Soil Quality, and Ecology

Calheiros et al. (Chapter 3), Figueiredo et al. (Chapter 4), and Ide et al. (Chapter 5) summarise the hydrology in the tributaries and the physicochemical conditions in the soils and water bodies of the Pantanal. The electrolyte content in the major tributaries of the Pantanal varies between 25 and 260 uS cm[-1], with considerable fluctuations between the dry and rainy seasons. The small upland streams are characterized by electrolyte contents as low as 3-5 uS cm[-1]. Inside the Pantanal, internal biogeochemical processes, local rainfall, and evapotranspiration considerably modify water quality. Salinas in the southern Pantanal have brackish water and their electrical conductivity can exceed 5,000 uS cm[-1] because of long-term hydrological isolation and increased evapotranspiration (Chapters 3 and 4).

The ecological processes in the Pantanal are described by the flood pulse concept (Junk et al. 1989, 2005, Junk and Wantzen, 2004). During rising and high water levels, the water bodies of the floodplain are dominated by aquatic macrophytes. At falling and low water levels, many water bodies present the highest concentrations of phytoplankton (Oliveira and Calheiros 2000, Loverde-Oliveira et al. 2011). The decomposition of organic matter together with bioturbation can dramatically increase electrical conductivity and nutrient loads, leading to blooms of cyanobacteria. Biological activity is documented by strong diurnal oscillations in oxygen content, with oxygen concentrations being lowest in the early morning because of high consumption and low production rates and by an explicit annual seasonality, expressed by lowest dissolved oxygen values during the rising water period, characterizing a phenomenon locally called as "*decoada*" (Oliveira and Calheiros 2000).

Although the water quality in the Pantanal has become increasingly affected by human activities, documentation has been insufficient. The lack of efficient collection and treatment systems for the solid and liquid waste produced in the major cities of the region has resulted in pollution of the Cuiabá River. Currently, industrial waste is a minor problem because of the absence of industries in the catchment areas. During the 1990s, gold mining led to mercury contamination, mainly in the Poconé, but with better regulation of mining activities further contamination has for the most part been prevented. The large agroindustries are becoming major sources of pollution by releasing large quantities of agrotoxics into the environment, thus threatening the water bodies of the Pantanal (Calheiros et al., Chapter 3, Figueiredo et al., Chapter 4). An additional major problem created by the agroindustrial activities is soil erosion. The eroded material enters the Pantanal via its tributaries, becomes deposited in the river channels, and leads to changes in the hydrology of major areas. At the beginning of the 1990s, soil erosion caused congestion of the Taquari River channel and the rupture of its natural levees. The river now inundates for much longer periods than before 5,000 to 11,000 km^2, with dramatic consequences not only for the environment but also for local ranchers (Ide et al., Chapter 5).

Political and Legal Aspects

In 1993, Brazil signed the Ramsar Convention, which requires a national policy for the protection and wise management of wetlands and their organisms. However, the conduction of inventories which is the scientific basis for a classification remains poor (Diegues 1994, 2002) and the government is slow to take effective actions. According to Junk (in press), about 20% of Brazil's land surface fulfills international criteria for wetlands but; unlike Europe and the USA (Ball, Chapter 10; Engel et al. Chapter 11), the Brazilian constitution does not consider wetlands as a specific landscape or vegetation category, nor as aquatic resources. Aquatic resources are rivers, lakes, and ground water for domestic and industrial use, irrigation, navigation, fishery, waste water purification, and hydroelectric power generation. This viewpoint is reinforced in the National Water Resources Plan and the State Plans of Water Resources of Mato Grosso and Mato Grosso do Sul (Figueiredo et al., Chapter 4). The distinction of water bodies in five quality classes, elaborated by the Ministry of Environment (MMA 2010), shows the same restricted view. The important and complex role of wetlands in the hydrological regime is therefore completely ignored. This holds true for swampy areas in the savannah catchments (i.e., *veredas*), small riparian wetlands along low-order streams, and large floodplains such as the Pantanal. The serious failure of the Brazilian legislation to protect the country's wetlands has lead to wetland destruction, with inevitably affect water quality and future availability.

Moreover, this will severely aggravate the predicted impacts of climate change on the Brazilian catchments in general and the Pantanal wetland more specifically (Junk, in press). Comparatively other developed and underdeveloped countries, are taking the impact of global climate change on wetlands much more seriously, e.g., such as Botswana, where climate change is considered as one of the major threats for the Okavango Delta (Vanderpost et al., Chapter 9).

As discussed by Ball (Chapter 10), the European Community has developed a very complex Water Framework Directive (WFD) to improve rivers, lakes, groundwater bodies, and transitional/ estuarine and coastal waters, with the aim of achieving their "good ecological status." Wetlands are to a certain extent implicated in the WFD. The Integrated Water Resource Management approach (IWRM) is shifting the narrow economic view of water towards one that considers ecological aspects, including the relationship to public well-being. Certainly, the problems related to wetlands in Western Europe are different from those in the Pantanal. However, in the industrialized areas of Brazil, the problems with water pollution and wetland destruction are similar to those encountered in many parts of Europe and the USA, but without a regulatory framework to deal with them. Therefore, the WFD and the "Clean Water Act" of USA can be considered interesting models for the improvement of the Brazilian National Water Resource Plan.

Modern wetland-friendly approaches, such as the Environmental Flow Assessment (EFA, King et al. 1999, Tharme 2003, Nel 2011), successfully established in Australia and several South African countries, have not yet been discussed in Brazil. The EFA requires well-defined amounts of water to be provided to the wetlands to maintain their principle structures and functions. With respect to the Pantanal, the applications of the principles of EFA would, for instance, force reservoir owners to consider the ecological requirements of the Pantanal and not only the maximization of energy production. The many reservoirs that have been proposed for the headwaters of the Pantanal's major tributaries must be integrated within a master plan that adjusts the release of water to the natural flood regime, thus combining energy production with the hydrological requirements of the Pantanal and minimizing the negative environmental impacts as well as leaving the reminiscent free flowing rivers without any barriers or dams (Calheiros et al., Chapter 3).

The Pantanal, despite being a floodplain with specific environmental conditions, is subjected to the same rules and regulations applied to all other Brazilian regions. The Brazilian Forest Code, established by law no. 4771/65, and actually under discussion in the Brazilian senate, regulates the protection and use of forests, with the aim of protecting their soils and water as well as the stability of timber markets. In 2002, the National Environmental Council (CONAMA) established, with resolution no. 302, parameters, definitions and limits of areas of permanent protection (APPs).

The use of these areas, whether public or private, is restricted in order to "preserve water resources, landscape, geological stability, biodiversity, and the gene flux of fauna and flora, to protect the soil and safeguard the well-being of human populations." The width of the gallery forest belts to be protected along streams and rivers, which are key to riverine protection, is currently under review by the Brazilian senate, within the forest code discussions. The old forest code considers the maximum flood level as basis for protection of the riparian zone. The proposal of the new forest considers a not well defined regular water level as basis. This would dramatically reduce wetland protection and result in massive wetland destruction.

However, both versions are not feasibly applicable to the Pantanal and other large Brazilian floodplains, as shown by Vieira (1997) for the Amazon River floodplains (*várzeas* and *igapós*). At their highest waters, most of these floodplains are inundated and could therefore be considered as APPs, which would not allow for human settlements. But, of course, the removal of local people and settlements from these wetlands is neither viable nor recommended, since they play an important role in the management of such vast areas. On the other hand, these wetlands also need a high level of protection guaranteed by specific legislation that harmonizes the requirements of local people and communities with those of environmental protection.

In 1988, the Brazilian Federal Constitution conferred upon the Pantanal the specific status of National Patrimony, although this status has yet to assume practical consequences. Law no. 6040 of 2007 confers upon the *pantaneiros* the specific status of a "traditional population," which guarantees them specific rights that can be used for the sustainable management of the Pantanal. However, these rights have yet to be specified and thus neither implemented in the Pantanal.

Ide et al. (Chapter 5) point out that the Upper Paraguay River Basin and the Pantanal belong to Brazil, Bolivia, and Paraguay, and that changes in hydrology will affect Argentina and Uruguay as well. National states cannot exploit their natural resources without bearing in mind the consequences for their neighbors. An integrated management concept requires the legal harmonization of the involved countries, but is still lacking for the Pantanal. Here, the experience provided by the Okavango Delta, under the Permanent Okavango River Basin Commission (OKACOM), is of interest. OKACOM is composed of representatives from Botswana, which owns the delta, and from Namibia and Angola, which own parts of the upper catchment area. OKACOM discusses management aspects and benefit-sharing through transboundary river basin management initiatives (Vanderpost et al., Chapter 9).

Benefits from and Threats to the Pantanal

The benefits of the Pantanal are discussed by Calheiros et al. (Chapter 3), Figueiredo et al. (Chapter 4), and Ide et al. (Chapter 5), Ioris (Chapter 8), and those of the Okavango Delta by Vanderpost et al. (Chapter 9), the European wetlands by Ball (Chapter 10), and the Everglades by Engel et al. (Chapter 11). They can be summarized as follows: (1) discharge buffering, (2) water purification, (3) groundwater recharging, (4) provision of water, (5) maintenance of biodiversity,

(6) interconnection of forest patches by riverine forests (thus promoting gene flow in forest plants and animals), (7) fish production, (8) provision of other renewable wetland products, (9) pasture for cattle ranching, (10) home to local human populations, (11) recreation for local people, and (12) ecotourism. The first six points are of high social but little, if any, commercial value such that each of these diverse wetland systems suffers the tragedy of the commons. Thus, for the Pantanal, while everyone enjoys its clean water, natural beauty, and rich biodiversity, these benefits are sustained only by the traditional *pantaneiros*, who have an economic interest in doing so.

Major threats to the Pantanal can be classified as internal and external. Internal threats include: (1) changes in vegetation cover (deforestation), (2) overexploitation of pasture areas (3) poaching of game animals and collecting of wild animals for pets, (4) local changes in hydrology by road and dike construction, and dredging of river channels and rectification of river curves

External threats are much more dangerous and they are increasing. They include: (1) deforestation in and sediment input from the upland, (2) input of pollutants (domestic, industrial, mining, and from agroindustry), (3) changes in hydrology (construction of reservoirs), (4) economic pressure on the ranchers, (5) construction of industries (e.g. gas pipeline, mining and smelting complex of Corumbá), and (6) the introduction of exotic plants, animals, and diseases. Several threats have abated, such as mercury release by gold mining. Also, the populations of endangered animal species have recovered because of intense control measures.

Less visible but nevertheless very powerful are indirect threats. These include intensive cattle ranching in the uplands on non-native pasture, which has made traditional low-density cattle ranching inside the Pantanal economically unattractive and, in turn, has drastically reduced the economic power of the local ranchers, who are beginning to be replaced by ranchers from outside, mostly from Sao Paulo and Paraná States, without any link to local knowledge and tradition. The political pressure in favor of the construction of the Upper Paraguay Waterway (*hidrovia*) came from soybean planters in the upland and not from the *pantaneiros*. The same is true in relation to hydroelectric power generation that is already produced at some tributaries and is in part distributed to consumers outside the basin. The detrimental impact of different pressure groups on the Pantanal are discussed by Safford (Chapter 7) and Ioris (Chapter 8).

The volume and distribution of the Pantanal's water, described by the concept of the flood pulse, is the driving force in the region. Major changes of the flood pulse will modify the environmental conditions in the affected areas (Junk et al. 1989). These modifications may occur slowly, over decades, e.g., by a decrease in the flood amplitude because of an increasing number of reservoirs in the catchment, or quickly, e.g., by changes in flooding behavior of rivers, such as the TaquariRiver, which inundates since a decade up to 11,000 km^2 in response to the increased sediment load. The experience in the Everglades, as described by Engel et al. (Chapter 11), shows clearly the disastrous effects of major changes in hydrology on a large wetland and the enormous costs to be paid by society for mitigation of the damage.

The above considerations highlight the vulnerability of the Pantanal and therefore the threat posed by the planned construction of numerous reservoirs at the tributaries of the upper Paraguay River. Beside altering discharge of water, nutrients and suspended sediments, reservoirs interrupt the longitudinal connectivity of rivers. Agreements between ecologists and reservoir owners are required to guarantee an ecologically sound discharge pattern of the rivers according to the hydrological cycle. Erratic discharge fluctuations may maximize economic benefits but do so by destroying the affected areas of the Pantanal, and should thus be prohibited. For example, some tributaries should be protected against reservoir construction, in order to maintain the migration routes of fish returning to their upriver spawning places (Calheiros et al., Chapter 3).

Pantanal protection efforts

There is a general agreement between most politicians, scientists, the interested national and international community, and the local population that the Pantanal is of high value and should be protected. But as noted by Ioris (Chapter 8), "*there are fierce disagreements over the interpretation of the value of ecosystem features.*" There is also a profound disagreement between the different stakeholders: Who should be involved in the decision-making process? What are the priorities, of protection? Which tools are to be used? And what are the time frames to reach the agreed upon goals? "*Through institutionalized forms of valuation, hegemonic approaches ignore the complex relations between social inequalities and environmental degradation (Scruggs 1998),*" This is shown by the Brazilian National Water Resources Plan, which is of little help in resolving problems between the riverine population and sport fishermen. For the former, fishery is an important source of dietary protein and of income. Sport fishing heavily competes for this resource, with the advantages of speed boats and modern fishing equipment, and the support of a powerful lobby.

Experience in the Okavango Delta, with its Community-Based Natural Resource Management (CBNRM), may provide a solution for this problem (Vanderpost et al., Chapter 11). CBNRM links the conservation of natural resources with rural development, which, since the 1990s, has become one of the leading themes in conservation. Its guiding principle is that the benefits to the community that arise from the sustainable use of resources must be shared among the stakeholders.

Protection proposals range from already-established private reserves (RPPNs), such as those of Fundação de Apoio à Vida nos Trópicos (ECOTROPICA), and of Social Service of Commerce (SESC) Pantanal, as well as national parks, to new proposals, such as the Biosphere Reserve Model, within the framework of the United Nations Educational, Scientific and Cultural Organization - Man and Biosphere Program (UNESCO-MAB program), which reports on ecological, social, and economic sustainability indicators. The systematically designed model using decision support software such as Marxan with Zones, presented by Lourival et al. (Chapter 6), is a good example of an approach that attempts to harmonize multiple objectives (i.e. economics, cultural and biodiversity issues) in a model aimed at identifying the best places for a network of protected areas. But the model also shows how far reality is distant from the best decision support science. On the one hand, it is common sense that protection measures will succeed only when they are accepted by the local population, and this acceptance can only be reached by its inclusion in the decision-making process. On the other hand, only a specialist is qualified to discuss Marxan with Zones and the importance of the parameters used in this model vs. those of other approaches. This gap must be bridged by improved communication among the parties involved.

Stakeholders and Pressure Groups of the Pantanal

Girard (Chapter 2) provides a good description of the historic role of traditional ranchers in the management of the Pantanal, the current economic pressures, and the need for additional wetland-friendly activities to maintain the competitiveness of low-density cattle ranching as a tool for long-term sustainable management of the area. The romantic view of a life in harmony with the beauty of nature quickly fades when one is confronted with reality. During the last few years, the program "Electricity For All" has brought significant comfort to the ranchers and their families, but television and the modern media have shown young people living on those ranches and riverside settlements the advantages of city life, such that many have left the rural Pantanal and moved to the cities with the hope of better education, jobs, and career opportunities. During the 1990s gold rush in the Poconé, many workers left the farms for the gold mines. Consequently, the gold rush destabilized both the structure and the economy of the local society (Callil and Junk 2011). Girard also provided examples of the conflicts between local fishermen and tourist activities in two major lakes. The chapter includes a discussion of the efforts of several owners of large ranches, who have banded together to create a Pantanal Natural Regional Park (PNRP) to protect the Pantanal Biome by traditional management methods, and the controversy arising from this approach between the different stakeholders. The owners of the large ranches consider themselves as the major stakeholders and thus entitled to speak for the *pantaneiros*, whereas neither the large majority of the *pantaneiros* (peons and fishermen) nor the region's indigenous people have political influence, nor a voice in the decision-making process. In the Brazilian Pantanal, there live about 150 remaining indigenous people of the Guató nation at the Indigenous Reserve Guató at Ínsua island, and about 250 people of the Bororo nation at the Indigenous Reserve Perigara. According to Girard (chapter 2), the current situation is plagued by significant problems limiting participative governance of the Pantanal.

His findings are consistent with the stakeholder analysis of Safford (Chapter 7), who also considers the ranchers as the strongest stakeholders in the Pantanal, because of their economic strength and good organization. But he also points to other major, outside players who influence development in the Pantanal. The State governments of Mato Grosso and Mato Grosso do Sul are responsible for the management of the Pantanal's natural resources. Highly efficient political pressure groups representing the business and manufacturing organizational sector use their direct influence on members of the government to obtain support for their clients' projects. On the other side are the well intentioned but diversified non-governmental organizations (NGOs) and civil society. (Safford, Chapter 7). Often the demands for strict conservation of NGOs are not helpful in furthering political discussions

Ioris (Chapter 8) offers a profound analysis of the socio-ecological disruption in the Brazilian part of the Pantanal. He points correctly to the fact that the Pantanal is part of a larger scenario in which globalization and global players have an increasing impact on decisions concerning regional development. He concludes that

> the differences between the past and present are mainly in terms of the magnitude and the speed of social and ecological impacts, but in the end it is the same overall model of development that has been responsible for the double exploitation of nature and society alike.

However, the impact of the overall model of development on nature and society changes dramatically with the increase in magnitude and speed. Adaptations by societies and nature itself can allow a smooth transition from old to new conditions, as long as the magnitude and speed of the imposed changes are sufficiently slow, as was the case in the Pantanal until approximately the 1950s. Since then, far-reaching political decisions have dramatically and rapidly altered the socio-economic environment of the Pantanal and its catchment. As a result, society and nature have not been able to respond appropriately. For example, while 200 years of low-density cattle ranching modified the Pantanal landscape without major losses of habitat and species diversity, intensive cattle ranching and soybean plantations during the last 40 years in the catchment area of the Taquari have caused enough soil erosion to transform an area of 5,000 to 11,000 km^2 of a well-structured floodplain landscape into a swampy monoculture of water hyacinth interspersed with dead trees and shrubs.

The economic problems of low-density cattle ranching in the Pantanal started 40 years ago, with the advent of high-density cattle ranching on non-native pastures in the surrounding uplands, without environmental safeguards. The transition from a low-impact ranching system dominated for over two centuries by large Pantanal ranches to a high-impact ranching and farming system in which industrial interests predominate is much more than a change from one elitist economic system to another; rather, it is a shift from an environmentally friendly system of inhabitants with a deep sense of local and regional socio-economic responsibilities and rooted for over 200 years in the traditions of the Pantanal wetland into a purely profit-oriented system of anonymous national and global players without any stake in local societies or in nature.

In the penultimate paragraph of Chapter 8, Ioris concludes that "Different than the conventional discourse about the management of the Pantanal floodplain, it is not accurate to say that traditional forms of economic production were wholly sustainable. ... It means that the floodplain was preserved after nearly three centuries mainly because of the low profitability of conventional cattle production." This statement neglects the social components and cultural achievements of traditional cattle ranching. Ranchers could have dramatically modified habitats and destroyed biodiversity during this period, but instead developed a great respect for the Pantanal's fauna and flora. Only in recent years habitat destruction and of overgrazing by intensification of cattle ranching is becoming a major problem. Without the still existing spirit of environmental protection among many traditional ranchers in the Pantanal, in contrast to the purely economic view of most newcomers from other regions, we would have little hope for the future of the area.

We agree with the statement by Ioris (Chapter 8), that official initiatives continue to subject socio-natural systems to economic exploitation and to an unfair distribution of opportunities. Newly formed decision-making forums have been dominated by the same political groups that have traditionally controlled economic and social opportunities. But a historical analysis shows that the current situation is transitional and is evolving rather quickly. Since the end of the military dictatorship, in the mid-1980s, Brazilian society has finally had the opportunity to develop civil structures to formulate and defend its interests. For many years, NGOs and other civil society associations were not accepted as serious partners by government institutions, and even today the level of acceptance is often low (Girard, Chapter 2). The process of developing efficient civil structures and integrating them in a solid, coordinated network ready to participate in development planning is still ongoing, in part with the encouragement of the Brazilian Government.

Resistance against the activities of international NGOs is often based not on rational arguments but on emotional nationalistic feelings, which do not contribute to resolving the problems at hand (Girard, Chapter 2). This situation will also change as the rising economic and political strength of Brazil necessitates the active integration of national projects in large international environmental programs, e.g., for the protection of biodiversity, in fulfilling the Ramsar Convention, and in regulating global climate change activities. This is mainly possible with an efficient cooperation between governmental and non-governmental organizations at national and international levels.

The Importance of Scientific Input into Environmental Planning in the Pantanal

Safford (Chapter 7) provides a detailed stakeholder analysis of the organizational actors who were involved in environmental planning in the Pantanal region between 1998 and 2002. The participants in that decision-making process came from a wide array of organizations representing the breadth of interests linked to the use and conservation of the Pantanal. It is interesting to notice that the scientific community does not appear as a specific stakeholder group. This is because the impact of scientists and scientific organizations is still diffuse, despite a rising number of universities, research organizations, and research programs in the area.

Mato Grosso has one federal and one state university with different campuses and two private universities, Mato Grosso do Sul one federal and one state university with different campuses and three private universities, respectively, with post-graduate courses and basic and applied research projects located in the Pantanal. The Brazilian Agricultural Research Corporation (EMBRAPA) has been involved in applied research in the Pantanal since 1984. Scientists have provided basic data on the ecology and sustainable management of natural resources, environmental impact analyses, and recommendations for sustainable development. They have also participated in multi-party environmental planning activities and knowledge transfer to stakeholder groups. Environmental research in Brazil has considerably increased during the last two decades and the government has supported this development. However, it is certainly true that scientists had, until now, only a small impact on the politics of the Pantanal.

But there is hope that this will change. From 1990 to 2002, a large cooperative project (Studies on Human Impact on Forests and Floodplains in the Tropics, SHIFT) between Brazil and Germany was realized at the Federal University of Mato Grosso (UFMT). In 2004, a research network (Center of Pantanal Research, CPP) was established at the university and cooperates with the majority of research institutions of the region. A group made up of the top scientists of the Universities of Mato Grosso and Mato Grosso do Sul was designated to run one of the new National Institutes for Science and Technology (INCTs): the INCT in Wetlands (INAU), hosted by the UFMT. In addition, a new National Institute of Pantanal Research, associated to the Ministry of Science and Technology is under construction also at the UFMT. Political pressure and increasing environmental problems will force the greater involvement of scientists in environmental development planning, environmental education, and environmental politics. Countries with a high educational level and significant environmental problems have recognized the need to foster an efficient scientific sector whose responsibilities include advising the governments and other decision makers in its efforts to deal with these problems, as discussed in Chapters 10 and 11 for the European Community's wetlands and the Everglades, respectively. Calheiros et al. (Chapter 3), exemplify a very recent case of direct influence of scientific opinion and action on the decision making processes related to the proliferation of hydroelectric power dams in the Basin.

Ioris (Chapter 8) points to the disparate values assigned to environmental services by the different stakeholder groups and to the fact that all stakeholders are lobbying for their own interests. As ecologists, we cannot remain neutral in this debate. As shown for the Everglades (Engel et al., Chapter 11), important general wetland priorities have already been determined and must be defended not only for the benefit of the Pantanal but also for the *pantaneiros*. We consider many economic arguments as extremely destructive. Increasing economic return by wetland-unfriendly management methods, e.g., the construction of the *hidrovia*, is certainly not on our priority list. The personal engagement of scientists with ecological and sociological backgrounds is required to make sure that their voices are heard in political discussions about the environment. They must compete for political space with technocrats who provide purely technical solutions to management problems without concern for either the environment or the well-being of traditional local societies. Many scientists from the Pantanal's surroundings and from abroad have already exposed their viewpoints in political discussions and the press. For some of them this has been a costly exercise, since they have confronted their institutions political position. Therefore scientists must organize themselves to ensure that their opinion will be taken into account.

In this context, mobilizing the population to participate in a discussion of the Pantanal's development remains a major challenge. This is, in part, a question of education, but medium- and long-term, progress is nonetheless expected, as the government increasingly recognizes that a lack of qualified people is the bottleneck of economic development. Universities will play a key role in providing teachers and administrators with sound ecological backgrounds.

The Pantanal as a Cultural Landscape

Girard (Chapter 2) reviews the different steps in the colonization of the Pantanal by European immigrants over a period of 250 years. However, indigenous nations had already inhabited the Pantanal for at least 6,000 years before the arrival of the Europeans. They constructed elevated sites for flood protection and managed fauna and flora according to their needs. Over time, the indigenous people accumulated a profound knowledge about the functions of the different landscape and vegetation units. This was of fundamental importance to survival. It allowed selective stimulation of the growth of useful trees on suitable habitats. For instance, the nearly monodominant stands of the palm tree *Scheelea phalerata* on many *capões* are probably the result of selective management in favor of this very useful palm species. The European immigrants partially incorporated this knowledge, as shown by the many specific names of indigenous origin which they adopted for plants, animals, and landscape and vegetation units. They amplified pasture areas and maintained specific tree species because of their high-quality wood, their medical importance, and/or their esthetic value, but they also protected those landscape units that had specific functions in the wetland. Consequently, large parts of the Pantanal slowly changed their vegetation cover but retained their characteristic habitat and species diversity. Exclosure experiments at Fazenda Pirizal showed that the esthetically very attractive park landscape of grasslands, single-tree savannahs, and forest islands along large stretches of the Transpantaneira Park Road is the result of careful management by the ranchers. Otherwise, most of the open grasslands and single-tree savannahs would, within a few decades, rapidly convert to unattractive scrubland with reduced biodiversity (Junk and Nunes da Cunha, in press). Accordingly, we have to consider large parts of the Pantanal not as pristine wetlands but as a carefully managed cultural landscape and one of the very few examples of the sustainable use of a tropical ecosystem by European immigrants (see also Girard, chapter 2).

Conclusions: Management Options and Solutions

The view of the Pantanal as a cultural landscape has far-reaching consequences for management options. Currently, only about 5% of the Pantanal is under full protection. This is certainly not enough, and we agree with Lourival et al. (2009) and Ide et al. (chapter 5) that at least around 20% of the area should be fully protected. A synthesis of threats and protecting efforts is provided by Machado et al. (2011). Even so, the majority of the Pantanal will be under private ownership, and the future of the Pantanal will depend on how these owners manage their properties.

From an ecological point of view, the only management options that can be advocated are those consistent with the following requirements: (1) They must take into account the natural hydrological cycle and sediment fluxes; (2) They should maintain, to the maximum possible extent, natural habitat and species diversity, including sufficient gene flux inside the Pantanal and between the Pantanal and the upland; (3) They must ensure the stability of the traditional human population and suitable living conditions; (4) They must avoid major infrastructure construction projects that alter the Pantanal's hydrology and sediment fluxes and thus create environmental disturbances and pollution.

These requirements exclude large-scale crop plantations in the floodplain, because the Pantanal's cyclical low-water and dry periods will necessitate large-scale and expensive irrigation. Furthermore, the regular floods would cause the run-off of fertilizers and pesticides from crop plantations directly into the aquatic environment, negatively affecting its plants and animals, including economically important fish species. Poldering, to control the floods, would modify the hydrological conditions not only inside but also outside the polders, by interrupting natural water flow and increasing the total amount of water in the adjacent areas. Moreover, the wetland system within the polders would be completely modified to terrestrial agro-ecosystems, inevitably leading to changes in the plant- and animal communities beyond them. Pollution by the run-off of agrochemicals used in cultivating the polders is another threat. The far-reaching consequences of altering wetland hydrology and nutrient status and the extremely high costs for mitigating the negative side effects have been well documented for the Everglades (Engel et al., Chapter 11).

Extended natural grasslands provide the basis for low-density cattle ranching as one of the natural vocation for human inhabitants of the Pantanal. The self-organization of the ranchers in the PNRP shows several deficits with respect to the definition of its goals and control by independent organizations, as indicated by Girard (Chapter 2). But these deficiencies can be overcome by specific, wetland-friendly legislation that fully protects key habitats and allows the management of other areas under the strict control of the Brazilian Institute for the Environment and Renewable Natural Resources (IBAMA). A basis for this approach is the habitat classification provided by Nunes da Cunha and Junk (2011). It differentiates the habitats of the Pantanal according to hydrological, soil-related, hydrochemical, and physical parameters and the characteristics of the vegetation. Most habitats have been given common names and are therefore familiar to the local population. This facilitates dialogue between *pantaneiros*, scientists, and decision-makers.

The economic constraints of this approach are obvious, thus raising the question, how to improve the economic situation of the traditional ranchers other traditional communities and support their wetland friendly management of the area. Additional, wetland-friendly activities were discussed by Girard (Chapter 2). These include tourism, which has also been shown to be of major importance for other large wetlands, such as the Okavango Delta (Vanderpost et al., Chapter 9) and the Everglades (Engel et al., Chapter 11).

Traditional ranchers in the Pantanal can be supported by the facilitation of pasture management, including modern technologies, as long as they do not place local hydrology, habitat, and species diversity at risk. The problem of pasture clearing was discussed by Junk and Nunes da Cunha (in press). Another option that merits discussion is the provision of economic stimuli, such as tax reduction for ranchers who practice wetland-friendly land management. Such measures can be justified by the fact that these ranchers provide a service to society, maintaining a wetland of national and international importance, and they should be compensated for income losses related to this service. Europe provides many examples for supporting the management of ecologically valuable cultural landscapes (Vos and Meeskes 1999), but these areas are very small compared to the Pantanal. Therefore, the development of Brazilian-specific solutions for the management of the Pantanal is required. An analysis of existing reserves, core areas of the Cerrado-Pantanal Priority Setting Workshop (CP-PSW), biosphere reserves, and 'core + corridors' CP-PSW led to the conclusion that "none of the four conservation scenarios met preferred areal targets for protection of habitats, nor did any protect all 17 biodiversity surrogates" (Lourival et al. 2009). These findings show that an alternative agreement with ranchers and other stakeholders, who own or inhabit most of Pantanal's lands and waters is required, which would support wetland-friendly management. Considering the transboundary nature of the Pantanal, discussions with the Bolivian and Paraguayan governments are required to harmonize protection efforts in the entire area.

As indicated by Calheiros et al. (Chapter 3), Figueiredo et al. (Chapter 4), and Ide et al. (Chapter 5), efforts to protect habitat and species diversity inside the Pantanal are vulnerable to serious threats from outside the area. Therefore, adequate protection of the Pantanal will be attained only by implementing an integrated management plan that includes the Upper Paraguay River Basin and the

Pantanal. This management plan should maintain the hydro-ecological processes in the entire basin, as determined by the Brazilian Federal Constitution and recommended by the Millennium Ecosystem Assessment (MEA 2005; Calheiros et al. Chapter 3).

The large financial efforts needed to maintain the Everglades (Engel et al., Chapter 11) and to recover the wetlands in Western Europe (Ball, Chapter 10) evidence the importance that industrialized societies have finally given to the preservation of intact wetlands. But the immense costs also make it clear that the maintenance of intact or moderately used wetlands is much cheaper than the recovery of deteriorated ones, a lesson that Brazil has yet to learn.

References

Brasil. 1983. *Projeto Radambrasil*. Folha SE.21 Corumbá e parte da Folha SE.20. Geologia, Geomorfologia, Pedologia, Vegetação e Uso Potencial da Terra. Levantamento de Recursos Naturais, 31. Rio de Janeiro: Ministério das Minas e Energia, Secretaria Geral.

Callil, C.T. and Junk, W.J. 2011. Gold mining near Poconé: environmental, social and economic impacts, in *The Pantanal: Ecology, Biodiversity and Sustainable Management of a Large Neotropical Seasonal Wetland*, edited by W.J. Junk WJ, C.J. da Silva, C. Nunes da Cunha and K.M. Wantzen. Moscow: Pensoft, Sofia, 695-717.

Diegues, A.C.S. 1994. *An Inventory of Brazilian Wetlands*. Gland, Switzerland: IUCN.

Diegues, A.C.S. 2002. *Povos e Águas*. 2nd edition. São Paulo: Núcleo de Apoio à Pesquisa sobre Populações Humanas e Áreas Úmidas Brasileiras.

Hamilton, S.K., Sippel, S.J. and Melack, J.M. 1996. Inundation patterns in the Pantanal wetland of South America determined from passive microwave remote sensing. *Archives of Hydrobiology*, 137(1), 1-23.

Eiten, G. 1983. *Classificação da Vegetação do Brasil*. Brasília: CNPq.

Junk, W.J. in press. Current State of Knowledge Regarding South America Wetlands and their Future under Global Climate Change. *Aquatic Sciences*.

Junk, W.J., Bayley, P.B. and Sparks, R.E. 1989. The flood pulse concept in river-floodplain systems. Special Publication of the *Canadian Journal of Fisheries and Aquatic Sciences*, 106, 110-127.

Junk, W.J. and Wantzen, K.M. 2004. The flood pulse concept: new aspects, approaches, and applications - an update, in *Proceedings of the Second International Symposium on the Management of Large Rivers for Fisheries*, Volume 2, edited by R.L. Welcomme and T. Petr, Bangkok: FAO Regional Office for Asia and the Pacific. Food and Agriculture Organization and Mekong River Commission. RAP Publication 2004/16, 117-149.

Junk, W.J. 2005. Flood pulsing and the linkages between terrestrial, aquatic, and wetland systems. *Proceedings. International Association of Theoretical and Applied Limnology*, 29(1), 11-38.

Junk, W.J. and Nunes da Cunha, C. in press. Pasture clearing from invasive woody plants in the Pantanal: a tool for sustainable management or environmental destruction? *Wetlands Ecology and Management*

Junk, W.J., da Silva, C.J., Nunes da Cunha, C. and Wantzen, K.M. (eds). 2011. *The Pantanal: Ecology, Biodiversity and Sustainable Management of a Large Neotropical Seasonal Wetland*. Moscow: Pensoft, Sofia.

King, J.M., Tharme, R.E. and Brown, C.A. 1999. *Definition and Implementation of Instream Flows: Thematic Report for the World Commission on Dams*. Cape Town, South Africa: Southern Waters Ecological Research and Consulting.

Lourival, R., McCallum, H., Grigg, G., Arcangelo, C., Machado, R. and Possingham, H. 2009. A systematic evaluation of the conservation plans for the Pantanal Wetland in Brazil. *Wetlands*, 29, 1189-1201.

Loverde-Oliveira, S.M., Adler, M. and Pinto Silva, V. 2011. Phytoplankton, periphyton, and metaphyton of the Pantanal floodplain: species richness, density, biomass, and primary production, in *The Pantanal: Ecology, Biodiversity and Sustainable Management of a Large Neotropical Seasonal Wetland*, edited by W.J. Junk WJ, C.J. da Silva, C. Nunes da Cunha and K.M. Wantzen. Moscow: Pensoft, Sofia, 235-256.

Machado, R.B., Harris, M.B., Silva, S.M. and Ramos Neto, M.B. 2011. Human impacts and environmental problems in the Brazilian Pantanal, in in *The Pantanal: Ecology, Biodiversity and Sustainable Management of a Large Neotropical Seasonal Wetland*, edited by W.J. Junk WJ, C.J. da Silva, C. Nunes da Cunha and K.M. Wantzen. Moscow: Pensoft, Sofia, 719-739.

MEA. 2005. *Millennium Ecosystem Assessment. Ecosystems and Human Well-Being: Wetlands and Water - Synthesis*. Available at: http://www.maweb.org/documents/document.358.aspx.pdf [accessed: 18 September 2009].

MMA. 2010. *Legislação Ambiental*. Available at http://www.mma.gov.br/conama [accessed: 20 November 2010]

Nel, J.L., Turak, E., Linke, S. and Brown, C. 2011. Integration of environmental flow assessment and freshwater conservation planning: a new era in catchment management. *Marine and Freshwater Research*, 62(3), 290-299.

Oliveira, M.D. and Calheiros, D.F. 2000. Flood pulse influence on phytoplankton communities of the south Pantanal floodplain, Brazil. *Hydrobiology*, 427, 102-112.

Scruggs, L.A. 1998. Political and economic inequality and the environment. *Ecological Economics*, 26(3), 259-275.

Silva, J.V.S. and Abdon M.M. 1998. Delimitação do Pantanal brasileiro e suas sub- regiões. *Pesquisa Agropecuária Brasileira*, 33, 1703- 1711.

Tharme, R.E. 2003. A global perspective on environmental flow assessment: emerging trends in the development and application of environmental flow methodologies for rivers. *River Research and Applications*, 19, 397-441.

Vieira, R.d.S. 2000. Legislation and the use of Amazonian floodplains, in *The Central Amazon Floodplain: Actual Use and Options for a Sustainable Management*, edited byW.J. Junk, J.J. Ohly, M.T.F. Piedade and M.G.M. Soares. Leiden: Backhuys Publishers, 505-533.

Vos, W. and Meeskes, H. 1999. Trends in European cultural landscape development: perspectives for a sustainable future. *Landscape and Urban Planning*, 46, 3-14.

Wantzen, K.M. 1998. Siltation effects on benthic communities in first order streams in Mato Grosso, Brazil. *Verhandlungen der Internationalen Vereinigung fur Theoretische und Angewandte Limnologie*, 26, 1155-1159.

Index